NUCLEIC ACID METABOLISM
CELL DIFFERENTIATION AND CANCER GROWTH

Participants in the Second International Symposium for Cellular Chemistry (1968).

NUCLEIC ACID METABOLISM CELL DIFFERENTIATION AND CANCER GROWTH

*Proceedings of the Second International Symposium for Cellular
Chemistry at Biwako Hotel, Ohtsu
October 17 to 21, 1966*

Members of Organizing Committee

H. Endo, T. Fujii, M. Hanaoka, S. Hibino, I. Honjo, T. Ishikawa,
N. Kamiya, I. Kawakami, R. Kinosita, S. Makino,
O. Midoriteawa, Y. Miura, H. Naora, T. Okada, B. Osogoe,
N. Shinke, M. Sugiyama, K. Takikawa, S. Tanaka, H. Terayama,
G. Wakisaka and T. Yamada

Chairman of Organizing Committee

S. Seno

Edited by

E. V. COWDRY AND S. SENO

*Department of Anatomy
Washington University
School of Medicine, St. Louis*

*Department of Pathology
Okayama University
Medical School, Okayama*

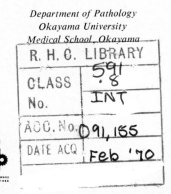

PERGAMON PRESS

OXFORD · LONDON · EDINBURGH · NEW YORK

TORONTO · SYDNEY · PARIS · BRAUNSCHWEIG

Pergamon Press Ltd., Headington Hill Hall, Oxford
4 & 5 Fitzroy Square, London W. 1
Pergamon Press (Scotland) Ltd., 2 & 3 Teviot Place, Edinburgh 1
Pergamon Press Inc., Maxwell House, Fairview Park, Elmsford, New York 10523
Pergamon of Canada Ltd., 207 Queen's Quay West, Toronto 1
Pergamon Press (Aust.) Pty. Ltd., 19a Boundary Street, Rushcutters Bay,
N. S. W. 2011, Australia
Pergamon Press S. A. R. L., 24 rue des Écoles, Paris 5e
Vieweg & Sohn GmbH, Burgplatz 1, Braunschweig

First edition 1969

Library of Congress Catalog Card No. 68-31342

PRINTED IN GERMANY
08 013252 9

CONTENTS

III. CELL MULTIPLICATION AND DIFFERENTIATION

IV. CONTROL OF CELL GROWTH, CELL TRANSFORMATION AND CANCER INDUCTION BY VIRUS

The late Dr. Seizo Katsunuma (1886–1963) the President of the First International Symposium for Cellular Chemistry (March, 1963). He had been also appointed the President of the Second International Symposium for Cellular Chemistry (1966) but he passed away from a cerebral hemorrhage in November, 1963, and he was succeeded by Dr. S. Akabori, President of Osaka University. Dr. Katsunuma, an authority in internal medicine, was President of Nagoya University from 1949 to 1959. He received numerous awards and prizes for his cytochemical studies of oxidase as early as in the twenties of this century, and is considered the Father of modern cell biology and cell chemistry.

Dr. Shiro Akabori, President of the Second International Symposium for Cellular Chemistry, Professor Emeritus of Biochemistry, Osaka University, formerly President of the same university (1960–1966), made a great contribution to research in amino acid and protein chemistry.

PREFACE

To write a preface to these Proceedings of the Second International Symposium for Cellular Chemistry is to relate the steps taken by our distinguished Secretary General, Dr. Satimaru Seno, Chairman of the Organizing Committee, that have made this Symposium an outstanding success. Although six steps, or stages, are recognizable in the development of this Symposium, these probably overlapped somewhat.

1. The selection of Dr. S. Akabori, President of Osaka University, himself a chemist of the highest reputation, to serve as President of the Symposium, constituted a long step in the right direction.

2. After consultation with the Organizing Committee, Dr. Akabori and others, Dr. Seno announced that the subjects of the Symposium would be *Nucleic Acid Metabolism, Cell Differentiation and Possible Ways to Control Cancer.* This proved a wise selection because it appealed to many organizations and individuals in Japan and abroad as including the most promising basic research made by the newest and most elaborate techniques fundamental to many aspects of human endeavor around the world.

3. In the difficult job of choosing those to be invited, the Executive Committee, the Secretary General, and President demonstrated intimate personal knowledge of leaders in these fruitful fields of enquiry in Japan and elsewhere. These individuals are to be congratulated on the scope and quality of the papers.

4. The subjects to be discussed and the names and reputations of the speakers were of great assistance to Drs. Akabori, Seno and every member of the influential Organizing Committee in obtaining financial support for the Symposium since they were acceptable as guarantees of quality.

5. Despite the fact that the Proceedings of the First International Symposium for Cellular Chemistry were published in Japan in flawless English by the Japan Society of Cell Biology, with illustrations second to none, in an attractive volume as a Katsunuma Memorial Issue, the decision of the Japan Society of Cell Biology to publish these Proceedings of the Second International Symposium by Pergamon Press Ltd. of Oxford was wise and far-sighted. To be consistently advertized with other notable books of the Pergamon Press will almost automatically lead to a large scale use of this

volume. This is important in placing them in the hands of all investigators in the basic and attractive fields of research that it covers.

6. I am convinced that the Officers of the Symposium and of the Japan Society of Cell Biology have carried out all matters as the late Seizo Katsunuma would have wished.

July 1967   E. V. COWDRY

ACKNOWLEDGEMENTS

THE Organizing Committee of the Second International Symposium for Cellular Chemistry gratefully acknowledges help and support from many sources. Among others, we wish to thank deeply.

1. Cooperation:

   Japan Science Council
   Japan Society for Cellular Chemistry
   Japan Society for the Promotion of Science

2. Support for the travelling expense of the American participants:

   National Science Foundation, U.S.A.

3. Financial support for congress, travel and printing:

   Japan Economic Association

   Contributors from U.S.A.:
   Parke, Davis & Co., Michigan; Schering Corporation, New Jersey; Smith Kline & French Overseas Co., Philadelphia; The Squibb Institute for Medical Research, New Jersey

   Brewery, Food and Chemical Manufacturing Companies:
   Sumitomo Chemical Co., Ltd., Osaka; Suntory Ltd., Osaka; Yamasa Shou Co., Ltd., Choshi

   Japan Electric Company Association:
   The Chubu Electric Power Co., Ltd., Nagoya; The Chugoku Electric Power Co., Ltd., Hiroshima; The Kansai Electric Power Co., Ltd., Osaka; The Tokyo Electric Power Co., Ltd., Tokyo

   Japan Medical Association

   Hospitals in Okayama and Hiroshima:
   Arichi Hospital, Hiroshima; Jikeikai Hospital, Okayama; Kawada Hospital, Okayama; Kawasaki Hospital and Cancer Institute, Okayama; Kobatake Hospital, Hiroshima; Okayama Chuo Hospital, Okayama; Sakakibara Hospital, Okayama; Shimotsui Hospital, Okayama

   The Life Insurance Association of Japan

   Medical and Biological Equipments Manufacturing Companies;
   Misuzu Sangyo Co., Ltd., Tokyo; Nichimen Co., Ltd., Osaka; Nippon Kogaku K.K., Tokyo; Olympus Optical Co., Ltd., Tokyo; Shimazu Seisakusho Ltd., Kyoto

   The Nagoya Chamber of Commerce and Industry:
   Matsusakaya Co., Ltd., Nagoya; Nagoya Railroad Co., Ltd., Nagoya; Toho Gas Co., Ltd., Nagoya; The Tokai Bank Ltd., Nagoya; Toyota Motor Co., Ltd., Nagoya

Personal contributors:
Kunishima, Kihachiro, Kyoto; Thuda, Ryotaro, Osaka

The Pharmaceutical Manufacturers of Tokyo, Osaka Pharmaceutical Manufacturers
Association and Related Companies:
Dainippon Pharmaceutical Co., Ltd., Osaka; Eisai Co., Ltd., Tokyo;
Fujisawa Pharmaceutical Co., Ltd., Osaka; The Green Cross Corporation,
Osaka; Kyowa Hakko Kogyo Co., Ltd., Tokyo; Nihon Shinyaku Co., Ltd.,
Kyoto; Nippon Merck-Banyu Co., Ltd., Tokyo; Otsuka Pharmaceutical
Factory, Tokushima & Okayama; Pfizer Taito Co., Ltd., Tokyo; Research
Association for Photosensitizing Dyes, Okayama; Shionogi & Co., Ltd., Osaka;
Takeda Che mical Industries, Ltd., Osaka; Tanabe Seiyaku Co., Ltd., Osaka;
Toshiba Pharmaceutical Co., Ltd., Kawasaki

Tokyo Bankers Association

LIST OF PARTICIPANTS

AKABORI, S., Institute for Protein Research, Osaka University, Osaka, Japan

*AMANO, M., Biology Division, National Cancer Center Research Institute, Tokyo, Japan

AOKI, I., Department of Physiology, Osaka City University, Medical School, Osaka, Japan

*BRAUN, A.C., The Rockefeller University, New York, U.S.A.

*BUSCH, H., Department of Pharmacology, Baylor University College of Medicine, Houston, U.S.A.

EBERT, J.D., Department of Embryology, Carnegie Institute of Washington, Baltimore, U.S.A.

ENDO, H., Research Institute of Cancer, Faculty of Medicine, Kyūshū University, Fukuoka, Japan

*FITZGERALD, P., Department of Pathology, State University of New York, Downstate, Medical Center, Brooklyn, U.S.A.

FUJIKI, N., Department of Internal Medicine, Kyoto Prefectural University of Medicine, Kyoto, Japan

FURUSAWA, M., Department of Biology, Faculty of Science, Osaka City University, Osaka, Japan

*HAGIWARA, A., Laboratory of Developmental Biology, College of Science, University of Kyoto, Kyoto, Japan

*HAMASHIMA, Y., Department of Pathology, Faculty of Medicine, Kyoto University, Kyoto, Japan

*HANAFUSA, H., The Public Health Research Institute of the City of New York, New York, U.S.A.

*HANAOKA, M., Institute for Virus Research, Kyoto University, Kyoto, Japan

*HARUNA, I., Department of Microbiology, University of Illinois, Urbana, U.S.A.

HIBINO, S., Department of Internal Medicine, Nagoya University, School of Medicine, Nagoya, Japan

HIRAOKA, T., Department of Botany, Faculty of Science, Kyoto University, Kyoto, Japan

HONJO, I., Department of Biology, Faculty of Science, Osaka University, Toyonaka, Japan

*HORI, S.H., Zoological Institute, Faculty of Science, Hokkaido University, Sapporo, Japan

HOTTA, Y., Department of Biology, University of California, San Diego, U.S.A.

ISHIDA, J., Zoological Institute, Faculty of Science, University of Tokyo, Tokyo, Japan

*ISHIKAWA, T., Department of Pathology, School of Medicine, Kanazawa University, Kanazawa, Japan

*ISHIZAKI, H., Laboratory of Development Biology, Zoological Institute, Kyoto University, Kyoto, Japan

ITO, N., Nara Technical College, Yamatokoriyama, Japan

*IWATA, S., Department of Anatomy, School of Medicine, Tokushima University, Tokushima, Japan

*IZAWA, M., Biology Division, National Cancer Center Research Institute, Tokyo, Japan

* Contributors.

xiii

Izutsu, K., Department of Pathology, Mie Prefectural University, School of Medicine, Tsu, Japan

Kaighn, M. E., Carnegie Institute of Washington, Department of Embryology, Baltimore, U.S.A.

*Kameyama, T., Laboratory of Cancer Research, School of Medicine, Kanazawa University, Kanazawa, Japan

Kamiya, N., Department of Biology, Faculty of Science, Osaka University, Toyonaka, Japan

*Kasten, F. H., Pasadena Foundation for Medical Research, Pasadena, U.S.A.

*Kato, S., Research Institute for Microbial Diseases, Osaka University, Osaka, Japan

Kawade, Y., Institute for Virus Research, Kyoto University, Kyoto, Japan

Kawai, T., Department of Pathology, Okayama University Medical School, Okayama, Japan

*Kawakami, I., Department of Biology, Faculty of Science, Kyūshū University, Fukuoka, Japan

Kawakami, M., Department of Microbiology, School of Medicine, Gunma University, Maebashi, Japan

Kawamata, J., Research Institute for Microbial Diseases, Osaka University, Osaka, Japan

*Kimoto, T., Department of Pathology, Okayama University Medical School, Okayama, Japan

Kimura, K., Department of Biophysics and Biochemistry, Faculty of Science, University of Tokyo, Tokyo, Japan

*Kinosita, R., City of Hope National Medical Center, Duarte, U.S.A.

Koshihara, H., Department of Zoology, Faculty of Science, Tokyo Kyoiku University, Tokyo, Japan

Kotani, M., Department of Biology, Faculty of Science, Osaka City University, Osaka, Japan

Kratochwil, W. K., Department of Biology, University of California, San Diego, U.S.A.

*Kuroda, Y., Department of Morphological Genetics, National Institute of Genetics, Mishima, Japan

*Marco, A. Di, Institute Tumori Nazionale, Milano, Italy

Matano, Y., Department of Anatomy, Osaka City University, Medical School, Osaka, Japan

Matsumoto, I., Department of Biochemistry, Faculty of Medicine, Kyūshū University, Fukuoka, Japan

Midorikawa, O., Department of Pathology, Faculty of Medicine, Kyoto University, Kyoto, Japan

*Mitsuhashi, S., Department of Microbiology, School of Medicine, Gunma University, Maebashi, Japan

*Miura, Y., Department of Biochemistry, School of Medicine, Chiba University, Chiba, Japan

*Miyahara, M., Department of Pathology, Okayama University Medical School, Okayama, Japan

Monden, H., Department of Pathology, Okayama University Medical School, Okayama, Japan

*Morikawa, S., Department of Pathology, Faculty of Medicine, Kyoto University, Kyoto, Japan

Murano, T., Department of Pharmacology, Wakayama Medical College, Wakayama, Japan

Nakazawa, T., Department of Neuropsychiatry, Keio University, School of Medicine, Tokyo, Japan

*Naora, H., Biology Division, National Cancer Center Research Institute, Tokyo, Japan

NISHI, K., Research Institute for Tuberculosis, Toneyama Hospital, National Sanatorium and Osaka City University, Toyonaka, Japan

NISHIDA, H., Biological Laboratory, Iwamizawa Branch, Hokkaido College of Education, Iwamizawa, Japan

OBUCHI, S., Department of Pathology, Okayama University Medical School, Okayama, Japan

OHTAKA, Y., Institute of Physical and Chemical Research, Tokyo, Japan

OKADA, S., Department of Pathology, Okayama University Medical School, Okayama, Japan

*OKADA, T.S., Zoological Institute, Faculty of Science, Kyoto University, Kyoto, Japan

OKIGAKI, T., Biology Department, Division of Natural Science, International Christian University, Tokyo, Japan

ROKUJO, T., Medical Division, Igaku Chosa-bu, Fujisawa Pharmaceutical Co., Ltd., Osaka, Japan

SATO, A., Technical Research Department, Nichimen Co., Ltd., Osaka, Japan

SATO, I., Department of Biology, College of General Education, Osaka University, Toyonaka, Japan

SEIJI, M., Department of Dermatology, Juntendo University School of Medicine, Tokyo, Japan

*SENO, S., Department of Pathology, Okayama University Medical School, Okayama, Japan

SHIBATA, T., Department of Pathology, Okayama University Medical School, Okayama, Japan

SHIMIZU, M., Dainippon Parmaceutical Co., Ltd., Osaka, Japan

*SHIRAKAWA, S., Department of Internal Medicine, Faculty of Medicine, Kyoto University, Kyoto, Japan

SOGABE, K., Department of Pathology, Okayama University Medical School, Okayama, Japan

*SPIEGELMANN, S., Department of Microbiology, University of Illinois, Urbana, U.S.A.

SUGINO, Y., Institute for Virus Research, Kyoto University, Kyoto, Japan

SUGIYAMA, M., Sugashima Marine Biological Station, Nagoya University, Toba, Japan

*SUYAMA, T., Department of Pathology, School of Medicine, Kanazawa University, Kanazawa, Japan

TAKAGI, Y., Department of Biochemistry, School of Medicine, Kyūshū University, Fukuoka, Japan

TAKAHASHI, T., Laboratory of Biochemistry, Aichi Cancer Center Research Institute, Nagoya, Japan

TAKAKI, R., Department of Internal Medicine, School of Medicine, Kyūshū University, Fukuoka, Japan

TAKEBAYASHI, J., Department of Pathology, Okayama University Medical School, Okayama, Japan

TAKEDA, S., Department of Pathology, School of Medicine, Mie Prefectural University, Tsu, Japan

*TAKEUCHI, I., Department of Biology, Faculty of Science, Osaka University, Toyonaka, Japan

TAKIKAWA, K., Department of Internal Medicine, Nagoya University School of Medicine, Nagoya, Japan

TANAKA, K., Central Research Laboratory, Sankyo Co., Ltd., Tokyo, Japan

TANAKA, S., Department of Biological Chemistry, Faculty of Science, Kyoto University, Kyoto, Japan

TANAKA, T., Laboratory of Tumor Biology, Aichi Cancer Center, Nagoya, Japan

*TERAYAMA, H., Department of Biophysics and Biochemistry, Faculty of Science, University of Tokyo, Tokyo, Japan

*THORELL, B., Department of Pathology, Karolinska Institute, Stockholm, Sweden

*VALLADARES, Y., Laboratory of Biology and Biochemistry of Cancer, Instituto Nacional de Oncologia, Madrid, Spain

*WAKISAKA, G., Department of Internal Medicine, Faculty of Medicine, Kyoto University, Kyoto, Japan

*WATANABE, R., Department of Pathology, School of Medicine, Kanazawa University, Kanazawa, Japan

*WATANABE, Y., Department of Pathology, National Institute of Health, Tokyo, Japan

*YAMADA, M., Department of Pathology, National Institute of Health, Tokyo, Japan

*YAMADA, M., Department of Anatomy, School of Medicine, Tokushima University, Tokushima, Japan

YAMAGATA, K., Department of Anatomy, Osaka City University, Medical School, Osaka, Japan

YAMAGATA, S., Department of Obstetrics and Gynecology, Osaka City University, Medical School, Osaka, Japan

YAMAMOTO, K., Faculty of Fisheries, Hokkaido University, Hakodate, Japan

*YAMAMOTO, T., Institute for Infectious Diseases, University of Tokyo, Tokyo, Japan

YAMANAKA, M., Department of Microbiology, Osaka Medical College, Takatsuki, Japan

YAMANE, I., Research Institute for TB, Leprosy and Cancer, Tōhoku University, Sendai, Japan

*YANAGITA, T., Institute of Applied Microbiology, University of Tokyo, Tokyo, Japan

YASUMASU, I., Department of Biology, College of Education, Waseda University, Tokyo, Japan

YOKOMURA, E., Department of Pathology, Okayama University Medical School, Okayama, Japan

*ZEUTHEN, E., Carlsberg Foundation Biological Institute, Copenhagen, Denmark

OPENING ADDRESS

S. AKABORI

Distinguished guests, ladies and gentlemen:

It is a great honor and pleasure for me to have the privilege of delivering a welcome address to you on the opening of the Second International Symposium for Cellular Chemistry.

It was at this same place in March 1963, when the First International Symposium was held with the theme of "Intracellular Membraneous Structure". The First Symposium was sponsored by the late President Seizo Katsunuma of Nagoya University with the collaboration of Dr. E. V. Cowdry of the United States. It was my privilege to attend this First Symposium, which was so very successful that President Katsunuma started planning a Second Symposium with the officers of the Japan Society for Cellular Chemistry, now called the Japan Society for Cell Biology. However, to our great regret, he suddenly passed away in the fall of 1963.

To cope with this difficult situation, the Japan Society for Cell Biology asked me to act as President of the Second International Symposium. Though I am not a cell biologist, I fully understand that cellular chemistry, binding together biochemistry and cell biology, is an exceedingly important and promising discipline for the understanding of the basic aspects of life and the causes of diseases like cancer at the cellular as well as the molecular level. It was indeed a great privilege for me to do what I could in organizing this Symposium.

The first Organizing Committee meeting of this Symposium was held at Osaka in March 1965. After repeated meetings, during which suggestions from many authorities were carefully considered, the Japan Society for Cell Biology decided the theme of this symposium should be "Nucleic Acid Metabolism, Cell Differentiation and Possible Ways to Control Cancer". The Science Council of Japan and the Japanese Association for the Promotion of Sciences have approved the preparation of this Symposium. Generous financial assistance from various sources, both domestic and foreign, has been thankfully received.

Taking this opportunity, it is a pleasure to express my sincere appreciation to Dr. Cowdry who has greatly encouraged the opening of this Symposium through his enthusiasm and suggestions to the officers of the Organizing Committee headed by Dr. Seno, who worked assiduously throughout

STUDIES ON THE SYNTHESIS
OF A VIRAL NUCLEIC ACID
WITH A PURIFIED ENZYME*

S. Spiegelman, I. Haruna and N. R. Pace†

Department of Microbiology, University of Illinois, Urbana, Illinois

THE ability of an RNA virus to complete its life cycle in a cell dominated by and designed for the requirements of a DNA genome poses a number of interesting biological problems. Molecular hybridization (Hall and Spiegelman;[1] Yankofsky and Spiegelman[2]) was used by Doi and Spiegelman[3] to determine whether a DNA complementary to the viral RNA could be found before or after infection. Under conditions in which complexes would have been readily detected, none were found.

The negative outcome of the hybridization test was taken to signify that the DNA to RNA pathway was not employed. This in turn implies that RNA viruses have evolved a mechanism of generating RNA copies from RNA. One predicts then the existence of an enzymatic mechanism involving an RNA dependent RNA polymerase which was named "replicase" for purposes of brevity and alliterative usefulness (Spiegelman and Doi[4]). It seemed desirable to begin an enzymological attack in the hope that the relevant enzyme system could be purified to the point where the mechanism of RNA synthesis could be examined in a simple system permitting hard inferences. Because of its obvious advantages, the biological system chosen was the class of RNA bacteriophages which specifically infect *E. coli* (Loeb and Zinder[5]).

THE SEARCH FOR THE MS-2 REPLICASE

The search for a unique RNA dependent polymerase is complicated by the presence of a variety of enzymes which, in addition to the DNA dependent RNA polymerase (transcriptase), can incorporate ribonucleotides either terminally or subterminally into preexistent RNA chains (Smellie[6]). In

* This investigation was supported by Public Health Service Research Grant No. CA-01094 from the National Cancer Institute and Grant No. GB-2169 from the National Science Foundation.

† Predoctoral trainee in Microbial and Molecular Genetics. Grant No. USPH-2-Tl-GM-319-06.

3

addition, there are others (e.g. RNA phosphorylase,[7] polyadenylate synthetase,[8] etc.) which can mediate extensive synthesis of polynucleotide chains. It is obvious that a claim for a new type of RNA polymerase must be accompanied by evidence for RNA dependence and a demonstration that the enzyme possesses some unique characteristic which differentiates it in one or more of its properties from previously known enzymes with which it can be confused.

In addition to these enzymological difficulties, we recognized a biological feature of the situation which influenced in at least one important detail the procedure we chose in the search for replicase. The point at issue may perhaps best be described in rather naive and admittedly somewhat anthropomorphic terms. Consider an RNA virus approaching a cell some 10^6 times its size and into which the virus is going to inject its only strand of genetic information. Even if the protein coated ribosomal RNA molecules are ignored, the cell cytoplasm still contains approximately 10,000 free RNA molecules of various sorts. If the new "replicase" was indifferent and replicated any RNA it happened to meet, *what chance would the single original strand injected have of multiplying?*

Admittedly, there are several ways out of this dilemma. One could, for example, segregate the new polymerase molecule and the viral RNA in some sequestered corner where they would be isolated and undisturbed by the mass of cellular RNA components. However, we entertained the unique possibility that the virus is ingenious enough to design a polymerase which would recognize its own genome and ignore all other RNA molecules.

At the outset, of course, we did not know which solution had been adopted by the virus to solve this dilemma, or even if the dilemma were real. However, the possibility that it did exist, and that replicase selectivity might be the chosen solution, required that its implications not be ignored; *for, if true, their disregard would guarantee failure.* In particular, this view meant that we could not afford the luxury of employing any conveniently available RNA in the search for replicase. It demanded the use of viral RNA in all steps of the purification. Further, one might perhaps push the selectivity property to its ultimate pessimistic conclusion. If the cleverness of the replicase extends further, it might well be true that even a fragment of its own genome would not be recognized and accepted for replication. This added possibility made it necessary to provide a guarantee that the RNA employed is not only homologous, but also intact. This in turn would introduce the complication that stages of purification preceding the removal of ribonuclease might well yield ambiguous or indeed false clues even with intact homologous RNA. Thus, one would have to "fly blind" initially and depend on very brief assays to provide the guides for the direction of the subsequent steps.

Despite all these potential obstacles, many of which were actually realized, our first success was achieved in 1963 (Haruna et al.[9]). A procedure involving negative protamine fractionation and column chromatography yielded what looked like the relevant enzyme for *E. coli* infected with the RNA bacteriophage, MS-2. Most important of all, the preparation exhibited a virtually complete dependence on added RNA, permitting a test of the expectation of specific template requirement. The response of MS-2 replicase to various kinds of nucleic acid revealed a striking preference for its own RNA. No significant activity was observed with either the host s-RNA or ribosomal RNA. Our guess was apparently confirmed. By producing a polymerase which ignores the mass of preexistent cellular RNA, a guarantee is provided that replication is focused on the single strand of incoming viral RNA, the ultimate origin of progeny.

CONFIRMATION OF SPECIFIC TEMPLATE REQUIREMENTS OF RNA REPLICASES

Our line of reasoning would lead to the expectation that RNA replicases induced by other RNA viruses would show a similar preference for their homologous templates. This was, however, not a foregone conclusion, since it was conceivable that other viruses might evolve different solutions to the problem of preferential synthesis. It seemed important to determine whether template selectivity could be observed in another virus unrelated to MS-2. The $Q\beta$ phage of Watanabe[10] was chosen because of its serological and other chemical differences (Overby et al.[11,12]).

TABLE 1. RESPONSE OF $Q\beta$ REPLICASE TO DIFFERENT TEMPLATES
(Haruna and Spiegelman[13])

As in all cases, assay for DNA dependent activity is carried out at 10 μg of DNA per 0.25 ml of reaction mixture. Control reactions containing no template yielded an average of 30 cpm. Numbers represent counts per minute (cpm) incorporated as detailed in Fig. 1.

Template	Input levels of RNA	
	1 μg	2 μg
$Q\beta$	4929	4945
TYMV	146	312
MS-2	35	26
Ribosomal RNA	45	9
s-RNA	15	57
Bulk RNA from infected cells	146	263
Satellite virus	61	51

DNA (10 μg) 36.

The isolation and purification of the $Q\beta$ replicase (Haruna and Spiegelman[13]) followed, with slight modifications, the procedures worked out earlier for the MS-2 replicase. The general properties of the $Q\beta$ replicase were similar to those observed with MS-2 replicase, including requirements

for all four triphosphates, and Mg^{++}. The abilities of various RNA molecules to stimulate the Qβ replicase to synthetic activity at saturation concentrations of homologous RNA, are recorded in Table 1. The response of
the Qβ replicase is in accord with that reported for the MS-2 replicase, the
preference being clearly for its own template. The only heterologous RNA
showing detectable activity is TYNV and it supports a synthesis corresponding to 3% of that observed with homologous Qβ RNA. Both of the heterologous viral RNAs, MS-2 and STNV, are completely inactive, and again, so
are the ribosomal and transfer RNA species of the host cell. It is important
to note that as in the case of MS-2 replicase, the absence of response to DNA
shows that our purification procedure eliminates detectable evidence of
transcriptase from our enzyme preparations.

RECOGNITION OF TEMPLATE INTACTNESS

We now focus attention on the discriminating selectivity displayed by the
two replicases in their response to added RNA. This is a phenomenon of
interest in itself, since it presents us with a system illustrating one of the
central unresolved issues of modern molecular biology, viz. the basis of
specific interaction between a protein and a nucleic acid. More immediately,
an understanding of this phenomenon is a necessary prerequisite to the
design of adequate experiments aimed at the mechanism of replicase action.

An obvious device to explain the specificity exhibited would invoke the
recognition of a beginning sequence, a possibility open to the simple test of
challenging the replicase with fragmented preparations of homologous RNA.
If the initial sequence is the sole requirement, RNA fragmented to half and
quarter pieces should serve adequately as templates. One would expect that
the initial rate with fragments would be the same as with intact templates,
although the reaction might terminate sooner. A different response would,
however, be predicted if a secondary structure requiring intact molecules
was involved in the recognition mechanism.

Fragments of Qβ RNA are readily obtained (Haruna and Spiegelman[14])
and fractionated for size on sucro segradients. The sedimentation profiles of
three such fragments are shown in the inset of Fig. 1. The first is the intact viral
RNA (28S), the second a half piece with a mean of about 17S, and the third
possesses a sedimentation coefficient of 7S. The response of replicase to the
three RNA preparations is shown in Fig. 1. It is obvious that the fragmented material is unable to stimulate the replicase to anywhere near its
full activity no matter how much is added. The half piece achieves approximately 10% the rate of the intact strand, whereas the rate attainable with
the quarter piece is only 2% of normal.

The inability of the replicase to properly employ fragments of its own
genome as templates argues against a recognition mechanism involving only
a beginning sequence. The enzyme can apparently sense when it is con-

fronted with an intact RNA molecule, implying that some element of secondary structure is involved. A plausible formal explanation can be proposed in terms of "functional circularity". Thus, a decision on the intactness of a linear heteropolymer can be readily made by examining both ends for the proper sequences. An examination of this sort would be physically aided by forming a circle using terminal sequences of overlapping complementarity.

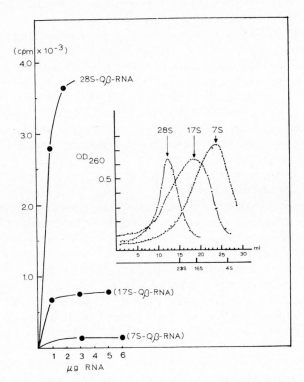

Fig. 1. Response of replicase to intact and fragmented $Q\beta$ RNA. Inset represents distribution in a linear sucrose gradient (2.5–15%) of RNA prepared and fractionated. Each preparation was run in a separate bucket for 12 hr at 25,000 rpm and 4°C in a Spinco SW-25 rotor with markers (ribosomal RNA) to permit S-value determinations (Haruna and Spiegelman[14]).

The enzyme could then recognize the resulting region of double-strands. This particular model is offered only as an example of how an enzyme could simultaneously distinguish both sequence and intactness. Whatever the details turn out to be, it appears that the RNA replicases are designed to minimize the futility of replicating either foreign sequences or incomplete copies of their own genome.

AUTOCATALYTIC SYNTHESIS OF A VIRAL RNA

The experimental analysis of a replicating reaction centers necessarily on the nature of the product. If, in particular, the concern is with the synthesis of a viral nucleic acid, data on base composition and nearest neighbors, while of interest, are hardly decisive. The ultimate issue is whether or not replicas are in fact being produced. To answer this question, information on the sequence of the synthesized RNA is required. Affirmative evidence of similarity between template and product would provide assurance that the reaction being studied is, indeed, relevant to an understanding of the replicative process.

The ability of a replicase to distinguish one RNA sequence from another can be used to provide information pertinent to the similarity question. Two sorts of readily performed experiments can decide whether the product is recognized by the enzyme as a template. One approach is to examine the kinetics of RNA synthesis at template concentrations, which start below those required to saturate the enzyme. If the product can serve as a template, a period of autocatalytic increase of RNA should be observed. Exponential kinetics should continue until the product saturates the enzyme, after which synthesis should become linear.

A second type of experiment is a direct test of the ability of the synthesized product to function as an initiating template. Here a synthesis of sufficient extent is carried out to insure that the initial input of RNA becomes a quantitatively minor component of the end product. The synthetic RNA can then be purified and examined for its template functioning capacities, a property readily examined by means of a saturation curve. If the response of the enzyme to variation and concentration of product is the same as that observed with the viral RNA, one would have to conclude that the product generated in the reaction is as effective a template for the replicase as is RNA of the mature virus particle.

Preliminary experiments established that $40\,\gamma$ of enzyme protein was saturated by approximately $1\,\gamma$ of Qβ RNA. An experiment was therefore performed (Haruna and Spiegelman[15]) in which the ratio of input template to protein was $\frac{1}{5}$ of the saturation value. The results are plotted in Fig. 2 arithmetically and semilogarithmically against time to permit ready comparison of kinetics. Exponential increase of RNA is evident over a period of approximately 3 hr. The arrows indicate the time at which the kinetics depart from the exponential and become linear. Extrapolation to the ordinate indicates that the change to linear synthesis occurs when approximately $1\,\gamma$ of RNA has accumulated.

The results just described are consistent with the implication that the product produced in the course of the reaction can serve to stimulate new enzyme molecules to activity. The enzyme is therefore able to recognize the product as being one which is homologous to its own genome.

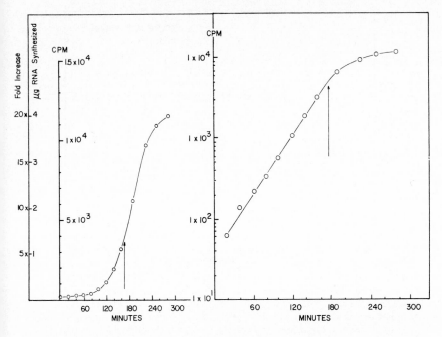

FIG. 2. Autocatalytic synthesis of viral RNA. A 2.5 ml reaction mixture contained 400 µg of enzyme protein and 2 µg of input Qβ RNA so that the starting ratio of template to enzyme was one-fifth of the saturating level. At the indicated times, 0.19 ml aliquots were removed and assayed for radioactive RNA. The ordinates for cpm and µg of RNA synthesized refer to that found in 0.19 ml samples. The data are plotted against time arithmetically on the right, and semilogarithmically on the left. The arrows indicate change from autocatalytic to linear kinetics (Haruna and Spiegelman[15]).

To carry out the more direct test of this conclusion, a 1 ml reaction mixture was set up and the synthesis allowed to proceed for 3.5 hr by which time a more than 60-fold increase of the input material was achieved. The reaction was then terminated and the RNA purified by the phenol method which yielded 55% of the synthesized product. Examination in a sucrose gradient showed that a majority of the product had the 28S size characteristic of Qβ RNA. Figure 3 illustrates the response of the replicase to various input levels of the product (triangles) compared to the original viral RNA levels (circles). It is evident that the RNA synthesized is as effective in serving as a template as the original viral RNA.

The data just summarized support the assertion that the reaction generates a polynucleotide of the same molecular weight (1×10^6) as viral RNA and which the replicase cannot distinguish from its homologous genome. Evidently, the enzyme is faithfully copying the recognition sequence employed by the replicase to distinguish one RNA molecule from another.

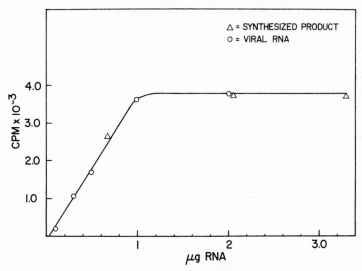

FIG. 3. Saturation of enzyme by synthesized RNA compared to viral RNA. The circles refer to the values obtained with RNA isolated from virus particles and the triangles to the rates obtained with the synthesized RNA. Since in the latter case, the template used was labeled with P^{32}, H^3-UTP at 1×10^6 cpm per 0.2 μmole was used to follow the synthesis (Haruna and Spiegelman[15]).

EVIDENCE FOR THE SYNTHESIS OF AN INFECTIOUS SELF-REPLICATING VIRAL RNA

The next question concerns the extent of the similarity between product and template. *Have identical replicas been in fact produced?* The most decisive test would be to determine whether the product contains all the information required to program the synthesis of complete virus particles in a suitable test system. The success we have just recorded encouraged an attempt at the next phase of the investigation which would subject the synthesized RNA to this more rigorous challenge. In the experiments to be described, all RNA preparations were first phenol treated prior to assay. Further, the phenol purified synthetic RNA was routinely tested for whole virus particles by assay on intact cells, and none were found in the experiments reported.

We now summarize experiments (Spiegelman *et al.*[16]) in which the kinetics of the appearance of new RNA and infective units were examined in two different ways. The first shows that the accumulation of radioactive RNA is accompanied by a proportionate increase in infective units. The second proves, by a serial dilution experiment, that the newly synthesized RNA is infective.

1. Comparison of the Kinetics of Appearance of RNA and Infectious Units

To compare the appearance of newly synthesized RNA with the presence of infectious units in an extensive synthesis, a reaction mixture was set up containing the necessary components at the concentrations required. Aliquots were taken at the times indicated for the determination of radioactive RNA and purification of the product for infectivity assay. Figure 4 shows the observed increase in both RNA and infectious units. The amount of RNA added is well below the saturation level of the enzyme present. Consequently,

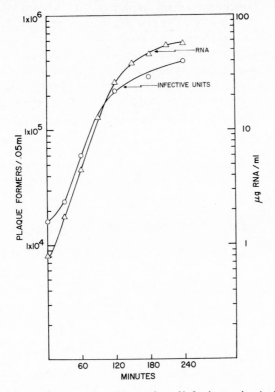

FIG. 4. Kinetics of RNA synthesis and formation of infectious units. An 8 ml reaction mixture was set up and samples were taken as follows: 1 ml at 0 time and 30 min, 0.5 ml at 60 min, 0.3 ml at 90 min, and 0.2 ml at all subsequent times. 20 λ were removed for assay of incorporated radioactivity. The RNA was purified from the remainder, radioactivity being determined on the final product to monitor recovery (Spiegelman et al.[16]).

the RNA increases autocatalytically for about the first 90 min followed by a synthesis which is linear with time. We note that the increase in RNA is paralleled by a rise in the number of infectious units.

2. Evidence for Infectivity of the Newly Synthesized RNA

The kind of experiment just described offers plausible evidence for infectivity of the newly synthesized radioactive RNA. It is not, however, conclusive, since the possibility is not eliminated that the agreement observed is fortuitous. One could argue, however, implausibly, that the enzyme is "activating" the infectivity of the *input* RNA while synthesizing new noninfectious RNA, and that the rather complex combination of exponential and linear kinetics of the two processes happen to coincide fortuitously.

Direct proof that the newly synthesized RNA is infectious can, in principle, be obtained by experiments which employ N^{15}–H^3 labeled initial templates to generate N^{14}–P^{32} product. The two can then, in principle, be separated in equilibrium density gradients of cesium sulfate. Such experiments have been carried out for other purposes and will be described elsewhere. However, the steepness of the cesium sulfate gradient makes it difficult to achieve a separation clean enough to be completely satisfying.

There exists, however, another approach which bypasses these technical difficulties by taking advantage of the biology of the situation and of the fact

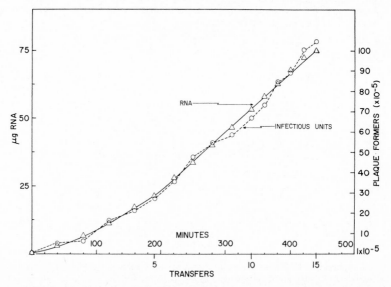

FIG. 5. RNA synthesis and formation of infectious units in a serial transfer experiment. Sixteen reaction mixtures of 0.25 ml were set up, each containing 40 γ of protein and the other components specified for the "standard" assay. 0.2 γ of template RNA were added to tubes 0 and 1; RNA was extracted from the former immediately, and the latter was allowed to incubate for 40 min. Then 50 λ of tube 1 were transferred to tube 3 and so on, each step after the first involving a 1 to 6 dilution of the input material. Every tube was transferred from an ice bath to the 35 °C water bath a few minutes before use to permit temperature equilibration. After the transfer from a given tube, 20 λ were removed to determine the amount of P^{32}-RNA synthesized, and the product purified from the remainder. Control tubes incubated for 60 min without the addition of the 0.2 γ of RNA showed no detectable RNA synthesis, nor any increase in the number of infectious units. All recorded numbers are normalized to 0.25 ml. The ordinates represent cumulative increases of infectious units and radioactive RNA in each transfer. The abscissa records elapsed time and the transfer number. Further details are to be found in Spiegelman *et al.*[16]

that we are dealing with a presumed self-propagating entity. Consider a series of tubes each containing 0.25 ml of the standard reaction mixture, but no added template. The first tube is seeded with 0.2 γ of Qβ RNA and incubated for a period adequate for the synthesis of several γ of radioactive RNA. An aliquot (50 λ) is then transferred to the second tube, which is in turn permitted to synthesize about the same amount of RNA, a portion of which is again transferred to a third tube and so on.

If each successive synthesis produces RNA which can serve to initiate the next one, the experiment can be continued indefinitely, and in particular until the point is reached at which the initial RNA of tube one has been diluted to an insignificant level. In fact, enough transfers can be made to insure that the last tube contains less than one strand of the input primer. *If, in all the tubes including the last one, the number of infectious units corresponds to the amount of radioactive RNA found, convincing evidence is offered that the newly synthesized RNA is infectious.*

A complete account of such a serial transfer experiment will be found in Spiegelman *et al.*,[16] and Fig. 5 describes the outcome. Apart from controls, fifteen transfers were involved, each resulting in a 1 to 6 dilution. By the eighth tube, there was less than one infectious unit ascribable to the initiating RNA and the fifteenth tube contained less than one strand of the initial input. Nevertheless, every tube showed an increment in infectious units corresponding to the radioactive RNA found.

FURTHER PURIFICATION OF REPLICASE ACCORDING TO ITS SIZE AND DENSITY

The experiments just summarized established that activation of the *input template* cannot be invoked to explain the results observed in the serial transfer experiment. They did not, however, eliminate arguments which involved "activation" of RNA, contaminating the enzyme preparation by an unknown reaction which requires both added template and new RNA synthesis. It seemed desirable to eliminate this possibility by further purification of the enzyme by procedures which would eliminate nucleic acid contaminants. In addition to this theoretical reason, there was a technical advantage to be gained. The *in vitro* serial transfer experiments which indicated that the newly synthesized RNA was a self-propagating and biologically competent entity, required infectivity assays of the reaction mixtures. A technical complication was introduced by the presence of viable virus particles in the replicase preparation. Their chemical contribution to the RNA content was trivial compared with the amounts synthesized. However, because complete viruses have a far greater infective efficiency than free RNA, even moderate contamination with intact particles cannot be tolerated in material being assayed for infectious RNA. To obviate this difficulty, all synthesized products were purified with phenol and checked for whole virus particles prior to measurement of infectivity. This made the assay both laborious and cumbersome, precluding its use in a variety of potentially interesting studies.

An obvious approach to the separation of virus particles from replicase would take advantage (Pace and Spiegelman[17]) of expected disparities in size and density. The Qβ virus has a molecular weight of 4.2×10^6 and a density of 1.43 g/cm^3. It was unlikely that the replicase would be as large or as dense.

1. Pycnographic Purification

The results of banding enzyme protein in a CsCl gradient are presented in Fig. 6. Virus particles are found at a density corresponding to 1.41 g/cm^3 and replicase bands at 1.26 g/cm^3, somewhat less dense than bulk protein. Approximately 10^{-6} of the original phage contamination was retained in the bulk protein peak, and repeated banding of the protein in CsCl failed to reduce the contamination substantially.

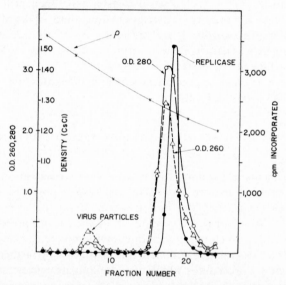

Fig. 6. Behavior of enzyme preparations during CsCl centrifugation. 1 ml of 10% ammonium sulfate in standard buffer (0.01 M tris, pH 7.4; 0.005 M MgCl$_2$; 0.0005 M 2-mercaptoethanol) containing 10–15 mg of post-DEAE enzyme protein was layered over 4 ml of CsCl solution adjusted to a density of 1.40 g/cm^3. The tube was centrifuged at 0 °C for 24 hr in the Spinco SW-39 rotor at 39,000 rpm. After centrifugation, the tube was pierced through the side, immediately below the visible protein band, and fractions were collected through a 20 gauge hypodermic needle. The needle was bent in a right angle to permit insertion with the long arm up, following which the latter is rotated through 180° to permit the contents to emerge. The lower portion of the tube was then removed by piercing the bottom and collecting fractions. In this manner, contamination of the protein band (ρ = 1.40) was minimized. Fractions were then diluted to 1 ml with 10% ammonium sulfate in standard buffer for optical density measurements and enzyme assays (Pace and Spiegelman[17]).

2. Sedimentation Purification

The residual virus contamination of the enzyme purified by banding was lowered to acceptable levels by sedimentation through linear gradients of sucrose. Figure 7 shows the sedimentation profile of replicase activity, with

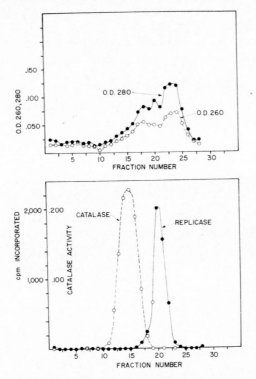

FIG. 7. Sedimentation through sucrose gradients. The peak tubes from the enzyme regions of CsCl gradients were pooled and precipitated from 50% saturated ammonium sulfate. The precipitate was dispersed in standard buffer to a protein concentration of 25–30 mg/ml and the suspension dialyzed for 2 hr against 250 ml of standard buffer at 0 °C. Between 0.2 and 0.3 ml of the dialyzed protein solution was layered onto a 5.4 ml 5% to 20% linear gradient of sucrose dissolved in standard buffer. When a reference protein was desired, 100 γ of catalase was included in each 0.2 ml of protein solution. Gradients were centrifuged at 0°C for 12–15 hr at 39,000 rpm in the Spinco SW-39 rotor. Again, to avoid contamination with pelleted virus particles and other material, the enzyme region was collected from the side as described for CsCl centrifugation (Pace and Spiegelman[17]).

catalase included as a reference. After 12 hr of centrifugation, infectious virus particles are found as a pellet in the tube. It will be noted that replicase activity sediments as a single peak. Employing catalase (molecular weight, 2.5×10^5) as a standard, the molecular weight of replicase may be estimated at 1.1×10^5.

The final residue of phage contamination is such that only one to ten plaque-forming units are introduced into a standard reaction mixture, a level which would not be detected by our usual sampling procedure. Synthesis routinely involves the appearance of 0.2 to 20 γ of new RNA, corresponding to 10^4 to 10^6 infectious units per reaction mixture under our assay conditions. The levels of mature phage contamination are clearly far below detectability.

It was now necessary to see whether an enzyme carried through the CsCl and sucrose purification steps could produce infectious RNA. A reaction

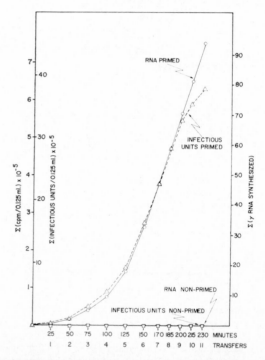

FIG. 8. Synthesis of RNA and infectious units by purified enzyme. Eleven reaction mixtures of 0.125 ml, each containing 22 μg of enzyme purified through CsCl and sucrose centrifugation and the other components of the standard reaction mixture were prepared. Specific activity of P^{32}-UTP was such that 8000 cpm signified 1 μg of RNA product. To the first tube was added 0.1 μg $Q\beta$ RNA, and the reaction was allowed to proceed for 25 min at 35 °C, whereupon 0.02 ml were withdrawn for counting and 0.01 ml used as template for the second reaction. The first tube was then frozen and stored at −70 °C. The second reaction product was used to initiate the third and so on. A second series of transfers were carried out in a manner identical to that described, save the *no* initial RNA template was added to the first reaction mixture. Aliquots of all the reaction mixtures were directly assayed for infectious units. In the case of the control transfer series, samples were diluted $\frac{1}{10}$ and then mixed with protoplasts. All other samples were adjusted to 0.2 to 0.8 μg RNA product/ml before mixing with protoplasts (Pace and Spiegelman[17]).

was initiated at less than primer saturation, and synthesis was allowed to proceed. After a suitable interval, 1/12.5th of this reaction was used to initiate a second one. Again, after permitting adequate synthesis, this reaction was used to initiate a second one. Again, after a period of synthesis, a similar aliquot of the second tube was employed as the template for a third reaction, and so on for ten transfers. Figure 8 gives the results in terms of cumulative increments. Since the reaction in the first tube was started with 6×10^{10} strands (0.1 γ) of Qβ RNA and each transfer involves a 1/12.5 dilution, the contribution of the initiating RNA to the infectious centers measured in the fourth tube would be below the level of detectability, and this tube contained 2.4×10^5 infectious units. Finally, the eleventh reaction contains *less than one strand of the initial primer*, and at the same time showed 2.5×10^5 infectious units as determined by plaques formed in the protoplast assay.

Clearly, the replicase purified by pycnography and sedimentation has retained its ability to produce biologically competent replicas.

There remains the task of a *rigorous* demonstration that the added RNA and not the replicase is the instructive agent in the replicative process. The discriminating selectivity of the replicase for its own genome as template makes it necessary to use mutants for this purpose. The required experiments are described in the next section.

A RIGOROUS PROOF THAT THE ADDED RNA IS THE SELF-DUPLICATING ENTITY

The central issue we now consider stems from the fact that two informed components are present in the reaction mixture, replicase and RNA template. None of the experiments thus far described *proved* that the RNA synthesized in this system is, in fact, a self-duplicating entity, i.e. one which contains the requisite information and directs its own synthesis. What is required is a rigorous demonstration that the RNA, and not the replicase, is the instructive agent in the replicative process. A definitive decision would be provided by an experimental answer to the following question: if the replicase is provided alternately with two distinguishable RNA molecules, is the product produced always identical to the initiating template?

A positive outcome would establish that the RNA is directing its own synthesis and simultaneously completely eliminate any remaining possibility of "activation" of preexisting RNA. Experiments to settle these issues were undertaken by Pace and Spiegelman.[18]

The discriminating selectivity of the replicase for its own genome as template makes it impossible to employ heterologous RNA in the test experiments and recourse was, therefore, had to mutants. For ease in isolation and simplicity in distinguishing between mutant and wild type, temperature sensitive (ts) mutants were chosen. Their diagnostic phenotype is poor growth at 41 °C as compared with 34 °C. The wild type grows equally well at both temperatures.

The data of Table 2 demonstrate that the ts-phenotype is easily recognized by parallel platings of intact virus particles at 34 °C and 41 °C on receptor cells. It remained, however, to see whether this difference would be retained when the corresponding purified mutant RNA preparations were assayed for infectivity in the protoplast system. This check is particularly necessary,

TABLE 2. RELATIVE EFFICIENCY OF PLATING AT 34 °C AND 41 °C
(Pace and Spiegelman[18])

Dilutions were plated with *E. coli* K 38 as the indicator organism, and duplicate plating series were incubated at 34 °C and 41 °C. The relative efficiency of plating (REOP) of 100 is defined relative to the plaque forming units (PFU) observed at 34 °C.

Virus		34°	41°
Qβ	REOP	100	100
	PFU	1.14×10^{13}/ml	1.16×10^{13}/ml
ts-Qβ	REOP	100	2.5×10^{-2}
	PFU	4.4×10^{7}/ml	1.1×10^{4}/ml

since one of the steps requires a 10-min incubation of the infected proto-plasts at 35 °C. During this interval "revertants" could be produced and add to the background of plaques developing at 41 °C. In addition, it was necessary to establish that the synthetic product of the replicase, primed by a normal Qβ RNA, behaves like the natural viral RNA in its behavior at 41 °C. Table 3 summarizes the results of the experiments performed to check these

TABLE 3. EFFICIENCY OF INFECTION OF PROTOPLASTS BY THREE RNA PREPARATIONS
(Pace and Spiegelman[18])

Infectious RNA assays were carried out on Qβ RNA, synthetic Qβ RNA, and ts-RNA. Duplicate pairs were incubated at 34 °C and 41 °C. Efficiencies at 34 °C are defined as 100. The synthetic Qβ RNA was the result of a 20-fold synthesis carried out by Qβ replicase purified through CsCl and sucrose centrifugation, using 0.1 µg Qβ RNA to initiate the standard reaction. REOP and PFU are as defined in Table 4.

RNA Species	REOP	34°	41°
Natural Qβ	REOP	100	93
	PFU	4.56×10^{5}/ml	4.24×10^{5}/ml
Synthetic Qβ RNA	REOP	100	92
	PFU	2.90×10^{6}/ml	2.66×10^{6}/ml
Natural ts-Qβ RNA	REOP	100	1.5
	PFU	1.86×10^{6}/ml	2.75×10^{4}/ml

points. It is evident that the synthetic wild type Qβ RNA behaves exactly like its natural counterpart at the two temperatures. On the other hand, the ts-Qβ RNA again shows the lower efficiency at 41 °C, although it will be noted that the background at 41 °C is higher than in the intact cell assay (Table 2) as was expected. The 65-fold difference at the two temperatures is, however, more than adequate for a clear diagnosis.

It is evident that the system available will permit us to determine whether the product produced by a normal replicase primed with ts-Qβ RNA is a mutant or wild type. As in previous investigations, this is best done by a serial transfer experiment to avoid the ambiguity of examining reactions containing significant quantities of the initiating RNA. Accordingly, seven standard reaction mixtures (0.25 ml) were prepared, each containing 60 μg of Qβ replicase isolated from cells infected with *normal* virus and purified through the CsCl banding and sucrose sedimentation steps. To the first reaction mixture was added 0.2 μg of RNA and synthesis allowed to proceed at 35 °C. After a suitable interval, $\frac{1}{10}$ of this reaction mixture was used to initiate a second reaction which, in turn, was diluted into a third reaction mixture, and so on for seven transfers. A *control* series was carried out in a manner identical to that just described, save that *no* RNA was added to the first tube. Aliquots from each reaction mixture were examined for radio-activity in TCA-precipitable material and assayed for infectious RNA at 34 °C and 41 °C.

Figure 9 summarizes the outcome of the experiment in a cumulative plot of the RNA synthesized and the plaque formers at the two test temperatures of 34 °C and 41 °C. It is clear that the RNA synthesized has the ts-phenotype, while plaque formation at 34 °C increases in parallel with the new RNA synthesized. No such increase is seen when the tests are carried out at 41 °C. It should be noted from the upper panel of Fig. 9 that no significant synthesis of either RNA or infectious units was observed in the unprimed control series of tubes.

The experiments described demonstrate that one and the same normal replicase can produce distinguishably different but genetically related RNA molecules. The genetic type produced is completely determined by the RNA used to start the reaction and is always identical to it. The following two conclusions would appear to be inescapable from these findings: (1) The RNA *is the instructive agent* in the replicating process and therefore satisfies the operational definition of a self-duplicating entity; (2) It is not some cryptic contaminant of the enzyme, but rather *the input RNA which multiplies.*

Finally, it should not escape the attention of the reader that the experiments described generate an opportunity for studying the genetics and evolution of a self-replicating nucleic acid molecule in a simple and chemically controllable medium. Of particular interest is the fact that such studies can be carried out under conditions in which the only demand made on the

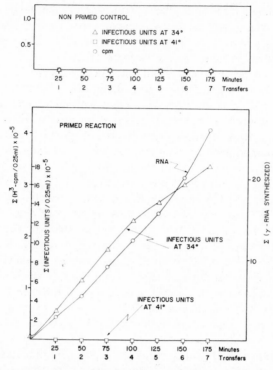

Fig. 9. Synthesis of mutant RNA. Each 0.25 ml reaction contained 60 μg of Qβ replicase purified through CsCl and sucrose centrifugation. The first reaction was initiated by addition of 0.2 μg of ts-RNA. Each reaction was carried out at 35 °C for 25 min, whereupen 0.02 ml was withdrawn for counting and 0.025 ml used to prime the next reaction. All samples were stored frozen at −70 °C until infectivity assays were carried out at 41 °C and 34 °C. A control series was carried out in which no initiating RNA was added (Pace and Spiegelman[18]).

molecules is that they multiply; they can be liberated from all secondary requirements (e.g. coding for coat protein, etc.) which serve only the needs and purposes of the complete organism.

AN ANALYSIS OF THE MECHANISM OF RNA REPLICATION WITH THE PURIFIED ENZYME

It is evident from the experiments described that the properties of the enzyme system have been brought to a stage which encouraged an attempt at an analysis of the mechanism of the synthesis. Since the reaction starts with viral RNA (plus strands) and ends with more of the same, every necessary intervening stage must be represented in the reaction mixture. A search was made (Haruna and Spiegelman[19]) during the early events of synthesis for evidences of intermediate forms involving complements (minus strands), a possibility implied by experiments with mutants (Lodish and Zinder[20]) and suggested by the presence of structures partially resistant to ribonuclease first observed with an animal virus (Montagnier and Sanders[21]) and also seen in cells infected with RNA bacteriophages.[22-28] The presence of complements was examined for by annealing experiments to plus strands. Advantage was also taken of the base composition of Qβ RNA in which A is not equal to U, making it possible to detect minus strands in terms of complementary A to U ratios of the early product. The base composition showed no tendency to linger at the complementary A/U value. Further, the extent of annealability to plus strands could be as readily explained in terms of partial copies of a template possessing some degree of self-complementarity. The data forced us to the conclusion that we could not provide unambiguous support for the intervention of a duplex composed of a plus strand and its complement.

However, such negative evidence cannot logically eliminate complementary copying mechanisms. More or less obvious complications could prevent precise interpretation of the data. Thus, only a minor fraction of the input templates might have been involved in the reaction. Further, the phenol method usually used to prepare samples rarely yields complete recovery and could introduce artifacts or selectively eliminate components of interest. Finally, base compositions and annealabilities of early products would not readily detect complementary copying involving transient complements. Similarly, they might not reveal complement sparing mechanisms in which a slow synthesis of a negative is accompanied by the immediate initiation of several positive strands.

We decided to invert our approach to the problem and focus attention on the early fate of the initiating templates. The pycnographic and sucrose density purification of the replicase made it possible to do direct infectivity assays of the reaction mixture, thus permitting a detailed analysis of any interval of synthesis. In addition we had developed[29] a simple treatment of the reaction mixture which avoids the phenol step and permits direct examination in sucrose gradients of templates and product with complete recovery of both.

With the aid of these technical advances, we were able to establish[30] the existence of a *latent period* in the reaction which precedes the appearance of new infectious RNA. A search during this latent period should maximize the chances of finding replicative complexes if they are mandatory intermediates. Further, at least some of these complexes should be non infectious and release infectious plus strands on heat denaturation as reported by Hofschneider and his colleagues.[28] Finally, the formation of these complexes should be accompanied by a disappearance of the input templates as infectious entities. The relevant experiments along these lines are now summarized.

TABLE 4. A COMPARISON OF THE SYNTHESIS OF RNA AND INFECTIOUS UNITS

A 2.00 ml standard reaction mixture contained in μmoles: Tris-HCl, pH 7.4; 160; $MgCl_2$, 24; ATP, CTP, GTP, UTP, 1.6 each; UTP-α-P^{32} to 4.05×10^7 cpm 0.2 μM UTP; 1.6 γ Qβ H^3-RNA (preparation 311); 400 γ enzyme protein was allowed to proceed at 35 °C. Zero time is represented by the addition of template to the prewarmed mixture. Column 1 indicates times at which samples were drawn. Column 2 presents P^{32} cpm in TCA precipitable product in each 0.25 ml. Column 3 gives acid precipitable H^3 cpm in template, and the calculated percent H^3-template remaining are given in column 4. Column 5 lists the amount of RNA product per 0.25 ml, calculated from column 2, where 1.62×10^5 cpm equals 1 γ RNA. Column 6 presents the number of infectious units per 0.25 ml observed in the spheroplast assay (Mills, Pace and Spiegelman[30]).

Time (min)	P^{32}-product	H^3-template	% H^3-template remaining	γ RNA synthesized 0.25 ml	Infectious units $\frac{}{0.25\,ml \times 10^{-5}}$
$\frac{1}{4}$	27	78,800	(100)	0	1.40
1	8530	82,100	104	0.053	0.73
4	28,700	74,100	94	0.177	0.35
6	81,550	81,500	103	0.504	0.45
8	139,200	73,600	94	0.859	1.32
12	343,000	75,300	96	2.118	6.22
16	540,000	78,400	99	3.332	9.35

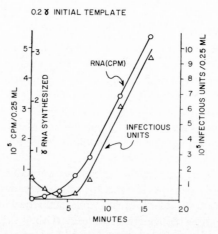

FIG. 10. Appearance of RNA and infectious units. Data are from columns 2 and 6 of Table 4 (Mills, Pace and Spiegelman[30]).

1. Kinetics of Appearance of RNA and Infectious Units

The possibility of performing direct assays for infectious RNA permits a detailed comparison of the appearance of mature plus strands with the amounts of RNA synthesized. To distinguish between initiating template and product, the former was labeled with H^3 and the latter with P^{32}. To maximize the detection of a complex involving template, the experiment was performed at low levels of initiating RNA. In addition, freshly prepared viral RNA was used for reasons which will be discussed later.

Conversion of templates into Hofschneider structures (HF) should result in their disappearance as plaque forming units (PFU). Thus, if the time required to complete the first plus strands is significant in the time scale of the experiment, one should observe a *latent period* accompanied by an *apparent eclipse* of measurable infectious RNA.

Table 4 and Fig. 10 record the data which show that these expectations are realized. It is evident that a considerable loss ($\sim 75\%$) of PFU is observed by the fourth minute of synthesis. Note further (Table 4, column 3) that the loss is not accompanied by disappearance of the H^3-template from the acid precipitable fraction. Figure 10 provides clear evidence of the eclipse and latent periods in PFU. Both end at about 6 min which appears to correspond to the time required to complete the first new mature strands. The 15-sec sample shows little evidence of synthesis of product or eclipse of template. At 2 min the RNA synthesized corresponds to 27% of the input and at that time 48% of the PFU present at 15 sec have disappeared. At 4 min an amount equivalent to 80% of the input RNA has been synthesized and 75% of the PFU are missing. By the sixth minute, when 25 times the input RNA has been formed, the reaction shows signs of emerging from the latent period.

2. Sucrose Gradient Analysis of the Reaction

If the disappearance of PFU is associated with the formation of HF, a peak containing H^3-template and P^{32}-product should appear in the 15S region of a sucrose gradient. Since, with time, the loss of PFU is extensive, the shift of H^3 to the 15S region should be considerable by 2 min but negligible at 15 sec. Further, this peak should yield infectious material only after heat denaturation. Accordingly, aliquots of the samples taken in the experiment of Fig. 10 at 15 sec, 2, 4, and 6 min were subjected to analysis in sucrose gradients.

The 15 sec sample (Fig. 11) shows that virtually all of the H^3-template still is found in the 28S position with little evidence of distortion. The number of PFU found at 15 sec agrees with the initial input of RNA and the amount of P^{32} incorporated is negligible, suggesting that the eclipse of PFU is associated with the synthetic reaction. The 2-min sample (Fig. 12) shows a dramatically different picture. A rather large proportion (62%) of the H^3-template has been shifted from the 28S to the 15S region, a movement which is accompanied by the appearance of P^{32}-product which is virtually confined

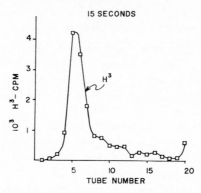

FIG. 11. Sedimentation analysis of the 15-sec sample. 0.20 ml of a 60-fold dilution of the 15-sec sample after digestion with pronase and SDS was submitted to sedimentation in sucrose as described in methods. All H³-cpm were adjusted to be equivalent to examining 0.05 ml of the reaction. No acid precipitable P³²-cpm were observed (Mills, Pace and Spiegelman[30]).

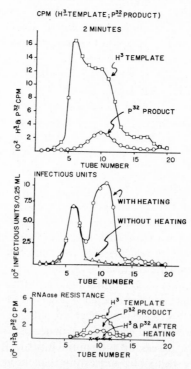

Fig. 12. Sedimentation analysis of the 2-min sample. 0.25 ml of reaction mixture was examined in a sucrose gradient as described in methods. 0.25 ml fractions were collected and diluted 5-fold, and aliquots were taken for P³²-and H³-cpm, for assay of infectious units, and for ribonuclease digestion. All cpm are per 0.25 ml of dilution. Infectious units are per 0.25 undiluted sample. In tube 11, H³ resistance is 31%, and P³² resistance is 49% (Mills, Pace and Spiegelman[30]).

to the 15S region. Assays for infectious RNA *without* prior heating yields PFU *only* in the region of mature plus strands (28S). However, heat denaturation uncovers a large peak of activatable PFU at 15S. Finally, it is evident that both the tritiated template and the P^{32}-product in the 15S region show resistance (about 50%) to ribonuclease. This resistance disappears completely when the samples are heated for 2 min at 100°C in 0.003 M EDTA.

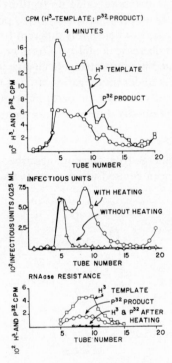

FIG. 13. Sedimentation analysis of the 4-min sample. Details are as in Fig. 3. Ribonuclease resistance in tube 10 is 41% for H^3 and 39% for P^{32} (Mills, Pace and Spiegelman[30]).

The 4-min sample (Fig. 13) shows the presence of product (P^{32}-RNA) which is almost all 28S. Again, we see the presence of both tritiated template and P^{32}-product in the 15S region and a corresponding peak of heat activatable plaque-forming units. Further, there is clear evidence again of resistance to ribonuclease of both template and product. The 6-min sample (Fig. 5) shows the same general pattern, the only difference being that some 28S-P^{32} product has clearly been produced.

3. A Requirement for Freshly Prepared Template

We ran into a surprisingly subtle demand on the nature of the template employed. We give as an example (Fig. 15) an experiment similar to that of

Fig. 10 carried out with another RNA preparation (#305) but using the same enzyme and the identical initiating input of 0.2 γ. A latent period exists which is slightly longer than that observed with the RNA preparation # 311 used in the experiment of Fig. 1. There is, however, almost no detectable eclipse. Examination of # 305 at an even lower level of template input (0.1 γ) again failed to yield a significant eclipse. Centrifugation of preparation # 305 in sucrose gradients revealed no features which distinguished it from # 311. Further, the infective efficiency (PFU per γ of RNA) of # 305 in the protoplast assay was not detectably different from that of # 311. The reason for this apparent decrease in ability to form non infectious complexes which is not accompanied by any detectable change in either sedimentation or biological efficiency is under investigation. It is of some practical interest to note that when originally isolated, preparation # 305 showed an excellent eclipse which was lost gradually during several months storage at −70°C.

4. Implications for Mechanism

The striking fact to emerge from the data described is that a set of experimental conditions has been devised which permits the *in vitro* synthesis of a

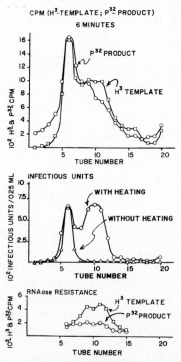

FIG. 14. Sedimentation analysis of the 6-min sample. Details were as in Fig. 3. Ribonuclease resistance in tube 11 is 48% for H^3 and 30% for P^{32} (Mills, Pace and Spiegelman[30]).

structure found in the infected cell. It has all the characteristics of HF including the key one of yielding infectious plus strands on denaturation. The relation of HF to the form recently reported by Franklin[31] remains to be elucidated. The latter is difficult to distinguish in sucrose gradients and other methods are being investigated to separate it from the 28S mature viral RNA.

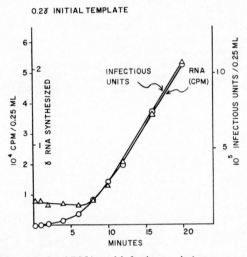

FIG. 15. A kinetic analysis of RNA and infectious unit Appearance with Qβ RNA ╪ 305. A 3-fold standard reaction as in Table 4 was initiated with 0.20 γ of ╪ 305 per 0.25 ml reaction. UTP-α-P³² was added to 6.58 × 10⁶ cpm 0.2 μM UTP. Samples were drawn at intervals and analyzed for cpm and infectious units after digestion with pronase and SDS as described in Methods (Mills, Pace and Spiegelman[30]).

It is clear from the results described (Figs. 10–15) that HF has the following relations to the evolution of the synthetic reaction. It contains both initiating H³-template and newly synthesized P³²-product. Further, the appearance of HF as a component in the latent period is accompanied by loss of PFU on direct assay. Finally, it is synthesized before any new 28S product is made and prior to emergence from the latent period. While these are all *necessary* characteristics of a replicative intermediate, *they are not logically sufficient to establish that HF is playing this role.*

Direct attempts to see whether HF can substitute for plus strands as an initiator of replicase activity have not yielded any signs of template activity until the plus strands are released by heat denaturation. However, such negative outcomes are again not decisive. We may be faced here with a problem analogous to the rethreading of a needle. Other approaches testing the functional capacities of HF in the enzyme reaction are being explored. At the moment firm conclusions are both premature and unnecessary. We can now synthesize this complex at will in any desired quantity and are in a position to design experiments to definitively illuminate both the structure and functions of HF.

It should now be possible to design and perform experiments which should definitively delineate the mechanism of RNA replication.

REFERENCES

1. B.D.HALL and S.SPIEGELMAN, *Proc. Nat. Acad. Sci.*, **47**, 137, 1961.
2. S.A.YANKOFSKY and S.SPIEGELMAN, *Proc. Nat. Acad. Sci.*, **48**, 1466, 1962b.
3. R.H.DOI and S.SPIEGELMAN, *Science*, **138**, 1270, 1962.
4. S.SPIEGELMAN and R.H.DOI, In Synthesis and structure of macromolecules. *Cold Spring Harbor Symposium on Quantitative Biology*, **28**, 109, 1963.
5. T.LOEB and N.D.ZINDER, *Proc. Nat. Acad. Sci.*, **47**, 282, 1961.
6. R.M.S.SMELLIE, In *Progress in Nucleic Acid Research*, vol. 1, p. 27. Ed. by J.N. Davidson and W.E.Cohn, Academic Press, N.Y., 1963.
7. M.GRUNBERG-MANAGO and S.OCHOA, *J. Am. Chem. Soc.*, **77**, 3165, 1955.
8. J.T.AUGUST, P.J.ORTIZ and J.HURWITZ, *J. Biol. Chem.*, **237**, 3786, 1962.
9. I.HARUNA, K.NOZU, Y.OHTAKA and S.SPIEGELMAN, *Proc. Nat. Acad. Sci.*, **50**, 905, 1963.
10. I.WATANABE, *Nihon Rinsho*, **22**, 243, 1964.
11. L.R.OVERBY, G.H.BARLOW, R.H.DOI, MONIQUE JACOB and S.SPIEGELMAN, *J. Bacteriol.*, **91**, 442, 1966.
12. L.R.OVERBY, G.H.BARLOW, R.H.DOI, MONIQUE JACOB and S.SPIEGELMAN, *J. Bacteriol.*, **92**, 739, 1966.
13. I.HARUNA and S.SPIEGELMAN, *Proc. Nat. Acad. Sci.*, **54**, 579, 1965.
14. I.HARUNA and S.SPIEGELMAN, *Proc. Nat. Acad. Sci.*, **54**, 1189, 1965.
15. I.HARUNA and S.SPIEGELMAN, *Science*, **150**, 884, 1965.
16. S.SPIEGELMAN, I.HARUNA, I.B.HOLLAND, G.BEAUDREAU and D.MILLS, *Proc.Nat.Acad. Sci.*, **54**, 919, 1965.
17. N.R.PACE and S.SPIEGELMAN, *Proc. Nat. Acad. Sci.*, **55**, 1608, 1966.
18. N.R.PACE and S.SPIEGELMAN, *Science*, **153**, 64, 1966.
19. I.HARUNA and S.SPIEGELMAN, *Proc. Nat. Acad. Sci.*, **55**, 1256, 1966.
20. H.LODISH and N.ZINDER, *Science*, **152**, 372, 1966.
21. L.MONTAGNIER and F.K.SANDERS, *Nature*, **199**, 664, 1963.
22. J.AMMANN, H.DELIUS and P.H.HOFSCHNEIDER, *J. Mol. Biol.*, **10**, 557, 1964.
23. C.WEISSMANN, P.BORST, R.H.BURDON, M.A.BILLETER and S.OCHOA, *Proc. Nat. Acad. Sci.*, **51**, 890, 1964.
24. R.B.KELLY and R.L.SINSHEIMER, *J. Mol. Biol.*, **8**, 602, 1964.
25. M.NONOYAMA and Y.IKEDA, *J. Mol. Biol.*, **9**, 763, 1964.
26. M.L.FENWICK, R.L.ERIKSON and R.M.FRANKLIN, *Science*, **146**, 527, 1964.
27. S.SPIEGELMAN and I.HARUNA, *J. Gen. Physiol.*, **49**, 263, 1966.
28. B.FRANCKE and P.H.HOFSCHNEIDER, *J. Mol. Biol.*, **16**, 544, 1966.
29. N.R.PACE and S.SPIEGELMAN, In preparation, 1966.
30. D.MILLS, N.PACE and S.SPIEGELMAN, *Proc. Nat. Acad. Sci.*, **56**, 1778, 1966.
31. R.M.FRANKLIN, *Proc. Nat. Acad. Sci.*, **55**, 1504, 1966.

THE MOLECULAR STRUCTURES AND FUNCTIONS OF RNA POLYMERASE OF *ESCHERICHIA COLI*

T. Kameyama, Y. Kitano, H. Kawakami, Y. Iida,
S. Murakami, Y. Tanaka and A. Ishihama

Biochemistry Division, Institute of Cancer Research, School of Medicine, Kanazawa University, Kanazawa, Japan

BASED on the experimental results reported by many investigators, it has been convincingly emphasized that DNA-dependent RNA polymerase may play a central role in the transcription of genetic information. Considering the molecular mechanism of the enzymic synthesis of RNA, catalyzed by DNA-dependent RNA polymerase, two unique properties, whereby genetic informations are transcribed, seem to be essential.

First, since genetic information is expressed by a certain unit, an operon or a cistron, an enzyme molecule must recognize a binding site on a primer DNA if the polymerase is the functional entity as transcriptase. Thus the first problem may arise: what is the molecular base of the recognizing mechanism? Second, the enzyme molecule bound on the site transcribes a linear sequence of nucleotides from DNA to RNA. Since an asymmetric selective reading has been proven to take place *in vitro*[1-4] as well as *in vivo*,[5] the enzyme molecule on DNA at the binding site must determine which strand should be selectively transcribed. Now the second problem may arise: what is the molecular base of the mechanism of asymmetric transcription? These two problems have been so far unsolved, but are essential.

Several hypotheses may explain these problems. However, any one of the hypotheses will not stand up against critical scrutiny, until more precise knowledge of the molecular properties of RNA polymerase is obtained.

We have attempted to investigate extensively the molecular properties of RNA polymerase and their relation to biological function. This paper deals first with a pertinent new device for the purification of RNA polymerase without any ambiguities in respect to multiple components of active polymerase and impurities. By this method we have succeeded in preparing, separately, 22S and 15S RNA polymerase from *E. coli* as highly purified proteins. The paper also reveals the molecular structure of the two types of RNA polymerase and the differences in kinetic profiles of the reactions catalyzed by both enzyme forms. Finally, the paper will propose a model whereby conformational change between the two types of enzyme molecules may be understood, and thus the regulation of transcription may be also considered.

(a) A New Device for Purification to Obtain Separately 22S and 15S RNA Polymerase

When RNA polymerase was prepared from *E. coli* according to the method of Chamberlin and Berg,[6] two active enzyme fractions were detected in the sedimentation pattern of a sucrose density gradient centrifugation, as shown in Fig. 1. The sedimentation coefficients of both active polymerase fractions were determined as 22S and 15S respectively, referring to catalase

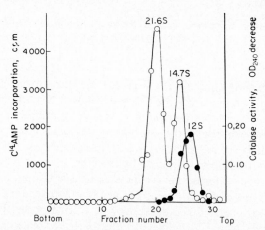

Fig. 1. Sedimentation pattern of RNA polymerase. 0.1 ml of DEAE-cellulose eluate according to the method of Chamberlin and Berg[6] was applied onto sucrose density gradient medium (5–20%) in 50 mM Tris-HCl (pH 7.6). 4.5 ml of total volume in a swinging roter (RPS-40) was centrifuged at 37,000 rpm for 4 hr. After centrifugation 34 fractions were collected; 0.14 ml/10 drops/tube. Using 0.01 ml of each fraction the polymerase assay was carried out. The volume of reaction mixture was reduced to 0.125 ml, thus containing a half of each ingredient as indicated in Table 2. 10-min incubation was used. ^{14}C-AMP incorporated plotted by —○—. For the assay of catalase, which was separately centrifuged as a reference under the same condition, the decrease in OD units at 240 mμ in a reaction mixture (3.3 ml), containing 10 mM potassium phosphate and 0.3% H_2O_2, was measured at 2 min by 0.01 ml of sample. The values are plotted by —●—.

activity (12S). However, the procedure did not consistently result in both enzyme forms being obtained. We have met with cases in which only one type of the enzyme was obtained by the original method. As summarized in Table 1, the results indicate the great differences in the distribution of active polymerase from one preparation to another.

Thus it is necessary to establish a new device for purification which makes it possible to obtain consistently both enzyme forms, and, further, to separate precisely individual enzymes from a single batch of cells. This may be useful to investigate the relationship between two enzyme forms and to answer the question: which is a true transcriptase?

TABLE 1. SEDIMENTATION COEFFICIENTS OF RNA POLYMERASE
(Chamberlin and Berg's Method)

Experiment	Active components				Inactive components	
I		22.2		14.6	7.3	
II				14.0	8.0	
III				14.4	8.9	
IV			19.7	12.4		4.9
V				13.2	6.5	
VI		21.8		13.7		3.8
VII				14.7	7.1	
VIII	30.4	22.8	18.0		6.8	

In the new device, the streptomycin-treatment in the original method has been omitted. And it must be pointed out that the most essential procedure is the stepwise elution from protamine precipitate with elevating concentrations of ammonium sulfate solution.

Frozen or freshly harvested cells of *E. coli* (wet weight: 180 g), grown up to an exponential phase in a glycerol-minimal salts medium, were suspended in an equal volume of buffer A (10 mM Tris-buffer, pH 7.9, 10 mM magnesium chloride, 10 mM β-mercaptoethanol and 0.1 mM EDTA) and disrupted through a French pressure cell. The extract was centrifuged for 20 min at 25,000 × g and the residue was washed with 70 ml of buffer A. After centrifugation, the resulting fluid was combined with the original supernatant. The combined fraction was adjusted to pH 7.8 with 1 M KOH solution and spun for 3 hr at 30,000 rpm in the RP-30 roter by the Hitachi model 40P preparative ultracentrifuge. The supernatant fluid was collected (crude extract).

To 220 ml of the supernatant were added 60 ml of 2% protamine sulfate (Yukigosei Co.) solution. This was neutralized with KOH just before use. After stirring for 30 min, the solution was centrifuged for 20 min at 25,000 × g. The precipitate was suspended in buffer A to 95 ml of the final volume. To the suspension were added 5 ml of 1 M ammonium sulfate solution with gentle stirring. After 30 min, the suspension was centrifuged and the resulting fluid was collected (0.05 M eluate). The precipitate, collected by centrifugation, was then resuspended in an adequate volume (60–80 ml) of buffer A and subjected to further stepwise elutions with 0.10, 0.15, and 0.20 M ammonium sulfate solution (at the final concentration).

Each eluate was subjected to purification by the usual fractionation steps with ammonium sulfate (0.42–0.53 saturation[7]), followed by DEAE-cellulose column (either linear gradient or stepwise elutions of KCl solution in buffer A), and finally by centrifugation in a glycerol gradient medium. The results are summarized in Table 2.

TABLE 2. PURIFICATION OF RNA POLYMERASE

	Method of Chamberlin and Berg		New method	
	Total activity (units)	Specific activity (units/mg protein)	Total activity (units)	Specific activity (units/mg protein)
1. Crude extract	232,000	8	310,000	10
2. Protamine eluate with				
0.05 M			37,000	20
0.10 M	36,000	110	190,000	140
0.15 M			170,000	170
0.20 M			58,000	90
3. Ammonium sulfate fractionation (0.42–0.53 sat.[7])	52,000	520	67,000[a] 86,000[b]	620 1360
4. Eluate from DEAE-cellulose column (0.25 M KCl)	40,000	2,400	53,000[a] 60,000[b]	1260 3100
5. Density gradient centrifugation				2830[a] 4050[b]

As a standard, the reaction mixture contained in 0.25 ml; ^{14}C-ATP (807 cpm/mμmole): 25 mμmoles, UTP/GTP/CTP: 100 mμmoles each, Tris-HCl (pH 7.8): 30 μmoles, magnesium acetate: 1.5 μmoles, manganese sulfate: 0.5 μmole, β-mercaptoethanol: 1.5 μmoles, E. coli DNA: 7.5 μg, and enzyme. The incubation was carried out at 37 °C for 5 min.

(a) and (b) are derived from 0.10 M eluate and 0.15 M eluate respectively.

0.1 M eluate and 0.15 M eluate were subjected to sedimentation analyses before further purification in order to determine whether the distributions of molecular species of RNA polymerase in both the eluates are identical of not. As shown in Fig. 2, almost complete separation of both active forms of RNA polymerase, 22S and 15S, was achieved by the stepwise elution at 0.10 M and 0.15 M, accompanying with a slight contamination of another active form. In addition, the sedimentation patterns do not essentially change even after the further purifications. The eluate with 0.05 M has given only 15S, whereas that with 0.20 M has given rather a broad peak at around 20–25S with much contamination of inactive proteins.

Consequently, it can be concluded that both molecular species of RNA polymerase, 15S and 22S, can be selectively eluted from the protamine precipitates with below 0.1 M and above 0.15 M ammonium sulfate solution respectively. The recoveries and specific activities of 15S and 22S are higher in 0.1 M eluate and 0.15 M eluate than in 0.05 M eluate and 0.20 M eluate respectively. From this reason 22S and 15S polymerase were usually prepared from 0.15 M eluate and 0.10 M eluate respectively. If all the eluates were pooled and followed by the further purification, the mixture of 22S and 15S polymerase was obtained.

22S and 15S preparations after final density gradient centrifugation in glycerol medium have exhibited 1.7–1.8 of OD ratio at 280 and 260 mμ and 3000–4000 units/mg protein of specific activity. (One unit is the enzyme

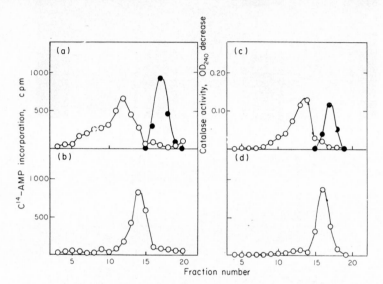

FIG. 2. Effect of stepwise elution on sedimentation patterns of RNA polymerase. 0.1 ml of each enzyme obtained by the new method was applied onto 4.4 ml of glycerol density gradient (10–35%) in Tris-HCl (10 mM, pH 7.8) containing 10 mM magnesium acetate and centrifuged at 37,000 rpm for 4 hr. After centrifugation, fractions were collected (0.2 ml/tube). Using 0.1 ml of sample, polymerase assay was carried out in the same reaction mixture as in Fig. 1. Catalase was also centrifuged and assayed by the same manner as in Fig. 1. ^{14}C-AMP incorporation: —○—. catalase: —●—. (a) 0.15 M eluate, (b) 0.1 M eluate were directly analyzed. (c) 0.15 M eluate, (d) 0.1 M eluate were followed by further steps and the eluates from the DEAE-cellulose column were analyzed.

activity which incorporates one mμmole of a nucleotide monophosphate per hour in the presence of a saturated amount of *E. coli* DNA as primer.) Contaminations of DNase and RNase were insignificant in both preparations judging by overnight incubations with either ^{32}P-DNA or RNA.

(b) Physical Properties and Molecular Structures of RNA Polymerase

22S and 15S polymerase were separately pooled, concentrated in dialyzing bags under negative pressure, and then subjected to various physico-chemical analyses.

The molecular weights of 22S and 15S were determined as 1.0×10^6 and

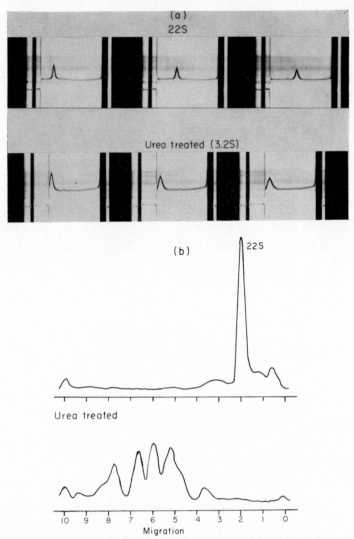

FIG. 3. Dissociation of 22S polymerase by urea treatment. 22S enzyme derived from 0.15 M eluate after serial steps was centrifuged in glycerol density gradient, and the fractions at 22S peak were pooled. The sample was minimized, the volume to be 5 mg protein/ml and then centrifuged in Spinco model E at 50,740 rpm. The photographs were taken at 8-min interval. On the other hand an aliquot of the concentrated 22S enzyme was treated by 6 M urea in Tris-HCl (10 mM, pH 7.6) containing 10 mM magnesium chloride. After dialysis for 2 hr the sample was centrifuged at 59,780 rpm and the photographs were taken at 8-min interval. (a) Schlieren patterns of 22S and the treated sample. (b) Electrophoretic patterns of 22S and the treated sample. According to the method of Davis[14] disc-electrophoresis was carried out. The patterns of protein bands were photographed on minicopy films (Fuji) from which density curves were scanned by a film densitometer.

4.5×10^5 respectively by equilibrium centrifugations in an analytical ultra-centrifuge (Spinco model E) equipped with Rayleigh interference optics.

Sedimentation coefficients were also determined more precisely as 21.6S and 14.7S by measurements of sedimentation velocities in the analytical ultracentrifuge equipped with Schlieren optics.

In order to obtain further information with respect to an interrelationship between the properties of the two enzyme molecules, a series of experiments was carried out. Treatments with urea (6 M) or sodium lauryl sulfate (0.03 %) resulted in a dissociation to smaller subunits from 22S and 15S. The urea-treated 22S shows a single component (3.2S), as shown in Fig. 3(a), whereas multiple components (6–7 subunits) are revealed by a discontinuous electrophoresis in polyacrylamide gel (Fig. 3(b)). These results indicate that 22S polymerase consists of multiple subunits which are similar in molecular shape

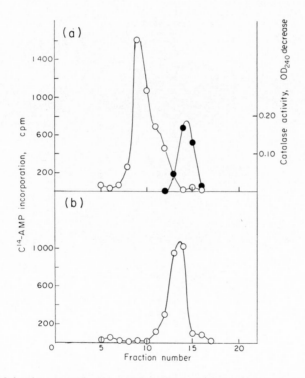

FIG. 4. Molecular conversion from 22S to 15S by increasing salt concentration. A preparation derived from 0.15 M eluate was subjected to sedimentation analyses. (a) 0.1 ml of eluate from DEAE-cellulose column was applied onto sucrose gradient medium (5–20%) in 50 mM Tris-HCl (pH 7.6). (b) The same eluate was adjusted KCl concentration at 0.3 M and applied onto sucrose gradient medium in the same buffer as in (a) containing 0.3 M KCl. The samples and catalase were separately centrifuged at 50,000 rpm for 2.5 hr. ^{14}C-AMP incorporation —○— and catalase activity —●— were measured in the same manner as in Fig. 2.

and size but different in ionic charge density. In addition it should be noticed that the dissociation of 22S mediated through 15S was not observed by urea.

22S polymerase has shown a dramatic conversion into 15S by addition of KCl at above 0.15 M, as shown in Fig. 4. And the resulted 15S could reversibly associate to active 22S by dialysis, while 15S polymerase prepared from 0.10 M eluate could never associate to form active 22S. The results may be interpreted to indicate that the 22S enzyme derived from 0.15 M eluate may contain two different components (designated as P and R, where P functions as a polymerizing unit and is identical with active 15S, and R functions as a regulating unit, which are essential to form 22S in a lower concentration of salt). On the other hand the enzyme fraction derived from 0.10 M eluate may consist of only one 15S component (P). This interpretation will be further supported by comparative analyses of 22S and 15S enzyme in connection with kinetic profiles.

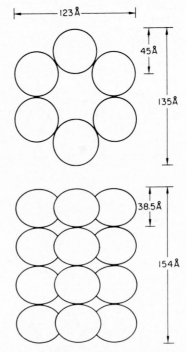

FIG. 5. Structure model of RNA polymerase.

In order to reveal the structural difference between both enzyme forms, a series of investigations was carried out using an electron microscope. 22S enzyme molecules, negatively stained with phosphotungustic acid according to the method of Brenner and Horne,[8] exhibited two regular distinct shapes as shown in Plates 1 and 2. One is a hexagon and the other is a rectangle.

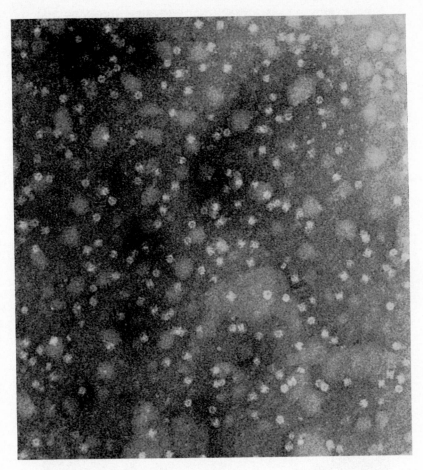

PLATE 1

PLATES 1 AND 2. *Electron micrographs of RNA polymerase* (22*S*): A JEM-7 (Japan Electron Optics Laboratory Co.) microscope was used, operating mainly at 80 KV. The astigmatism of the objective lens was readily corrected and reduced to less than 0.1 μ. All the high resolution micrographs were taken using liquid-nitrogen cold stages and anti-contamination devices (below 0.1 Å/min). According to the method of Brenner and Horne,[13] the negative-staining technique was used. After density gradient centrifugation the enzyme fractions were pooled, concentrated in a dialyzing bag under a negative pressure and dialyzed against Tris-HCl (10 mM, pH 7.6) in order to remove glycerol or sucrose. The protein concentration was about 1.0 OD unit at 280 mμ. Magnification; Plate 1: $3.0 \times 10^5 \times \frac{1}{2}$, Plate 2: 4.8×10^6.

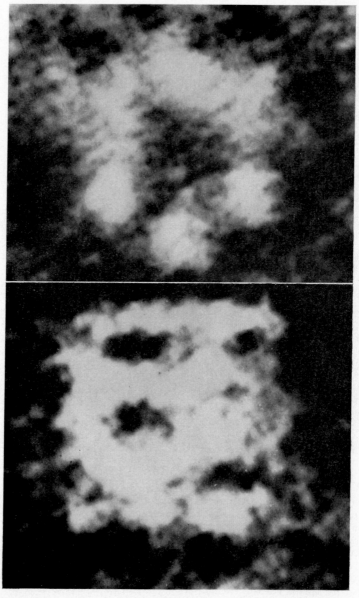

PLATE 2

At a higher magnification in Plate 2, it can be recognized that these two distinct shapes may be reflections of two profiles of a single entity as a 22S enzyme molecule. After measurements of a thousand molecules on the plates, a tentative molecular model of 22S was proposed as shown in Fig. 5. The model consists of 24 subunits, each of which is an oblate spheroid with a short diameter of 39 Å and a long diameter of 45 Å, and thus it is different from the model proposed by Fuchs et al.[9] which is a hexagonal assembly of six cylindrical units. Based on our model, the molecular weight is 9×10^5 if the partial specific volume of the protein is 0.72. This assumption agrees well with the molecular weight (1.0×10^6) by the physical measurement.

On the other hand 15S polymerase showed several irregular shapes different from those of 22S. This might be due to an unstable molecular state in the condition used. In order to obtain the precise molecular structure of 15S, further investigations are required.

(c) Kinetic Studies on RNA Polymerase Reaction Catalyzed by 22S and 15S

Now we turn our attention to another important problem. The problem is which component is essential in the transcription of genetic information. Both enzyme molecules may operate as machinery in vivo. If so, why does transcriptase exist in two molecular forms? And also it may be considered that one of these forms may be at a premature stage of functional development, or on the contrary one may be at a stage of decomposition or of reconstruction in the process of cellular metabolism.

These views will be clarified by an accumulation of experimental results and especially by the comparison of functional activities of both enzyme forms.

Since Hayashi et al.,[1] Geiduschek et al.,[2] Green[3] and Luria[4] have reported that the asymmetric reading might depend more or less on a degree of intactness of DNA used as primer, the kinetic studies on the polymerase reactions catalyzed by two types of enzyme should be carried out systematically by different preparations of DNA with respect to the chain length. The DNA preparations, obtained from cells or phages according to the usual method,[10] are not intact since artificial random breakages have occurred during the procedures.

On this aspect the number of residual operons as a whole in a given DNA preparation is statistically calculated as follows;

$$S = N \left(1 - \frac{1}{N}\right)^n$$

where N is a total number of operons in an intact DNA within a cell, n is a number of random scissions during the procedure, and S is a number of residual intact operons in a given preparation of DNA after random scissions (n times). According to the equation it can be calculated that the molecular weight of *E. coli* DNA must be more than 1.0×10^7 in order to certificate 90 % of the residual intact operons in a given DNA preparation, on the assumption that *E. coli* DNA consists of a thousand operons per 1.0×10^9 molecular weight.

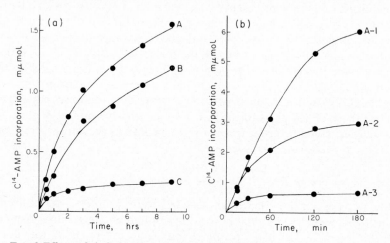

FIG. 6. Effects of chain length and relative amount of DNA on persistence of RNA polymerase reaction. 22S enzyme (3840 units/mg protein) derived from 0.15 M eluate was used. The reaction mixture was the same as in Table 2 except ^{14}C-ATP (20 mμ-moles, 1488 cpm/mμmole). (a) The reactions primed by *E. coli* DNA (35 μg) —A: 2.3×10^7, B: 1.4×10^7, C: 2.0×10^6 M.w.—were carried out using 0.5 unit of enzyme. (b) The reactions primed by DNA A—A-1: 60 μg, A-2: 6 μg, A-3: 0.6 μg— were carried out using 3.5 units of enzyme. At the times indicated each reaction was stopped and assayed.

On this aspect the relation of the chain length of DNA to the priming activity was tested in the reactions primed by different DNAs processed through serial breakages from the same preparation of *E. coli* DNA. In the presence of a constant amount of 22S enzyme, different preparations of DNA (molecular weight; A: 2.3×10^7, B: 1.4×10^7, C: 2.0×10^6) were added to prime the reactions. The reactions primed by DNA A and B are proceeding over 9 hr, as shown by curves A and B in Fig. 6 (a), while the reaction primed by DNA C stops at 60–90 min as shown by curve C.

Therefore it can be concluded that the higher intactness of primer DNA is required for the longer continuation of RNA synthesis *in vitro* as well as for asymmetric reading. However, even if DNA has a higher degree of intactness, the reaction does not persist for a longer period in the system containing a relatively lower concentration of DNA (enzyme excess). As shown in Fig. 6(b), increasing the relative ratio of DNA to enzyme a longer linear-proceeding of the reaction was observed.

Consequently, these results clearly indicate that the persistence of polymerase reaction is a function of chain length and relative amount of DNA added in the system, and that the requirement of two factors for this effect may reflect a cellular situation in which the genetic information is transcribed *in vivo*.

However, the reactions primed by 15S enzyme stopped early at around 60–90 min, even if enough excess amount of DNA having higher intactness was used.

Now we intend to discuss the molecular interactions of both enzyme forms with different DNAs. As already stated at the beginning of this paper, since the binding of RNA polymerase to the sites on DNA is the most important feature of transcriptase, it is interesting to test whether any significant differences can be detected in the interactions of two enzyme forms with phage T_6 DNA and *E. coli* DNA.

Using constant units of enzyme, the reactions were carried out and the activities for 6 min were plotted as a function of DNA concentration, as shown in Fig. 7. There can be seen a great difference in the amounts of DNA needed for saturation of the two enzyme preparations. The ratios of DNA

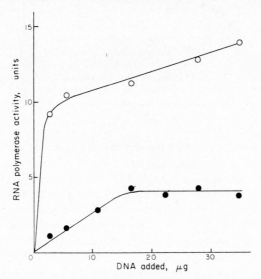

Fig. 7. Amounts of DNA for saturation of 22S and 15S enzyme. 14 units of 22S enzyme (3840 units/mg protein) and 4.2 units of 15S (1560 units/mg protein) were used. The reaction mixture was the same as in Table 2 except for the DNA. T_6DNA (M.W. 1.3×10^8) was used. 22S: —○—; 15S: —●—.

to enzyme at the saturation points, 0.2 µg/unit for 22S and 4 µg/unit for 15S, represent the minimum amounts of DNA required for the maximum expression of enzyme activity by 22S and 15S enzyme respectively. The results thus obtained may suggest that 22S can be utilized more effectively than 15S by the same DNA molecule and therefore 15S binds only to a limited number of sites of DNA. This interpretation is further supported by the next experiment.

2a NAM

With constant amounts of T_6 DNA (0.56 µg) the reactions were carried out by increasing the units of both enzyme preparations. Figure 8 shows that T_6 DNA is saturated by 22S enzyme at a higher level than by 15S enzyme. A similar result was obtained with *E. coli* DNA.

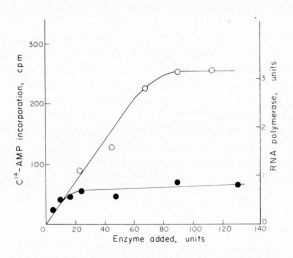

Fig. 8. Amounts of 22S and 15S enzyme for saturation of DNA. 0.56 µg. of T_6DNA (M.w. 1.3×10^8) was used. The reaction mixture was the same as in Table 2. ^{14}C-AMP incorporated for 6 min was plotted against units of enzyme added. 22S: —○—; 15S: —●—.

Thus it can be concluded that the specific sites on either T_6 DNA or *E. coli* DNA recognized by 22S molecules are more than those by 15S molecules. Moreover, it has been indicated that the two types of sites were not competitive between both types of enzyme molecules because the reaction carried out by a mixed preparation resulted in an additive activity of the individual reactions. From these results one can easily calculate the average number of enzyme molecules bound on a DNA molecule at each saturation level; 70 molecules of 22S and 19 molecules of 15S on T_6 (M.w. 1.4×10^8) and 20 molecules of 22S and 3–4 molecules of 15S on *E. coli* (M.w. 1.4×10^7) on the assumption that the specific activities of 22S and 15S are 10,000 and 20,000 units/mg protein respectively.

These assumptions have suggested that the average length of DNA transcribed by one enzyme molecule of 22S might correspond to the average size of the genetic functional unit, the operon or cistron.

Since it has been assumed that RNA polymerase operates along the primer at a constant velocity, which can be determined by measurements of chain lengths of product RNAs as a function of time, RNAs synthesized by 22S and 15S polymerase in the reactions primed by *E. coli* DNA were centrifuged in sucrose density gradient medium. As shown in Fig. 9(a), no differences were observed in the sedimentation patterns obtained by both enzyme forms. Since the chain length of RNA is about a linear function of S^2, the values obtained by different DNAs are plotted against the reaction time in Fig. 9(b). The rates of chain growth are classified into two groups. In the reactions primed by DNAs of *E. coli*, T-odd phages, calf thymus, and other bacteria, the rates are 10 nucleotides/sec, whereas in those of T-even phages, the rates are 4–5 nucleotides/sec. In addition the fact that there is no difference in the rate of chain growth between 22S and 15S even in the reaction primed by T_6 DNA has been observed. These results may indicate that two types of polymerase molecules can operate equally well on DNA when they bind on it.

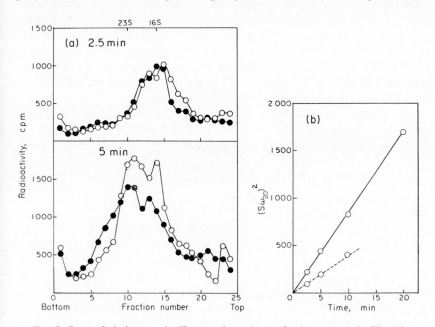

FIG. 9. Rate of chain growth. The reaction mixture is the same as in Fig. 1 but ^{14}C-ATP was replaced by 12.5 mμmoles of ^{3}H-ATP (10,000 cpm/mμmole). (a) 4 units of 22S enzyme, 3 units of 15S enzyme, and 13 μg of *E. coli* DNA (M.w. 1.4×10^7) were used. The reactions were stopped at 2.5 min and 5 min by addition of 0.01 ml of 4% sodium lauryl sulfate. After 3 min the sample was layered onto 4.4 ml of sucrose gradient medium (5–20% in 10 mM Tris-HCl, pH 7.8), containing 1 mM magnesium chloride and centrifuged at 37,000 rpm for 4 hr. Fractions were collected and precipitated in the presence of 500 μg of BSA as carrier by 5% TCA. The washed precipitates were dried and counted. 22S: —○—; 15S: —●—. (b) From the sedimentation patterns the values of Sw_{20} were estimated referring to *E. coli* ribosomal RNAs. Square values of the sedimentation coefficients obtained by the reactions primed by different species of DNA were plotted; *E. coli*, T-odd phage, *Micrococcus lysodeikticus*, and calf thymus: —○—; T-even and T_6^*: ⋯○⋯.

DISCUSSION

The new purification method has been established in obtaining consistently two types of RNA polymerase and also given us satisfaction not only with the almost complete separation of one from the other type of RNA polymerase in the early stages of purification but also with the final yields. It is obvious that these effects are due to the stepwise elution from the protamine precipitate, and thus two types of RNA polymerase are different in the molecular entities as DNA complexes.

On the other hand Richardson[11] and Stevens et al.[12] have recently considered that 22S enzyme molecule may be a dimer of two active 15S molecules, based on the reversible conversion from 22S to 15S by alterations of salt concentration in accordance with our results. However, the interpretation may not elucidate the fact that 15S enzyme obtained by the new method could not form the active 22S molecule, although an inactive dimer (about 22S) appeared when the salt concentration was lowered.

Considering these results in connection with the difference in the patterns observed at the stepwise elution, it might be conceivable that the events in the purification and the conformational change could be expressed as follows;

$$\text{DNA} - \text{RP} \xrightarrow{\text{low salt}} \text{DNA} - \text{R} + \text{P} \tag{1}$$

$$\text{DNA} - \text{R} \xrightarrow{\text{high salt}} \text{DNA} + \text{R} \tag{2}$$

$$\text{P} + \text{R} \underset{-\text{salt}}{\overset{+\text{salt}}{\rightleftharpoons}} \text{PR} \text{ (Active 22S)} \tag{3}$$

$$\text{P} + \text{P} \xrightarrow{-\text{salt}} \text{PP} \text{ (Inactive 22S)} \tag{4}$$

Within a cell, RNA polymerase may exist as a DNA-RP complex which can be precipitated by protamine from the extract. Reaction (1) may partially take place at the low concentration of salt, while increasing the salt concentration reaction (2) may take place at above a critical point to release R as well as P. Consequently whether one of two types of molecular conversion (3) or (4) does occur in a preparation, may completely depend on the purification procedures applied.

Electron microscopic studies enabled us to visualize the shape and the surface structure of 22S polymerase. The discrepancy in the models proposed by Fuchs et al.[9] and the authors may be derived mainly from an attainment of higher resolution. In addition the fact that 22S dissociates into homogeneous 3.2S ingredients in the Schlieren pattern may give much support to the model proposed in this paper since the molecular weight of a subunit can be calculated as 4.2×10^4 which is sizable to be around 3S.

The results obtained from the kinetic studies have strongly suggested that 22S polymerase might be a true functional entity to be transcriptase with respect to the number of sites recognized on DNA and the persistence of the reaction. Since one could assume that the polymerase reaction consisted of four steps; (a) specific binding, (b) polymerization, (c) termination and (d) reutilization from one to another genetic region, 22S molecule could accomplish all the steps whereas 15S molecule could more or less steps do (b) and (c) but would be defective partially in (a) and completely in (d).

That the rates of RNA growth are classified into two types by the species of DNA used can be explained by hydroxymethyl cytosine (HMC) as a constituent of T-even phage DNA, because the rate of chain growth on T_6^* DNA is much the same as that on T_6 DNA.

Moreover it should be pointed out that a discrepancy in the rates obtained independently by Bremer and Konrad[13] (2–3 nucleotides/sec for T-even DNA) and the authors (4–5 nucleotides/sec for T-even DNA) might be due to the difference of preparations especially in a contamination of RNase.

All the results presented in this paper seem to suggest that the 22S molecule may be a true transcriptase. Thus in the next step our efforts will be concentrated on the problem of how 22S molecule can recognize the specific site on DNA. On this aspect it is at first necessary to obtain information on whether 22S polymerase molecules are homogeneous with respect to the recognizing function. Thus an attempt along this line will afford a clue to investigate the regulation mechanism of genetic transcription in connection with the hypothesis which 22S polymerase molecule consists of P and R components, as described.

ACKNOWLEDGEMENTS

The authors wish to thank the U.S. Public Health Service for supporting this research with Grant GM-10635. This work was also partly supported by grants from the Ministry of Education of Japan.

The authors also want to express their appreciation to Mrs. K. Kameyama, Miss M. Kanno and Mr. T. Mizukami for excellent technical assistances.

REFERENCES

1. M. HAYASHI, M. N. HAYASHI and S. SPIEGELMAN, *Proc. Nat. Acad. Sci. U.S.*, **51**, 351, 1964.
2. E. P. GEIDUSCHEK, G. P. TOCCHINI-VALENTINI and M. T. SARNAT, *Proc. Nat. Acad. Sci.*, *U. S.*, **52**, 486, 1964.
3. M. H. GREEN, *Proc. Nat. Acad. Sci.*, *U.S.*, **52**, 1388, 1964.
4. S. E. LURIA, *Biochem. Biophys. Res. Comm.*, **18**, 735, 1965.
5. J. MARMUR, C. M. GREENSPAN, E. PALECEK, J. LEVINE and M. MANDEL, *Cold Spring Harbor Symp. Quant. Biol.*, **28**, 191, 1963.
6. M. CHAMBERLIN and P. BERG, *Proc. Nat. Acad. Sci.*, *U.S.*, **48**, 81, 1962.
7. A. ISHIHAMA and T. KAMEYAMA, *Biochem. Biophys. Acta*, **138**, 480, 1967.
8. S. BRENNER and R. W. HORNE, *Biochem. Biophys. Acta*, **34**, 103, 1959.
9. E. FUCHS, W. ZILLIG, P. H. HOFSCHNEIDER and A. PREUSS, *J. Mol. Biol.*, **10**, 546, 1964.
10. J. MARMUR, *J. Mol. Biol.*, **2**, 208, 1961.

11. J. P. RICHARDSON, *Proc. Nat. Acad. Sci., U.S.*, **55**, 1616, 1966.
12. A. STEVENS, A. J. EMERY and N. STERNBERGER, *Biochem. Biophys. Res. Comm.*, **24**, 929, 1966.
13. H. BREMER and M. W. KONRAD, *Proc. Nat. Acad. Sci., U.S.*, **51**, 801, 1964.
14. B. J. DAVIS, *Annals New York Acad. Sci.*, **121**, 404, 1964.

DISCUSSION

S. SPIEGELMANN: Have you determined by molecular hybridization how much of the DNA has been transcribed?

T. KAMEYAMA: Only a few percent.

S. SPIEGELMANN: This is important to know with your preparation. It has been our experience, and that of others, that transcription in the test tube always results in messages representing only a small fraction of the genome.

PROTEIN SYNTHESIS
IN THE NUCLEAR RIBOSOME SYSTEM

HIROTO NAORA

Biology Division, National Cancer Center Research Institute,
Tsukiji, Chuo-ku, Tokyo, Japan

IN ALL the animal and plant cells that have been examined so far, an extensive protein synthesis takes place in the cytoplasm in which usually no DNA exists except for the chloroplastic and mitochondrial DNA.[1] It seems quite reasonable to suppose that in the cytoplasm, an information transfer from DNA to proteins is mediated through messenger RNA.

There are also independent pieces of evidence indicating that there does exist a complete machinery for protein synthesis in the nucleus.[2-7] A little is known about protein synthesis in nuclei. It is evident that this synthesis proceeds, probably for the most part, but not entirely, in a manner quite similar to that in the cytoplasm.[3,8] Mediation through mRNA appears to be an important step of this process. It has not yet been established, however, whether this process takes place in chromosomes or at the site where DNA is absent within the nucleus.

Biochemical and autoradiographic studies have evidently demonstrated that some of protein synthesis in the nucleus indeed takes place in chromosomes.[4,9-13] The finding that protein synthesis takes place in chromosomes immediately raises a question concerning the molecular mechanism of protein synthesis at the site where DNA exists. Is such a mechanism similar to that at the site where DNA does not exist? Unfortunately, the molecular mechanism of protein synthesis in chromosomes has not been investigated.

This paper describes recent results from our laboratory which examine the interaction of DNA with the nuclear ribosomes, and also a model of protein synthesis in chromosomes in which the mechanism of transcription and translation are integrated and which may account for or predict various phenomena.

EFFECT OF DNA ON AMINO ACID INCORPORATION *IN VITRO*

Nuclear ribosomes and pH 5 fractions have been prepared from calf thymus nuclei isolated in 0.25 M sucrose–0.003 M MgCl$_2$–0.006 M mercaptoethanol.[14] As an approach to understanding the molecular events in chromosomes, an attempt was made to investigate the *in vitro* effect of DNA on the activities of ribosomes.

When nuclear ribosomes were incubated *in vitro*, these ribosomes were quite active in incorporating labeled amino acids into proteins. One of the most interesting results was that in a nuclear ribosome system *in vitro*, labeled amino acid incorporation into proteins is stimulated by an addition of DNA.[14] This is shown in Fig. 1, demonstrating that addition of heat-denatured calf thymus DNA to the incubation mixture results in stimulation of ^{14}C-leucine incorporation into proteins. Though the amount of amino acid newly incorporated is much less than the value obtained in the bacterial cell-free extract supplemented with RNA polymerase and native DNA, the stimulation observed here is certainly significant and reproducible.

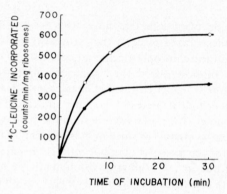

FIG. 1. Stimulation of ^{14}C-leucine incorporation by addition of DNA. 1.9 mg nuclear ribosomes and 1.1 mg of pH 5 fraction were present in the incubation mixture. The mixtures were incubated at 37°C with (○——○) or without (●——●) 400 μg of heat-denatured calf thymus DNA.

Characteristics of the amino acid incorporation stimulated by DNA in the nuclear ribosome system *in vitro* have been extensively investigated.[14,15] Results are summarized:

1. Addition of the DNA prepared from a variety of sources, such as phage T2, phage ΦX174, *E. coli*, *M. lysodeikticus*, salmon sperm, rat hepatoma and calf thymus, to the incubation mixture results in stimulation of labeled amino acid incorporation. Maximal stimulation of ^{14}C-leucine incorporation above the basal level ranged from 50 to 110% in 30 min of incubation.

2. Actinomycin D has no effect in these experiments on stimulation of amino acid incorporation by addition of DNA.

3. UTP and CTP are not necessary for stimulation. But stimulation was largely dependent upon the presence of ATP and GTP.

4. Addition of DNA to the reaction mixture did not result in appreciable stimulation of ^{14}C-GTP and ^{14}C-UTP incorporation into the alkali-labile material of the acid-insoluble fraction; nor of ^{14}C-ATP and ^{14}C-CTP in-

corporation into the acid-insoluble fraction. Marked stimulation of ^{14}C-GTP incorporation by DNA was due to the terminal addition of ribonucleotide to the DNA molecule. Furthermore, analysis of the reaction product indicated that the incorporation into RNA is mainly due to the terminal exchange of soluble RNA and the attachment of AMP to the terminal AMP residue in RNA molecules. Therefore, these results clearly suggest that RNA synthesis was not stimulated by DNA, while amino acid incorporation was being stimulated in this system.

5. An interesting observation was that double stranded DNA has less or no stimulatory activity compared with single stranded DNA.

6. The added DNA, and not the contaminating RNA, is indeed responsible for the stimulatory effect upon amino acid incorporation.

7. Stimulation is certainly due to the specific effect of DNA added and the labeled products are characterized by the base composition of the DNA added.

8. If we assume that the DNA code-words for each amino acid are the same as those in mRNA, except that T is equivalent to U, the values of the relative amino acid composition of the products newly formed *in vitro* by the DNA added were in good agreement with those calculated theoretically from the base composition of the DNA added with certain assumptions.

9. Nuclear ribosomes are capable of binding DNA and forming the nuclear ribisome-DNA complex which gives the electrophoretic and centrifugal patterns distinct from either of the components.[16] The formation of such a ribosome-DNA complex can be generalized among different organisms.[17-21]

$$\text{Lys}\left\{\begin{array}{l} \text{A - T} \\ \text{A - T} \\ \text{A - T} \\ \text{C - G} \end{array}\right. \rightarrow \begin{array}{l} \text{A} \\ \text{A} \\ \text{A} \\ \text{C} \end{array} \rightarrow \left.\begin{array}{l} \text{A} \\ \text{A} \\ \text{A} \\ \text{C} \end{array}\right\}\text{Lys}$$

DNA   mRNA

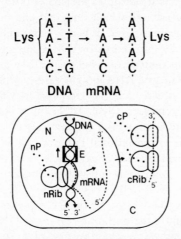

FIG. 2. Schematic representation of the model of protein synthesis in chromosomes and in the cytoplasm. N: nucleus; C: cytoplasm; nP: proteins synthesized in chromosomes; cP: proteins synthesized in the cytoplasm; nRib: nuclear ribosomes; cRib: cytoplasmic ribosomes; E: DNA-dependent RNA polymerase; and 5′ or 3′:5′ or 3′ terminal of DNA or mRNA strands.

The results obtained here suggested that genetic information from DNA is directly transferred to the proteins in the nuclear ribosome system without any mediation through messenger RNA, although DNA added might affect protein synthesis in some other way unknown at present.

A TENTATIVE HYPOTHESIS

As mentioned above, a direct transfer of genetic information from DNA to proteins, and the formation of the nuclear ribosome-DNA complex, were only demonstrated under *in vitro* conditions. Assuming that these reactions take place within an intact cell nucleus, a question to be answered concerns the relationship between the syntheses of proteins and RNA in chromosomes. Is protein synthesis in chromosomes independent of RNA synthesis?

Among the possible models that account for the syntheses of RNA and proteins in chromosomes, we now present the model which is based on the data, mainly resulting from stimulation of amino acid incorporation by addition of DNA *in vitro*, from the formation of the ribosome-DNA complex and from the known feature of the *in vivo* transcription catalyzed by a DNA-dependent RNA polymerase.[22]

The model is concerned with the simultaneous read-out of the genetic information from both strands of DNA molecules, that is the transcriptive read-out from one strand and the translation process on another strand. This model is schematically shown in Fig. 1.

The chromosomes consist of duplex molecules of DNA and of substances that reinforce the structure and which surround and/or associate with DNA as a repressor of the gene action. When a particular region in repressed chromosomes is once derepressed through the physical and chemical alteration of the complicated interaction between DNA and the repressor substances, the DNA dependent RNA polymerase is attached to such a derepressed region. We conceive that the attached enzyme is able to detect a specific site of DNA template which is the initiating point of a certain operon. Around this site, a sequential alignment of DNA bases of one strand should at least contain a translational initiator codon, like ATG, in the direction of 5′ to 3′ in the sequence of polynucleotide linkage. Consequently, another strand should contain an anti-initiator codon, like TAC, in the direction of 3′ to 5′. The specific effect of ATP and GTP on initiation of transcription and not on polymerization by the RNA polymerase might correlate with this code alignment in the initiating regions of DNA molecules.[23] The enzyme moves from this site in the direction of 3′ to 5′ along the strands which contain the anti-initiator codon but never in the opposite direction. Therefore, this model may imply the possibility of the specific codon alignment and hence the strand selection for transcription.[24-29]

Significance of the movement of the RNA polymerase along the DNA strand during the transcription process deserves mention here. In addition to the polymerization, we conceive that an important function of this enzyme is strand separation of the DNA duplex molecule. When the enzyme moves forward, two strands of DNA will be pulled apart, and hence the hydrogen bonds which unite the two strands will be broken. The complementary base-pairing in the strand on which RNA is being formed is responsible for specific alignment of ribonucleotides, though the mechanism rather than the complementary base-pairing might also conceivable be in the process of *in vivo* transcription.[21] However, if the former mechanism is the case, the formation of a hybrid between newly formed RNA and the template strand behind the enzyme leads to the appearance of the free single strand in the other chain, which contains the initiator codon in the direction of 5' to 3'.

An important mechanism in this model is the appearance of the free single strand at a certain point of the DNA duplex molecule during a short period and the movement of the single strand region along the duplex molecule during the transcription process. When the single strand, which contains the initiator codon, appears by the forward movement of the enzyme, a nuclear ribosome immediately attaches to the initiator codon or nearby, and thus forms the nuclear ribosome-DNA complex.[16] When once the transcription process by the RNA polymerase is initiated in one strand, the translation process by the attached ribosome on another strand, which is free from RNA synthesis, is initiated about at the same time. During this translation process, the genetic information encoded in DNA strands is directly transferred to proteins without any mediation through mRNA. Two processes, translation and transcription, proceed concomitantly; when the transciption is terminated at the end of the operon, translation may be likewise terminated.

We now conceive the possibility of the presence of the terminator codon alignment[30,31] in one DNA strand, on which the translation process takes place, at or near the end of the cistron. The terminator codon in this strand and anti-terminator codon in another strand give the newly formed proteins, the nuclear ribosomes and mRNA an opportunity to liberate from the DNA strands. When the nonsense or some particular code alignment is followed by the terminator codon in the DNA strand, while the polymerase moves forward further and reads out the genetic messages continuously, a polycistronic mRNA synthesis would take place in chromosomes. Failure of liberation of the newly formed proteins and/or the nuclear ribosome from the DNA strand might lead to the interruption of the repetitive transcription even if the operon is derepressed.

It should be mentioned here that the reaction of protein synthesis on the DNA strand and even the attachment of ribosomes to the DNA strand may not be necessarily involved in the mechanism of the RNA synthesis itself catalyzed by the polymerase, but may considerably affect it in several ways. In fact, RNA is formed in the *in vitro* incubation mixture containing only the DNA template, purified RNA polymerase and ribonucleoside triphosphates, but addition of ribosomes to the mixture results in stimulation of RNA synthesis and in the formation of higher molecular weight RNA.[32] A possibility may exist that when the reaction of protein synthesis is inhibited by some factors while ribosomes attach to the DNA strand, the RNA synthesis alone takes place on chromosomes without concomitant protein synthesis but with the co-operation of the attached ribosome. The puffed regions of the polytene chromosomes of certain insects, which exhibit the active RNA synthesis but practically no protein synthesis, may eventuate.[33-36]

This tentative hypothesis does not exclude the possibility of a nuclear ribosome-mRNA complex, in the nucleus.[37] Although it might be conceivable that the ribosome-mRNA complex is the majority of the functioning structure in the whole nucleus, such a function was not discussed as the protein synthesis in chromosomes in the model described here. Since the protein synthesis in the nuclear ribosome-mRNA complex can be also interpreted in the way quite similar to the case of the protein synthesis in the cytoplasmic polysomes, conventionally particular attention is not focussed on this process.

It is stressed that the model presented here is highly tentative and does not exclude alternative possibilities. But this model accounts for several data and also predicts certain future data which await experimental confirmation. Among them, one of the most interesting predictions is the synthesis of identical protein molecules in chromosomes and in the cytoplasm.

We assume in the present model that two separate molecules of proteins are independently synthesized in chromosomes and in the cytoplasm. It will be predicted by this model that these protein molecules are identical in chemical structure and also in chemical polarity. The upper part of Fig. 1 schematically shows the transfer of genetic information from DNA to the proteins synthesized in chromosomes and in the cytoplasm. These points will be discussed in detail elsewhere.[38] This prediction is of significance in nuclear-cytoplasmic interaction within a whole cell.

It seems possible that protein synthesis in chromosomes may regulate protein synthesis in the cytoplasmic machinery by interfering with the repetitive transcription of the operon. The dissociation of nascent protein-nuclear ribosome-DNA complex may result in the inhibition of the forward movement of the succeeding RNA polymerase. Thus, the repetitive transcription may be prevented even if the operon is derepressed. Detailed exploration is certainly needed for better understanding of the molecular events of nuclear-cytoplasmic interaction. Such a study is now in progress.

SUMMARY

1. In a nuclear ribosome system *in vitro*, the incorporation of labeled amino acids into proteins was stimulated by addition of DNA. It was previously demonstrated that nuclear ribosomes are capable of binding DNA and

forming a nuclear ribosome-DNA complex. It was then suggested that genetic information from DNA is directly transferred to proteins without any mediation through mRNA.

2. This paper described a model of protein synthesis in chromosomes into which the mechanisms of transcription and translation are adequately integrated. Such a model is concerned with the simultaneous read-out of the genetic information from both strands of DNA molecule: transcriptive read-out from one strand and translation process on another strand.

ACKNOWLEDGEMENTS

The author wishes to extend his sincere thanks to Dr. W. Nakahara, Director of the Institute, for advice and encouragement in this work and to Dr. Sibatani, Hiroshima University, Hiroshima, for stimulating discussion and helpful advice. Particular thanks are due to Mr. M. Shimizu and Miss. H. Osawa for expert assistance. This work was in part supported by a grant from the U.S. Public Health Service (CA-06986).

REFERENCES

1. H. CHANTRENNE, *The Biosynthesis of Proteins*, Pergamon Press, Oxford, 1961.
2. V. G. ALLFREY, A. E. MIRSKY and S. OSAWA, *J. Gen. Physiol.*, **40**, 451, 1957.
3. V. G. ALLFREY, *Exptl. Cell Res.* Suppl., **9**, 183, 1963.
4. V. G. ALLFREY, V. C. LITTAU and A. E. MIRSKY, *J. Cell Biol.*, **21**, 213, 1964.
5. T-Y. WANG, *Biochim. Biophys. Acta*, **68**, 52, 1963.
6. W. G. FLAMM and M. L. BIRNSTIEL, *Biochim. Biophys. Acta*, **87**, 101, 1964.
7. K. S. McCARTY, J. PARSONS, W. A. CARTER and J. LASZLO, *J. Biol. Chem.* **241**, 5489, 1966.
8. T-Y. WANG, *Biochim. Biophys. Acta*, **68**, 633, 1963.
9. J. H. FRENSTER, V. G. ALLFREY and A. E. MIRSKY, *Proc. Nat. Acad. Sci.*, *U.S.*, **50**, 1026, 1963.
10. G. PATEL and T-Y. WANG, *Biochim. Biophys. Acta*, **95**, 314, 1965.
11. J. G. GALL and H. G. CALLAN, *Proc. Nat. Acad. Sci.*, *U.S.*, **48**, 562, 1962.
12. J. G. GALL, *Cytodifferentiation and Macromolecular Synthesis*, p. 119. Ed. by M. Locke, Academic Press, New York, 1963.
13. M. IZAWA, V. G. ALLFREY and A. E. MIRSKY, *Proc. Nat. Acad. Sci.*, *U.S.* **50**, 811, 1963.
14. H. NAORA, *Biochim. Biophys. Acta*, **123**, 151, 1966.
15. H. NAORA, *Biochim. Biophys. Acta*, **114**, 437, 1966.
16. H. NAORA, *Biochim. Biophys. Acta*, **61**, 588, 1962.
17. T. KAKEFUDA, *Fed. Proc.*, **23**, 498, 1964.
18. H. NAORA and H. NAORA, *Biochim. Biophys. Acta*, **134**, 277, 1967.
19. R. BYRNE, J. G. LEVIN, H. A. BLADEN and M. W. NIRENBERG, *Proc. Nat. Acad. Sci.*, *U.S.*, **52**, 140, 1964.
20. H. A. BLADEN, R. BYRNE, L. G. LEVIN and M. W. NIRENBERG, *J. Mol. Biol.*, **11**, 78, 1965.
21. G. S. STENT, *Proc. Roy. Soc.*, B**164**, 181, 1966.
22. A. SIBATANI, *Progr. Biophys. Mol. Biol.*, **16**, 15, 1966.
23. D. D. ANTHONY, E. ZESZOTEK and D. A. GOLDTHWAIT, *Proc. Nat. Acad. Sci.*, *U.S.*, **56**, 1026, 1966.
24. M. HAYASHI, M. N. HAYASHI and S. SPIEGELMAN, *Proc. Nat. Acad. Sci.*, *U.S.*, **50**, 664, 1963.

25. M. HAYASHI, M. N. HAYASHI and S. SPIEGELMAN, *Proc. Nat. Acad. Sci., U.S.*, **51**, 351, 1964.
26. B. D. HALL, M. GREEN, A. P. NYGAARD and J. BOEZI, *Cold Spring Harbor Symp. Quant. Biol.*, **28**, 201, 1963.
27. E. K. F. BAUTZ, *Cold Spring Harbor Symp. Quant. Biol.*, **28**, 205, 1963.
28. E. P. GEIDUSCHEK, G. P. TOCCHINI-VALENTINI and M. T. SARNAT, *Proc. Nat. Acad. Sci., U.S.*, **52**, 486, 1964.
29. M. H. GREEN, *Proc. Nat. Acad. Sci., U.S.*, **52**, 1388, 1964.
30. M. TAKANAMI and Y. YAN, *Proc. Nat. Acad. Sci., U.S.*, **54**, 1450, 1965.
31. M. C. GANOSA and T. NAKAMOTO, *Proc. Nat. Acad. Sci., U.S.*, **55**, 163, 1966.
32. D. SHIN and K. MOLDAVE, *J. Mol. Biol.*, **21**, 231, 1966.
33. W. BEERMANN, *American Zoologist*, **3**, 23, 1963.
34. W. BEERMANN, *Differentiation and Development*, p. 49, Little, Brown and Co., Boston, 1964.
35. U. CLEVER, *Brookhaven Symposia in Biology*, No. 18, p. 242, Univ. of Tokyo Press, Tokyo, 1966.
36. C. PAVAN, *Brookhaven Symposia in Biology*, No. 18, p. 222, Univ. of Tokyo Press, Tokyo, 1966.
37. M. HILL, A. MILLER-FAURÉS and M. ERRERA, *Biochim. Biophys. Acta*, **80**, 39, 1964.
38. H. NAORA, *J. Theoret. Biol.* (in press).

DISCUSSION

H. BUSCH: Ro has found in our laboratory that high doses of actinomycin D (650 μg/kg) produces blocks in synthesis of nuclear but not cytoplasmic proteins. It seems that this selective inhibition of nuclear protein synthesis is a support for Dr. Naora's concept and may provide a tool for further studies.

H. NAORA: Yes, it does. I think that the characterization of proteins synthesized on chromosomes is of a great importance.

S. SPIEGELMAN: You may recall Zalokar's elegantly simple experiments which showed that the polarity of new protein synthesis indicated that the principal site of protein synthesis was the cytoplasm. These experiments suggest that your model can account for only a minor fraction of protein synthesis. This situation makes it accordingly more difficult to experimentally verify your interesting model.

H. NAORA: It is not easy to determine how much of the proteins synthesized within the nucleus was really synthesized on chromosomes.

ULTRASTRUCTURE
OF MAMMALIAN CHROMOSOMES*

R. KINOSITA

City of Hope Medical Center, Duarte, California, U.S.A.

CYTOLOGICAL studies of mitotic as well as meiotic chromosomes, especially those at metaphase, have brought to light many basic problems on cell biology and genetics, for instance chromosome replication, splitting, gene alleles, crossing-over, recombination, rearrangement, etc. Most of the contributions were initially made by studies of cells of plants and lower animals. Studies of mammalian chromosomes followed later, mainly facilitated by the introduction in the early 1950's of an improved technique, employing treatment with hypotonic solution and squashing, which enabled reasonably precise analysis of the complements of metaphase chromosomes.

Since the electron microscope became available, many investigators have employed this new tool in an effort to uncover submicroscopic structures of chromosomes. The approach to the aimed objectives has proved to be very hazardous. Even for the framework of chromosomes, miscellaneous theories have been proposed, suggesting multistranded formation,[1] codes and lateral loops,[2] three-dimensional irregular folding,[3] etc. These differences may be due to the fact that materials from varying species of animals have been used. Moreover, there are technical limitations in electron microscopy. The objects must be as thin as about 500 Å or less. There is also difficulty in the reconstruction of serial images. Most of the examinations, therefore, have been focused on fine fibrillar structures. The descriptions reported are fairly divergent, probably confused as a result of failure in exact orientation of the fine objects in the chromosomes. The techniques must also be extremely delicate, in order to preserve the chromosome construction and essential components in the course of fixation and other processing and to achieve effective staining.

At the beginning of the 1960's, an epoch-making development took place in biochemical studies on the chromosomal materials of micro-organisms and viruses, eloquently elucidating, at the macromolecular level, the genetic codes on DNA, replication of DNA, transcription to RNA, and translation to proteins. It is, of course, an ambitious move to translate these principles onto microscopic chromosomes of higher animals and, if possible, elucidate at the macromolecular level the microscopically acquired learning on chromosome

* Supported by U.S.P.H., N.I.H. Grant No. C-6019.

biology. It is most urgent, therefore, to bridge our knowledge on chromosome structures from the microscopic level to the macromolecular level.

Some efforts have been in progress in this line at our laboratories. Our earlier studious endeavors, simply depending on light microscopes, have been emphatically complemented, in these later years, by submicroscopic examinations by the use of electron microscopes, combined with adequately modified techniques. In the present paper, some significant results that have been obtained are described and commented on.

MICROSCOPIC ORIENTATION

First, some short remarks are made about the microscopic framework of chromosomes, in which submicroscopic studies are to be orientated, and also about certain terms which need clarification.

By light microscopy, chromosomes are recognized as thread- or rod-like bodies in a cell during the process of mitosis, but not as such in interphase. Instead, numerous irregularly shaped particles of blue staining or condensed material, namely chromatin, are discernible, which often are called karyosomes or chromocenters. In the interphase nucleus of a mouse spermatogonial cell, 40 chromocenters are usually found, of which 38 are identified as the condensed or heterochromatic regions of the 38 autosomes in the subsequent mitotic phase, while the other two are identified as the totally heterochromatic X- and Y-chromosomes, associated with the nucleolus.[4]

When chromosomes are duplicated during interphase, each chromosome will have two rods, called chromatids. Each carries one full set of the genes of this chromosome. The genes are stored in DNA, and represent the ultimate units that direct the nature and activities of the cell, such as reproduction, differentiation, modulation, growth, secretion, etc.

On entering prophase, chromosomes become visible. Each chromosome evidently consists of two chromatids, lying parallel and firmly attached to each other only at a heterochromatic region, adjacent to the kinetochore locus of the respective chromatid. In the mouse spermatogonial cell at prophase, it is found that 38 autosomes have heterochromatic regions adjacent to the kinetochore loci, while the X- and the Y-chromosomes are entirely heterochromatic and are concerned with nucleolus formation.[4]

At metaphase, condensation of the autosomes advances with the result that the 38 autosomes become equally condensed with the X- and the Y-chromosomes.

Another structural feature of chromosomes is spiralization.[5] This was noted first in plant cells and then in protozoa and insects. It was quite recently, however, that the helical structure was successfully visualized on mammalian chromosomes by the following method.[6] Pretreated with distilled water and fixed with acetic acid, squash preparations were prepared from mouse seminiferous tubules, immersed in 5% acetic acid solution, kept in a refrigerator for an over-night period, and then stained by the Feulgen method. The spiral structure was conspicuous on spermatogonial anaphase autosomes, especially so on their non-condensed or euchromatic portions. Lately, Ohnuki[7] showed spiralization on the metaphase chromosomes of human lymphocytes taken from peripheral blood. The lymphocytes were cultured in a colchicine containing medium, treated with a 4:2:1 mixture of equal 0.055 M solutions of KCl, $NaNO_3$, and CH_3COONa, and fixed with Carnoy solution. A drop of the resulting cell suspension was spread on a slide, air-dried, and stained with Giemsa stain.

For example, Fig. 1 supplied by Ohnuki shows three human No. 2 chromosomes from a preparation thus prepared. The chromosome on the left is from an early metaphase complement. Both chromatids are almost equally spiral, but at several portions are narrower or more distended and faintly stained. The heterochromatic regions adjacent to the kinetochore loci are firmly attached together. The chromosome on the right is from a late metaphase complement. Both chromatids are tightly wound in a wider gyre-width (about 1 micron), becoming considerably thicker, shorter, and heterochromatic all over. The chromatids are only slightly connected at the kinetochore regions. The chromosome in the middle is from a complement at the middle stage of metaphase. It shows intermediate features between those of the other two chromosomes. The gyre-width is about 0.7 micron.

At anaphase, the two chromatids of each chromosome separate to become independent chromosomes, which move to respective poles. Thus, one complete set of chromosomes group at each pole, which, at telophase, goes to form the nucleus of one daughter cell. Larger portions of the chromosomes then start to unwind and stretch, which at interphase become fibrillar, namely euchromatic, and in part microscopically invisible, while the heterochromatic regions at prophase remain condensed all the way, and are then called chromocenters, as stated previously. Some additional portions of chromosomes occasionally are found condensed to form granules and patches. This condensation varies to a great extent according to the types and activities of the cells. The large condensed patches are usually found in the nuclear periphery and on the nucleolus.

Thus, no doubt, the mammalian mitotic chromosomes have orderly coiling, and are chromatid wide at pro- and meta-phases and chromosome wide at ana- and telo-phases. The coiling becomes gradually condensed at pro- and metaphases until completely condensed at the end of metaphase, and starts to unwind at telophase. The heterochromatic regions and some additional portions remain condensed through interphase, which are known as chromocenters and patches. The microscopic features of chromosomes thus undergo considerable changes in the course of the whole mitotic cycle. These changes evidently are intimately involved with varying degrees of the spiralization. The discernible spiral structure in the mitotic phase, therefore, will be a good target to start with for the submicroscopic investigation.

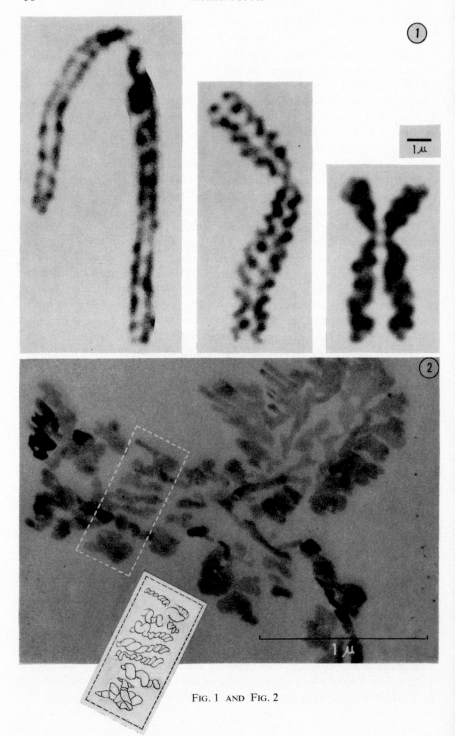

FIG. 1 AND FIG. 2

SECTIONS OF ISOLATED MITOTIC CHROMOSOMES

As previously mentioned, the spiral structure of chromosomes was clearly visualized by microscopes in metaphase cells in stretch preparations. Accordingly, a similar approach was chosen for the first submicroscopic study.

Short-term cultures of human blood lymphocytes were treated for 2 hr in media containing colchicine to arrest mitoses at metaphase. From these, a suspension of mitotic chromosomes was obtained, which was fixed in Carnoy solution. A drop of it was spread on a cover-slip and air-dried. A metal ring, 1 cm in diameter and 2 mm in height, was placed on this area, filled with partially polymerized methacrylate, and covered with another cover-slip. After complete polymerization of metacrylate, the preparation was refrigerated in dry-ice to remove the methacrylate cast. This was turned over and examined by a microscope to locate a small area with horizontally stretched metaphase chromosomes. This portion was marked, cut out, and mounted horizontally on the top end of a methacrylate column. After careful trimming, ultra-thin sections, about 500 Å in thickness, were sectioned in parallel, as much as possible with the horizontal plane. Chances were not frequent to strike flattened chromosomes longitudinally. However, this did happen.

Figure 2 is from a slice of a metaphase chromosome of the human lymphocyte thus prepared. It includes a large piece of chromatid, on the middle portion of which a small piece of chromatid lies. It appears that the large one is superimposed on the small one, the sister chromatid. The large one is about 0.7 micron wide. Its right portion is obliquely cut, and the left portion is almost longitudinally sectioned, although the end is sharply bordered, from which, apparently, some portion is lost. It is noted that the left portion consists of rectangular segments, about 0.7 micron wide and 2000 Å long, arranged longitudinally in series These apparently are cross-sections of repeats of the chromatid coiling, about 0.7 micron in gyre-width, visible by light microscope. Further, it is conspicuous, especially at the marked area, that the segment is represented by transversally parallel threads, about 2000 Å long and about 400 Å wide on average. These threads apparently are repeats of another coiling, about 2000 Å in gyre-width. Each thread shows a further lower order of coiling. Its gyre-width generally is about 400 Å, but at stretched portions about 150 Å.

Thus, the 0.7 micron chromatid coiling is superimposed upon the 2000 Å coiling, which is further superimposed on the 400 Å coiling. This is schematically illustrated in Fig. 3A. Nebel[8] on the basis of electron microscopic study of mammalian spermatogonial cells, has formulated a model which consists of a hierarchy of three levels of coiling, including the minor one, with the axial core. Cole[9] also has developed a similar model with three orders of coiling. To the contrary, DuPraw[3] has proposed a model showing a three-dimensional irregular folding rather than the orderly chromatid and thread coilings, based on whole-mount electron microscopy of metaphase chromosomes from honey-bee cells. Nevertheless, in our studies of mammalian metaphase chromosomes, as described previously, the chromatid coiling of 0.7 micron gyre-width has been microscopically demonstrated without doubt.

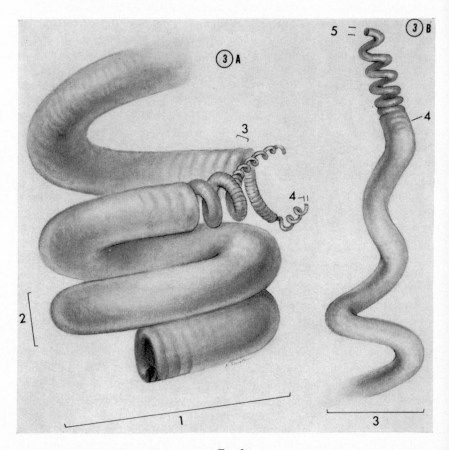

FIG. 3

In addition, two more lower levels of coiling, about 2000 Å and 400 Å in gyre-width, are now recognized in the 500 Å section of the metaphase chromatid. It is especially important that the 400 Å minor coiling is brought to light, because, according to Wilson,[5] the question of the reality of the minor coil is still open. The success in demonstration of these two submicroscopic gyres in the 500 Å section is largely due to the preparation procedures employed, especially chromosome isolation, washing, distention, flattening, and careful sectioning.

Any of these normally rough procedures could cause, to some extent, disruption of peripheral fringes, distortion of the natural constitution, and obscuration of fine structures. Therefore, confirmatory and complementary examinations are desirable, approaching from different angles.

SECTIONS OF MITOTIC CHROMOSOMES IN CELLS

To minimize the handicaps pointed out, ultrathin sections were studied of the least roughened mitotic cells, especially by emphatically observing finer structures after they had been adequately fixed, stained, and embedded.

Cell pellets, prepared from cultures of human lymphocytes, L-4946, C-1498, 4NQO-induced lymphoma, etc., were fixed with glutaraldehyde and OsO_4, stained with lead hydroxide and uranyl acetate at pH 7.4, embedded in epoxy resin, and sectioned. Since colchicine was not used to arrest mitosis at metaphase, there were cells at various phases of mitosis as usual.

Figure 4A, is a cross-section of a 4NQO-induced mouse lymphoma cell at late anaphase. Two groups of anaphase chromosomes are almost longitudinally sectioned. The chromosomes in each group lie parallel. At a high magnification, fine structures are discernible. Contrary to some previous reports, but agreeing with Ham,[10] neither limiting membrane nor central cores are found at all. It is observed in this 500 Å slice that the chromosomes are massing together, partially overlapping. At a higher magnification, for instance in the marked area of Fig. 4B, a segment of about 2000 Å is seen across the chromosome. Apparently, this is a section of the series of repeats of the microscopic chromosome coiling, about 0.7 micron in gyre-width. In this segment, about 2000 Å long parallel threads are conspicuous, which apparently represent sections of repeats of another order of coiling, about 2000 Å in gyre-width (Fig. 3A). Each thread itself is helical, for the most part about 400 Å in gyre-width, but partially becoming distended to a twisting fibril.

Figure 5A is a 500 Å slice of a lymphoma cell of the same origin at early metaphase. There are electron-dense patches of chromatid masses, of which the boundaries of individual chromosomes and chromatids are not distinguishable. Accordingly, neither the chromatid coiling nor the 2000 Å coiling can easily be traced. At the peripheries of the compact patches, however, there are large loosened portions. Especially at the marked areas, the 400 Å thread coiling, extending to form the 100 Å fibril, is conspicuous. Although vaguely, there is occasionally a noted distention of the 100 Å fibril itself, demonstrating a twisting structure, less than 100 Å in gyre-width. This is further stretched to form an extemely fine filament, about 30 Å in diameter (see the structures pointed to by the arrows). Figure 5B is a pretty well focused picture at a very high magnification. The mentioned fine structures, especially those from the 100 Å coiling to the 30 Å filament, are precisely discernible. These fine structures are schematically illustrated in Fig. 3B. Since the filament is the finest structure ever identified as such in chromosomes, it is tentatively called "the

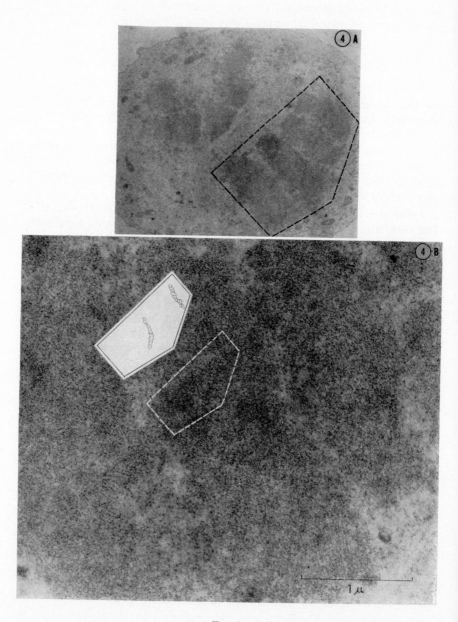

FIG. 4

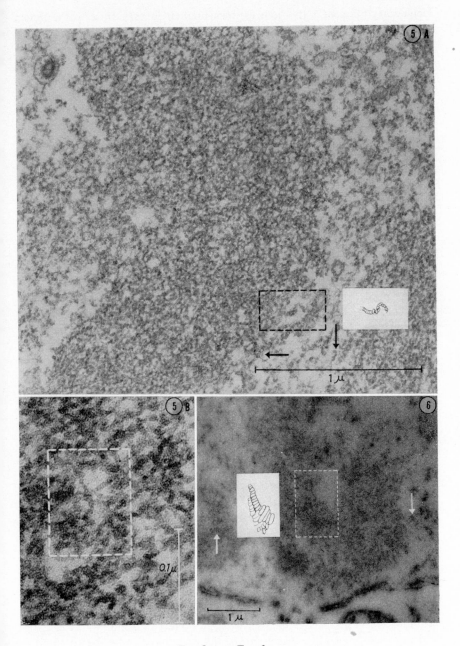

FIG. 5 AND FIG. 6

elementary filament". Its diameter measurement is very close to the figure given for the width of the DNA molecule.

Figure 6 is a section of a cell at prophase. Broken pieces of the nuclear membrane still remain, partially enveloping patches of chromatid masses, which in parts are loose and electron-luscent, and in other parts are condensed and electron-dense. The luscent patches and the peripheries of dense patches are fibrillar in structure, in which the formation of 400 Å are seen, for instance at the places pointed to by arrows. In dense patches, for instance in the marked area, the 400 Å threads are seen becoming wound at a high level, to gradually form the 2000 Å coil. It is occasionally found that two of the 2000 Å coils are being formed side by side.

SECTIONS OF INTERPHASE CHROMOSOMES IN NUCLEI

During the studies of mitotic cells, many sections of interphase cells were found as well. In fact, the same preparations also served for examinations of interphase nuclear contents. As previously mentioned, cells which are low in metabolic activities and seldom divide have plenty of condensed granules and patches in their nuclei. These nuclei should be the most favorably accessible specimens for the study following the investigation of condensed mitotic chromosomes. Kuppfer cells are one example, which are known to be scarcely divisible and not high in metabolism.

Figure 7 is a picture from a section of liver tissue of a rabbit, showing a Kuppfer cell at interphase. Although individual chromosomes or chromatids are not distinguishable as separate forms, a large portion of chromosome material is condensed to form electron-dense masses and patches, distributed in three regions: peripherally on the inner surface of the nuclear membrane, centrally associated with the nucleolus, and scattered in the intermediate zone. The remaining, not very spacious areas are electron-luscent and fibrillar.

In the peripheral dense area, a slightly twisting configuration of bands, on average about 0.7 micron long and 2000 Å wide, are conspicuous. Apparently, these are repeats of the 0.7 micron chromatid coils of the mitotic chromosomes. They, for the most part, are attached to the inner membrane of the nuclear envelope with one end, and run in parallel towards the interior almost vertically but slightly slanted, as seen in the marked area. It is suggested, therefore, that the chromosomes are lying on their side on the nuclear membrane. Each band consists of transversally parallel 2000 Å long and 400 Å wide threads. These correspond to the 2000 Å coil of the mitotic chromosome. The threads themselves show coiling of about 400 Å in gyre-width, as seen especially clearly on the thread pointed to by the arrow. It is remarkable that the 400 Å coil at the interior portion of a 2000 Å coil often is unwound to become less electron-luscent, and that this portion sometimes further transits to a condensed portion, or a microscopic chromatin granule.

The electron-dense patches associated with the nucleolus demonstrate similar features to the peripheral dense patches. The 2000 Å coils usually lie centrifugally from the nucleolus. Their central portions are electron-luscent, and are often found penetrating deep into the nucleolus (Fig. 7). In fact, the positive labeling has been noted in the nucleolus in autoradiography preparations from cultures of human amnion cells and HeLa cells, incubated with H^3-thymidine (20 μc/ml, 30 min) prior to fixation with calcium containing OsO_4 solution.[11] These loosened central portions apparently represent the nucleolus organizers.

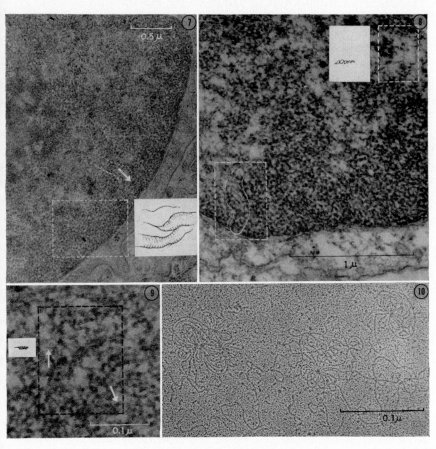

FIG. 7, FIG. 8, FIG. 9 AND FIG. 10

The study was then extended to more functional cells, for comparison. Figure 8 is a picture from a section of a 4NQO-induced lymphoma cell. The chromosomes are loosened in large portions. Only a few of the 2000 Å wide bands are distinguishable. One in the left marked area stands on the nuclear

3 NAM

membrane and shows repeats of the 2000 Å gyre. In the right marked area, the 400 Å coil is found stretching to form the 100 Å fibril. At a very high magnification it is observed, as shown in Fig. 9, two 100 Å fibrils pointed to by the arrows are uncoiled to form the elementary filament of about 30 Å in diameter.

The additional submicroscopic formations, found besides the fibrillar structures, are electron-dense granules, from about 150 to 500 Å in diameter. They are distributed singly, in chains, or in clusters, which are probably nuclear ribosomes. This problem will be discussed later.

Thus, either in mitosis or in interphase the essential chromosomal material is the elementary filament which is wound in four orders of coiling, the only difference is in the degree of spiralization. According to the literature, however, views on the basic structure are considerably manifold and without agreement. For instance, several recent publications are quoted. Gall[12] reports a single 150 Å core structure lying within a 400 to 600 Å fiber. Ris and Chandler[13] interpret a 200 to 500 Å fiber as being composed of two twisted 100 to 120 Å subunits. Ham[10] suggests that 50 Å filaments become aggregated to form filaments of greater diameter. Wolfe[14] sees chromosomes as being made up of 250 Å cylindrical fibers, interconnected by 80 Å fibers. Sotelo and Trujillo-Cenóz[15] propose a structure consisting of ribbon-like groups of three threads: two external, dense and thick (500 Å wide) and one medial, light and thin (120 Å wide). Wettstein and Sotelo[16] consider that chromosomes are formed as a mass of 100 Å fibrils, interwoven with no appearance of order. Hay and Revel[17] describe a meshwork of interconnected filaments, from 50 to 75 Å in diameter. In this confusion, it is significant to know from the present objective observations, where and how the elementary filament stands within the microscopic chromosome.

ISOLATED CHROMOSOME FILAMENTS

Studies were further extended in depth to the elementary filament itself. Some part of the results was published at the Sixth Annual Meeting of the American Society for Cell Biology.[18]

Cultures of human amnion cells were incubated for 1 hr in media with H^3-thymidine or H^3-uridine, 20 $\mu c/ml$. A pooled preparation of the cellular elements were trypsinized, gently homogenized in 0.1 M potassium citrate solution, and centrifuged at 1000 g for 30 min. The sediment was resuspended in 0.1 M potassium citrate solution and centrifuged similarly. This was repeated two more times. The final sediment, almost entirely consisting of nuclei, was gently homogenized in 2 M sodium chloride solution and left overnight in a refrigerator. Then, the same volume of absolute ethanol was added. The resulting flocculant precipitate, mainly containing DNA, RNA, and proteins, was dissolved in 2 M sodium chloride solution and kept at −60 °C. To one tenth ml of this were added 2.5 ml of 2 M sodium chloride solution and 2.5 ml of 2×10^{-4} g/ml cytochrome C solution. A small drop of the final solution was spread to a thin film on the surface of distilled water. A portion of the film was taken on a collodian coated grid, air-dried, and shadowed with platinum carbon. For autoradiography, the grids were coated with Ilford 1–4 photographic emulsion. After 14 to 42 days of exposure, the preparations were developed, stained with uranyl acetate, and then examined by an electron microscope. In some experiments, preparations were prepared from films spread on the surface of water containing RNase or DNase, and after being taken on grids, incubated for 30 min.

Figure 10 is a picture of spread chromosomes from one of the preparations. A winding filament is seen forming loops. The loops are often in a group and radially arranged in a rosette pattern, having one point of each loop close together in the center. A portion of filament forming each loop is about 0.3 to 0.7 micron long. Apparently, this is produced by the unwinding and extension of a small part of the 100 Å fibril, with its two ends not far apart, probably due to enclosure in relatively stable proteins. A similar extension of serially adjoining parts results in rosette formation. There is no evident sign of ending or branching at least within the limited area of this picture. The filament is apparently continuous, and for the most part is uniform in diameter, about 30 Å, corresponding to the measurement of the elementary filament observed in mitotic as well as interphase cells. Occasionally, however, it is two or three times thicker at small portions especially near the rosette centers. The thick portions often show periodical repeats, suggesting helical twisting. In addition to the filament, there are numerous spherical particles, about 120 to 200 Å in diameter, often located at or near the rosette centers and associated with the thick portions of the filament.

Figure 11 is from one of the DNase treated preparations. The 30 Å filament is almost out of sight, whereas the 120 to 200 Å spherical particles are still visible. In addition, extremely fine fibrils, previously not noticed become evident. Figure 12 is from one of the RNase treated preparations. The 120 to 200 Å spherical particles are markedly decreased in number, whereas the 30 Å filament shows little change in morphology. Some thick portions are still visible, but the loop formations become less orderly. The extremely fine fibrils recognized in Fig. 11 are not distinguishable.

Figure 13 is from one of the preparations processed for autoradiography with H^3-thymidine. The labeling, indicating sites of H^3-thymidine incorporation or the DNA replication, is recognized by the appearance of intensely electron-dense grains of reduced silver. All grains are found on or close by the filament. The large ones are worm-like in shape. Often two are seen not far apart from each other. The smaller ones are numerous, spherical, and irregularly distributed. These in form are like the afore-mentioned 120 to 200 Å particles, but much more electron-dense than the latter. Figure 14 is from one of the preparations processed for autoradiography with H^3-uridine. The labeling, indicating sites of H^3-uridine incorporation or the RNA syn-

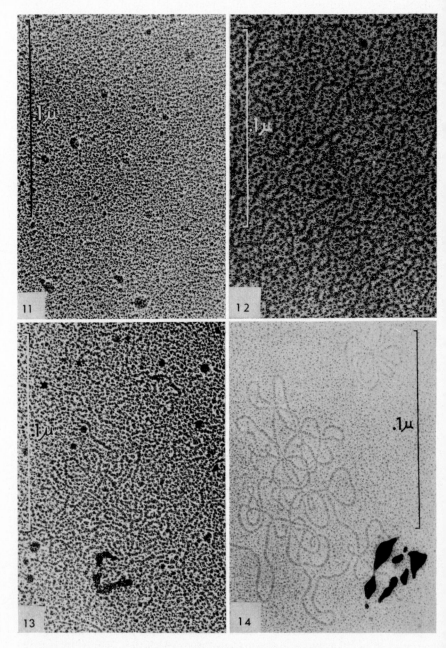

Fig. 11, Fig. 12, Fig. 13 and Fig. 14

thesis, is found on the 30 Å filament, especially frequent at the thick portions near the rosette center.

These findings suggest that the 30 Å filament is mainly made up of a DNA molecule, on which the DNA replication takes place at many positions, and that many of the less electron-dense spherical particles are closely associated with RNA which is usually synthesized on the filament near the rosette centers. The extremely fine fibrils noted in the DNase treated preparations are probably the units of RNA.

Following the work of Kleinschmidt and Lung,[19] who had obtained excellent preparations of fibers by spreading bacteria and viruses on a liquid surface, Gall[12] spread a drop of newt blood on a water surface, picked up the resulting thin layer of erythrocyte components with carbon-coated grids and processed with platinum–palladium for shadowing. The nuclear material of disrupted nuclei were found to entirely consist of long fibers, approximately 400 to 600 Å in diameter, as quoted previously. He further claimed that HeLa cell chromosomes at metaphase and grasshopper spermatogonial chromosomes at meiotic prophase also consisted of fibers similar to those from the newt erythrocytes. Since, the method has been improved. DNA molecules have been obtained from tumor viruses and various phages, of which several types of circular form were demonstrated by electron microscopy.[20] However, a similar approach to DNA in the far more complicated mammalian chromosomes is very difficult. The isolated metaphase contains for each mg of DNA, 0.66 mg of RNA, 2.0 mg of acid-soluble proteins and 2.7 mg of acid-insoluble proteins.[21] The RNA bound to the isolated chromosome consists mainly of ribosomal RNA, but there is also a significant amount of 45S RNA. Many non-histone proteins are acid soluble. Various histone fractions have a high lysine content.[22] Some lysine-rich histones are removed from chromosomes by the preparation processes.[21] Ham[10] concluded that the precise distribution of DNA and histones is not yet well established. As for the present preparations of isolated chromosomal materials, certainly, RNA and some proteins are retained associated with the elementary filament and the spherical particles. The elementary filament itself represents the DNA molecule.

Cairns,[23] by autoradiography, has demonstrated the semi-conservative nature of DNA replication in *E. coli*. Duplication of the daughter DNA strands starts by forming a loop or fork and advances at the rate of 20 to 30 microns/min until completed, reaching the other end of the chromosomes in over 30 min, covering altogether 1 to 1.5 mm. In chromosomes of higher cells, however, the apparently continuous DNA molecule within a single chromosome is considerably long. To explain the replication of such a long DNA molecule, Freese[24] and then Taylor[25] have proposed a model of DNA with many intermediate linkers. There is, however, little morphological evidence for the existence of linkers. De Robertis[26] has suggested that the synaptinemal complexes, which appear during meiotic prophase and are probably involved in the pairing and interchange of fragments of chromosomes during crossing-over, could be related to the concept of linkers. The multi-centric H^3-thymidine incorporation and the formation of loops and rosettes, observed in the present experiments, if not artefact, support a possibility of some structural provision for functional segmentation.

CONCLUSION

The analysis of the results from a series of reasonably planned experiments lead to the logical understanding that a long string of the elementary filament, mainly of a DNA molecule, is wound at four orders to form a piece of chromosome, of which some portions, especially large in number and range at interphase, are unwound to leave them loose in the surrounding foundation for the functional operations.

ACKNOWLEDGEMENTS

The author wishes to acknowledge the assistance of Dr. T. Kakefuda and Dr. J. Ito in the execution of experiments and of Mr. Kurt Smolen and Mr. Richard Ray in the preparation of illustrations.

REFERENCES

1. D. M. STEFFENSON, Chromosome structure with special reference to the role of metal ions. *Int. Rev. Cytol.*, **12**, 163, 1961.
2. B. R. NEBEL and E. M. COULON, The fine structure of chromosomes in pigeon spermatocytes. *Chromosoma (Berl.)*, **13**, 272, 1962.
3. E. J. DU PRAW, Macromolecular organization of nuclei and chromosomes: A folded fibre model based on whole-mount electron microscopy. *Nature*, **206**, 338, 1965.
4. S. OHNO, W. D. KAPLAN and R. KINOSITA, Heterochromatic regions and nucleolus organizers in chromosomes of the mouse, *Mus musculus. Exptl. Cell Research*, **13**, 358, 1957.
5. G. B. WILSON and J. H. MORRISON, The chromosome coiling cycle. In *Cytology*, p. 164. Reinhold Publ. Corp., New York, 1961.
6. S. OHNO, W. D. KAPLAN and R. KINOSITA, Demonstration of bi-partite spiral structure on spermatogonial anaphase chromosomes of *Mus musculus. Exptl. Cell Research*, **15**, 426, 1958.
7. Y. OHNUKI, Demonstration of the spiral structure of human chromosome. *Nature*, **208**, 916, 1965.
8. B. R. NEBEL, On the structure of mammalian chromosomes during spermatogenesis and after radiation with special reference to cores. In *Fourth Intern. Conf. on Electron Microscopy*, p. 228. Springer-Verlag, Berlin, 1960.
9. A. COLE, A molecular model for biological contractility: Implications in chromosome structure and function. *Nature*, **196**, 211, 1962.
10. A. W. HAM, *Histology*, 5th ed., p. 97. Lippencott Co., Phil., 1965.
11. T. KAKEFUDA, Electron-microscopic autoradiography of nuclei. *Fed. Proc.*, **24**, No. 2, 258, 1965.
12. J. G. GALL, Chromosome fibers from an interphase nucleus. *Science*, **139**, 120, 1963.

13. H. RIS and B. L. CHANDLER, The ultrastructure of genetic systems in prokaryotes and eukaryotes. *Cold Spring Harbor Symp. Quant. Biol.*, **28**, 1, 1963.
14. S. L. WOLFE, The fine structure of isolated chromosomes. *J. Ultrastruct. Res.*, **12**, 104, 1965.
15. J. R. SOTELO and O. TRUJILLO-CENÓZ, Submicroscopic structure of meiotic chromosomes during prophase. *Exptl. Cell Research*, **14**, 1, 1958.
16. R. WETTSTEIN and R. SOTELLO, Fine structure of meiotic chromosomes: The elementary components of metaphase chromosomes of *Gryllus argentinus*. *J. Ultrastruct. Res.*, **13**, 367, 1965.
17. E. HAY and H. P. REVEL, The fine structure of the DNP component of the nucleus: An electron microscopic study utilizing autoradiography to localize DNA synthesis. *J. Cell Biol.*, **16**, 29, 1963.
18. T. KAKEFUDA and R. KINOSITA, Ultrastructure of elementary fibrils of chromosomes and RNA in mammalian cell nuclei. *J. Cell Biol.*, **31**, 55A, 1966.
19. A. KLEINSCHMIDT and D. LUNG, Intrazelluläre Desoxyribonucleinsäure von Bakterien. *Fifth Intern. Congr. Electron Microscopy*, **2**, 1962.
20. J. VINOGRAD and J. LEBOWITZ, Physical and topological properties of circular DNA. *J. Gen. Physiol.*, **49**, part 2, 103, 1966.
21. J. A. HUBERMAN and G. ATTARDI, Isolation of metaphase chromosomes from HeLa cells. *J. Cell Biol.*, **31**, 95, 1966.
22. J. A. V. BUTLER, Fractionation and characteristics of histones. In *The Nucleohistone*, p. 36. Ed. by J. Bonner and P. O. P. T'so. San Francisco, Holden-Day, Inc., 1964.
23. J. CAIRNS, The bacterial chromosome and its manner of replication as seen by autoradiography. *J. Molec. Biol.*, **6**, 208, 1963.
24. E. FREESE, The arrangement of DNA in chromosome. *Cold Spring Harbor Symp. Quant. Biol.*, **23**, 13, 1958.
25. J. H. TAYLOR, The replication and organization of DNA in chromosomes. In *Molecular Genetics*, vol. 1, p. 65. Ed. by J. H. Taylor. New York, Academic Press Inc., 1963.
26. E. DE ROBERTIS, Advances in ultrastructure of the nucleus and chromosomes. *N.C.I. Monograph*, **14**, 33, 1964.

DISCUSSION

F. H. KASTEN: The elegant light microscope slide of coiling in leucocyte chromosomes is from a preparation of Dr. Y. Ohnuki and needs to be considered with regard to its manner of preparation. He uses a special hypotonic salt solution, which contains certain critical concentrations and proportions of three different salts in order to demonstrate the spirals. This recipe is shown on the first slide which I would like to project. On the second slide, we see that in seat-treated interphase nuclei of cultured salamander cells, there is a loss of nuclear granules, leaving an optically empty and homogeneous nucleoplasm while the normal nucleolar density is unaffected. Since the salt treatment most likely extracts some chromosomal histone, I believe it would be inappropriate to compare this type of preparation with standard electron microscope images where no such extraction has been employed.

RNA OF NUCLEOLI OF WALKER TUMOR
AND LIVER CELLS*

Harris Busch, Joe Arendell, Ken Higashi,
Leroy Hodnett, Toru Nakamura, Rajat Neogy,
Steven M. Schwartz and Steven J. Smith

Department of Pharmacology, Baylor University College of Medicine,
Houston, Texas

Studies on the nucleoli of cells have been made by morphologists ever since Fontana described nucleoli in the 1770's.[1] The relationship between nucleolar size and protein synthesis was documented by the intensive studies of Caspersson and his associates,[2,3] who attempted to correlate aspects of the structure and function of this nuclear organelle when they established the relationship of nuclear size and RNA content to protein synthesis in specialized cells.

The development of autoradiographic techniques enables cytologists to observe the dynamic functions of the nucleus and the nucleolus, with the aid of radioactive precursors. With these techniques, evidence was obtained that the nucleolus is a site of synthesis of ribosomal RNA.[4–10] Conclusive evidence has now been obtained that the nucleolus is the site of synthesis of the 28S RNA of the ribosomes (Table 1) from experiments with isolated nucleoli,[11–22] and studies with mutants having differences in their nucleolus-organizers.[23,24] Some evidence has shown that nucleoli are active in protein synthesis.[16,18,25–31] Most nucleolar proteins strongly resemble ribosomal proteins,[19] but the source of these nucleolar proteins is still unknown. It is possible that ribosomal proteins[16] or histones are synthesized in the nucleoli, but this is not proven.

Maggio, Siekevitz and Palade[38,39] re-investigated the procedure of Monty, Litt, Kay and Dounce[40] and found that the sonication procedure could be markedly improved, provided the container was cooled and Ca^{++} was added to the medium. Unfortunately, the procedure they reported was not found to provide a satisfactory nucleolar preparation from rat liver nuclei although a satisfactory preparation was apparently obtained from

* These studies were supported in part by grants from the American Cancer Society, the Jane Coffin Childs Fund, the National Science Foundation and the U.S. Public Health Service (CA-08182).

3a NAM

TABLE 1. EVIDENCE THAT THE NUCLEOLUS IS THE SITE OF SYNTHESIS
OF RIBOSOMAL RNA

1. The nucleolus contains RNA which is very similar in composition to cytoplasmic 28S RNA.[20]
2. In vivo treatment of cells with actinomycin D results in a rapid loss of 45S RNA initially, 35S later and 28S finally. If the nucleolar RNA is labeled with a pulse of radioactive precursor, the label is transferred from 45S to 35S to 28S RNA giving rise to the equation 45S RNA → 35S RNA → 28S RNA. Since, in some preparations, the 55S RNA is labeled as rapidly as 45S RNA, it may be that 55S RNA is a precursor of 45S RNA.[20, 32,33]
3. Isolated nucleoli contain a DNA-dependent RNA polymerase which is capable of biosynthesis of RNA and other enzymes of RNA metabolism.[34,35]
4. Inhibition of biosynthesis of ribosomal RNA is caused by U.V. microbeam irradiation of nucleoli.[9,10,36]
5. Selective inhibition of biosynthesis of nucleolar RNA by actinomycin D.[33]
6. Failure of synthesis of ribosomes or ribosomal RNA in mutants containing no nucleolus organizers.[23]
7. Hybridization of 28S RNA with nucleolus organizers or nucleolar DNA.[17]
8. Quantitative relationships between hybridization of ribosomal RNA with DNA in proportion to the number of nucleolus organizers present.[24]
9. Presence in nucleoli of proteins with amino acid compositions very similar to those of ribosomal proteins.[18,19,22]

guinea pig nuclei. However, they generously provided details of their procedure at a time when efforts were under way for isolation of nucleoli from tumors and precancerous livers in this laboratory.

THE SONICATION PROCEDURE FOR ISOLATION OF NUCLEOLI

A re-investigation of the variables of the procedure employing sonication led to a method[30,34,41,42] which has been successfully employed in a number of laboratories for the isolation of nucleoli of various tumors and non-tumor tissues.[43]

For the purposes of the isolation of nucleoli, it is essential that a divalent ion be present in the medium for isolation from the nuclei. The 'sucrose-calcium procedure' is generally employed. After the tissues are perfused or excised and chilled in 0.15 M NaCl or 0.25 M sucrose, they are homogenized in 0.25 M sucrose containing 3.3 mM Ca^{++} or 10 mM Mg^{++}. Sucrose solutions of higher concentrations such as 2.0–2.2 M have also been employed. Isolated nuclei from 1 g of liver or nuclear preparations from tumors are suspended in 0.34 M sucrose in a ratio of 1 and 5 ml per g of original liver and tumor, respectively.

The Raytheon sonic oscillator is adjusted in this laboratory to provide 1.0–1.1 A of output current for periods of 15–30 sec. At each time interval, a drop of the solution is used to determine the extent of destruction of the nuclei and the sonic oscillation is discontinued when the nuclei are virtually

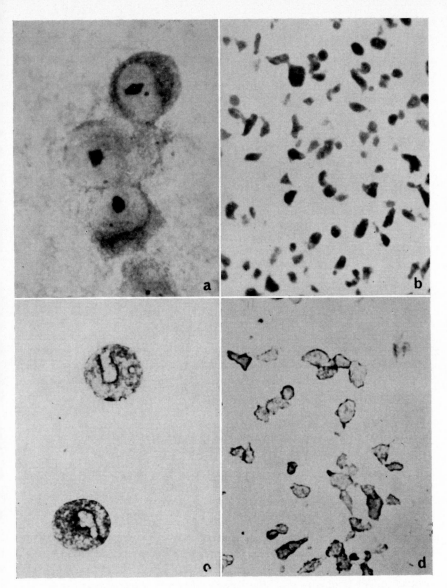

Fɪɢ. 1a. Smear of isolated nuclei from the Walker tumor, stained with toluidine blue. The nucleoli and the cytoplasmic basophilic structures are stained intensely. × 1800.

Fɪɢ. 1b. Smear of isolated nuclei from the Walker tumor, stained with toluidine blue after hydrolysis with 1N HCl. Chromatin-containing structures are stained. The nucleolus is not stained, but the perinucleolar nucleolus-associated chromatin is stained. × 1800.

Fɪɢ. 1c. Smear of isolated nucleoli from the Walker tumor, stained with toluidine blue at pH 5.0. Note the similar morphology and staining properties of isolated nucleoli and nucleoli in situ. × 1800.

Fɪɢ. 1d. Isolated nucleoli from the Walker tumor, stained with toluidine blue after hydrolysis with 1N HCl. The nucleolus-associated chromatin is stained, whereas the rest of the nucleolus is not stained. Septa penetrating from the nucleolus-associated chromatin into the nucleolus are visible in both the isolated nucleoli and the nucleoli in situ. × 1800.

completely destroyed.[30] The destruction of the nuclei does not ordinarily require longer than 45–60 sec in the Raytheon sonic oscillator but with the Branson sonifier it generally requires 30–60 sec.

The modification introduced by Ro et al.[34,42] is used in which the whole sonicate in a volume of 20 ml is layered over 20–25 ml of 0.88 M sucrose and the mixture is centrifuged for 20 min at 2000 × g in a swinging bucket rotor. The sedimented nucleoli may be resuspended in 0.25 M or 0.88 M sucrose and centrifuged once more to remove co-precipitated chromatin. This preparation constitutes the type of preparation shown in Fig. 1, which is a satisfactory nucleolar preparation.

NUCLEOLAR PREPARATIONS OBTAINED BY CHEMICAL EXTRACTIONS

After nuclei are extracted with 0.15 M NaCl and 1 or 2 M NaCl, a residue fraction is left in which the labeling of RNA is significantly greater than that of the extract.[44–51] A similar result is obtained when the nuclei are initially extracted with 0.1 to 0.2 M phosphate buffer. The residue fraction has been referred to as the 'nucleolar fraction' because of the presence of structures that stain with the density of nucolei when the samples are stained with pyronine. The basis for the chemical technique for preparation of the nucleolar fraction is successive extraction of the nuclei with dilute and concentrated saline solutions.[32,34,52–54]

RELATIONSHIP OF RNA TO NUCLEAR PARTICLES

Recent studies[55] have indicated that, in all probability, 6S nucleolar RNA is present in the fibrillar elements and the 28S nucleolar RNA is present in the granular elements of the nucleoli. The data which support these conclusions arose from studies indicated in Fig. 2, in which gradient sedimentation patterns of nucleolar RNA of control and actinomycin D-treated liver nucleoli of untreated animals were compared along with control and actinomycin-D-treated nucleoli obtained of rats pretreated with thioacetamide. In the control (upper gradient) all of the RNA components were present that have been previously indicated. In the actinomycin D-treated nucleoli, segregation of the nucleolar components occurred (Fig. 3) and a marked loss of the rapidly sedimenting RNA was observed such that only 28S RNA was present. By planimetric measuremnets, the amount of 6S RNA increased to approximately equal amounts of that of the 28S RNA, and this change corresponded to an average increase in the amount of fibrillar components from 5 to almost 50% of the total nucleolar area. In the thioacetamide-treated livers, the amount of fibrillar component in the nucleoli was markedly reduced. However, the fibrillar elements still comprised approximately 5% of the total nucleolar area. The next to the bottom gradient in Fig. 2 shows that under these conditions, the amount of 6S RNA was small, i.e., about 5% of the total RNA.

As shown in Fig. 4, administration of actinomycin D to animals pretreated with thioacetamide resulted in a coalescence of the fibrillar element into

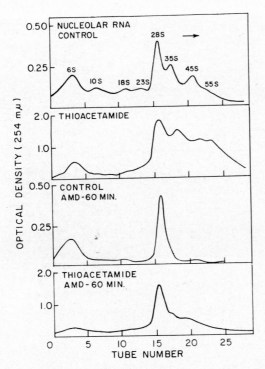

FIG. 2. Sucrose density gradient patterns for RNA of isolated nucleoli. The top figure is the control. The next figure is that for RNA of nucleoli of animals treated with thioacetamide. The figure next to the bottom is the pattern following treatment with actinomycin D, and the bottom figure is that for RNA of animals treated with both thioacetamide and actinomycin D.

small spherules which then were apparently extruded from the nucleolar surface. Since this unexpected effect produced nucleoli which consisted almost entirely of granular elements (Fig. 5) it was interesting to find, as shown in the bottom gradient of Fig. 2, that 6S RNA was virtually absent from these nucleoli. Accordingly, it seems most likely that 6S RNA is a component of the fibrillar element of the nucleolus, and 28S RNA is a component of the granular element. Whether the more rapidly sedimenting RNA fractions are present in the granular element only, or are part of the hazy substructure of the nucleolus is not apparent from these data.

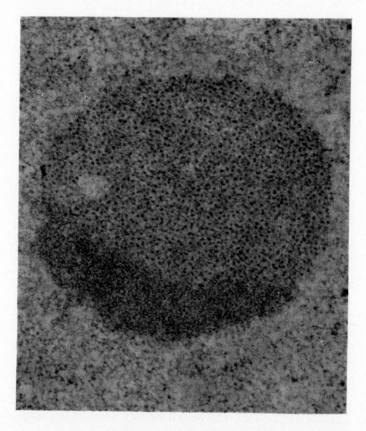

FIG. 3. Segregation of fibrillar and granular components of nucleoli of rats treated with actinomycin D (50 mg/kg body weight). × 60,000.

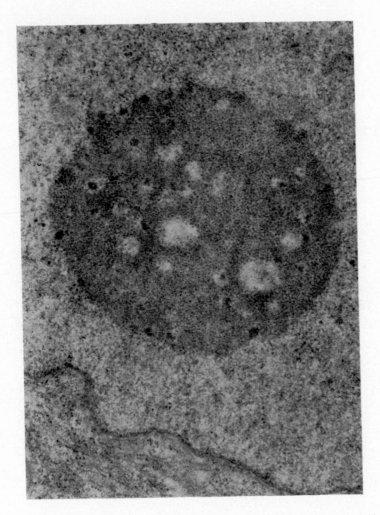

FIG. 4. Early effect (10–30 min) after injection of combined actinomycin D and thio-acetamide. Small spherules of fibrillar elements are apparently formed within the nucleoli. ×30,000.

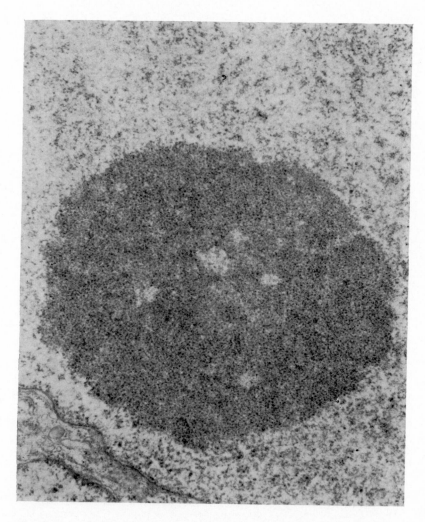

FIG. 5. Late effect (90 min) after injection of combined actinomycin D and thio-acetamide. The small spherules of fibrillar elements are apparently extruded from the nucleolar surface. ×30,000.

ROLE OF THE NUCLEOLUS IN RNA SYNTHESIS

Using highly purified nuclear preparations, it was found[54,60] that the 45S and 35S RNA existed in liver nuclear RNA in quantities readily demonstrable by U.V. patterns on the sedimentation profile. Evidence that these RNA components derive from nucleoli could be demonstrated by experiments with thioacetamide[54] and with regenerating rat liver.[56] Large increases in 45S and 35S RNA in the nuclear RNA, as demonstrated on the sedimentation profile, were correlated with large increases in nucleolar size and RNA.

Nucleolar RNA was analyzed on sucrose density gradients, as shown in Fig. 2. The nucleolar RNA contained large peaks of 45S, 35S and 28S RNA and a smaller peak of 6S RNA. Minor peaks were found in the 10S, 14S, 18S and 23S regions. In contrast to this, when the RNA of the extranucleolar fraction (S-1) was analyzed, three major peaks of 28S, 18S and 6S were present. However, it seems likely from other data that this fraction contains 45S RNA which is degraded in the course of sonication.

The sucrose density gradient profile of nucleolar RNA of the Walker 256 carcinosarcoma was essentially similar to that of the normal liver. Recent studies have been devoted to large-scale isolation of nuclear and nucleolar components, especially the RNA.[57-61]

COMPOSITION OF NUCLEAR AND NUCLEOLAR RNA

The initial studies on the base composition of the RNA of isolated nucleoli revealed that it was very GC-rich, as determined by optical density. The values for nuclear aqueous (a) and interphase (i) RNA (isolated according to Muramatsu and Busch[62]) are given in Table 2, which also shows the composition of the a- and i-RNA of whole nuclei, nucleoli, and extranucleolar nuclear RNA of the Walker tumor and the liver. Table 3 presents the analogous determinations on ^{32}P base compositions.[62]

Although the base compositions of the nucleoli of the Walker tumor and the liver were quite similar when analyzed by optical density, the ^{32}P base compositions differed remarkably in the high content of adenylic acid in the newly synthesized RNA of liver nucleoli by comparison with the low content of adenylic acid in these nucleolar fractions of the Walker tumor. This difference from the Walker tumor exists not only with respect to the nucleoli of normal liver but also with the nucleoli of regenerating liver (Table 4). It was initially believed that the difference between the ^{32}P-labeled RNA of the Walker tumor and the liver may have reflected the high growth rate of the Walker tumor. However, it would now appear that the difference was one of tissue type rather than rate of growth.

Recent studies have demonstrated that the base composition of newly synthesized nucleolar RNA differs markedly in tumors as compared to other

TABLE 2. BASE COMPOSITION (ULTRAVIOLET DETERMINATION)

Tissue	Adenine (A)	Uracil (U)	Guanine (G)	Cytosine (C)	$\dfrac{A + U}{G + C}$
		Nuclear a-RNA			
Walker tumor	18.6	19.0	33.6	28.8	0.60
Liver	19.2	23.2	32.2	25.2	0.74
		Nuclear i-RNA			
Walker tumor	23.4	21.6	29.0	26.0	0.82
Liver	23.3	28.2	25.5	23.0	1.06
		Nucleolar a-RNA			
Walker tumor	15.2	20.4	35.6	28.8	0.55
Liver	16.4	21.8	34.1	27.7	0.62
		Nucleolar i-RNA			
Walker tumor	14.8	21.3	36.2	27.7	0.57
Liver	14.6	20.0	34.9	30.5	0.53
		S-1 a-RNA			
Walker tumor	20.3	19.0	32.8	27.9	0.65
Liver	20.6	23.8	30.7	24.9	0.80
		S-1 i-RNA			
Walker tumor	24.9	25.6	27.5	22.0	1.02
Liver	23.6	29.3	27.2	19.9	1.12

tissues (Table 4). For each of the tumors studied, the content of adenylic acid in the nucleolar RNA was 12–14% of the total nucleotides, and the content of uridylic acid ranged from 17.5–22.5% of the total nucleotides. In liver, the adenylic acid content was higher, but in monkey kidney (tissue culture) the adenylic acid content was similar to that of the tumors. In the latter case, however, the content of uridylic acid was significantly higher than that of the livers or tumors studied.

Recent evidence has indicated that actinomycin D markedly suppresses the biosynthesis of the very GC-rich nucleolar RNA. In an effort to establish whether or not other species of 45S RNA were present in nucleoli, studies were made on the base composition of nucleolar RNA after injections of increasing amounts of actinomycin D.

Table 5 shows that the nucleoli do biosynthesize relatively small amounts of rapidly labeled AU-rich 45S RNA. Further efforts have been made to subfractionate the nucleolar 45S RNA. Although a number of problems

TABLE 3. BASE COMPOSITION (^{32}P)

Tissue	Adenylic acid (A)	Uridylic acid (U)	Guanylic acid (G)	Cytidylic acid (C)	$\dfrac{A + U}{G + C}$
Nuclear a-RNA					
Walker tumor	18.1	20.9	32.9	28.2	0.64
Liver	24.6	23.0	29.8	22.6	0.91
Nuclear i-RNA					
Walker tumor	19.1	24.3	29.9	26.8	0.76
Liver	25.6	24.3	27.1	23.1	0.99
Nucleolar a-RNA					
Walker tumor	13.0	21.1	35.5	30.4	0.52
Liver	20.5	17.0	38.2	24.3	0.60
Nucleolar i-RNA					
Walker tumor	12.9	21.1	33.8	32.3	0.51
Liver	21.5	17.5	36.4	24.7	0.64
S-1 a-RNA					
Walker tumor	21.9	28.6	26.5	23.0	1.02
Liver	26.2	25.1	25.3	23.4	1.05
S-1 i-RNA					
Walker tumor	23.6	28.3	24.4	23.7	1.08
Liver	30.6	21.6	25.8	22.0	1.09

TABLE 4. BASE COMPOSITION OF NUCLEOLAR RNA—^{32}P

	A	U	G	C
Tumors				
Ehrlich ascites	12.8	21.2	34.3	31.6
Novikoff hepatoma	12.0	22.1	32.9	32.9
Walker carcinosarcoma	13.0	21.1	34.7	31.3
HeLa	12.7	17.7	36.6	33.1
H-50	13.4	22.4	35.4	28.8
Liver				
Normal	21.0	17.2	37.3	24.5
Regenerating (6 hr)	20.6	16.1	36.9	26.2
Regenerating (18 hr)	19.7	16.8	37.7	25.6
Kidney				
Monkey	12.6	31.2	34.4	21.8

exist, including degradation of RNA, the results of the fractionation by means of partition chromatography on Sephadex have been encouraging. Highly purified nucleolar 45S RNA from the Novikoff hepatoma was chromatographed on Sephadex G-25 with the aid of the Kirby solvent system. A number of RNA fractions were obtained with base compositions

TABLE 5. EFFECT OF ACTINOMYCIN D ON BASE COMPOSITION
OF RAPIDLY LABELED 45S RNA

	A	U	G	C	r
Nucleolar 45S	14.6	22.6	34.0	28.7	0.59
Nucleolar 45S					
Actinomycin D	23.0	26.3	25.7	25.0	0.98
S-1 Walker	22.8	28.5	25.5	23.3	1.08
S-1 Novikoff	26.2	26.4	29.7	17.8	1.11

shown in Table 6. The most rapidly sedimenting fractions emerged in the last major peak which contains fractions 8–10. The A + U/G + C ratios in these fractions ranged from 0.45 to 0.49. The adenylic acid content was very low.

TABLE 6. NUCLEOLAR 45S RNA—NOVIKOFF HEPATOMA CHROMATOGRAPHIC
FRACTIONS (^{32}P BASE COMPOSITION)

Fraction number	Adenylic acid (A)	Uridylic acid (U)	Guanylic acid (G)	Cytidylic acid (C)	$\dfrac{A + U}{G + C}$
1	18.2	22.0	31.8	28.0	0.67
2	14.9	22.5	34.0	28.5	0.60
3	14.9	22.5	31.6	30.9	0.60
4	13.0	20.4	33.6	33.0	0.50
5	13.4	22.5	32.0	32.1	0.56
6	11.8	20.8	34.4	33.0	0.48
7	14.0	20.9	32.7	32.4	0.54
8	10.7	20.2	35.7	33.3	0.45
9	11.1	21.6	34.3	33.0	0.49
10	9.5	21.5	34.9	34.1	0.45

These studies indicate that either nucleolar 45S RNA consists of chains that are in part aggregated, or that a multiplicity of species of RNA may be present, of which some are extremely rich in guanylic and cytidylic acids. Recently, suggestions have been made that the ribosomal RNA consists of long strands of GC-rich RNA, to which AU-rich RNA is linked. These results would support such a conclusion.

TURNOVER OF NUCLEOLAR RNA

Studies with actinomycin D have shown that, at various time intervals after a pulse of actinomycin D, a rather marked shift of the distribution of nucleolar RNA occurs such that there is a relatively rapid loss of 45S and 55S RNA, a slower loss of 35S RNA and, finally, a 28S RNA peak remains in the nucleoli, as a residue of the rapidly sedimenting RNA. It has thus become possible to determine the half-lives or turnover times of the rapidly sedimenting RNA of the Walker tumor and other tissues.[59] In the Walker tumor, the half-life-time of 45S RNA was only 6.8 min; that of regenerating liver was 4–5 min. In normal liver nucleoli, the half-life of 45S RNA was 8–10 min. Since there is so much more 45S RNA in nucleoli of the Walker tumor than there is in the nucleoli of the regenerating liver, it was of interest to determine the total turnover.

On the basis of the previously determined amounts of RNA and the half-lives found by analysis of the actinomycin D-treated samples, the rate of synthesis of nucleolar 45S RNA in the Walker tumor was approximately 32 femtograms of RNA per min per nucleolus by comparison with approximately 16 in the regenerating liver. In normal liver nucleoli, the rate was much slower, i.e. 4 femtograms of RNA per min.

The role of the rapidly sedimenting nucleolar RNA is apparently to serve as a precursor for the ribosomal RNA, although it is possible that other functions exist. The series of reactions established is 45S to 35S to 28S RNA (Table 1).

From our present data, it seems that the rapidly sedimenting RNA represents either a hydrogen bonded precursor or a larger molecular weight precursor which must be subjected either to changes in tertiary structure or to cleavage in order to yield the 28S RNA of the ribosomes.[20]

THE NUCLEAR RIBONUCLEOPROTEIN NETWORK (NRN)

A variety of experiments in this laboratory has demonstrated that the product which remains after nuclei are extracted with 0.14 M NaCl and 2 M NaCl contains elements which are present in the nucleolus, but also a variety of elements that are absent in the nucleolus, i.e. 6S RNA is absent from this "residue fraction", and 18S RNA is present in substantial amounts.

Both the NRN and the nucleolar fractions contain substantial amounts of RNA that sediment in the 35S to 55S regions.[32,54] The base compositions of the RNA of the NRN is in many respects similar to that of the nucleolus, but contains in addition some RNA richer in adenylic acid. Both by ^{32}P base composition (Table 7) and U.V. base composition, there is a higher content of adenylic acid than is found in the nucleolar fraction (Table 3). What is of greater significance, however, is the fact that the NRN of the tumors contains newly synthesized RNA which differs in base composition

TABLE 7. BASE COMPOSITION OF NUCLEAR RESIDUE RNA—^{32}P

	A	U	G	C	r
Liver, normal	23.4	20.4	32.2	23.9	0.78
Liver, TA	23.5	19.2	33.8	23.5	0.75
Walker tumor	15.8	22.4	32.3	29.1	0.62
Novikoff hepatoma	15.7	20.2	35.2	29.0	0.56

from that of the liver to the same extent as that of the nucleolar RNA (Table 7) and nuclear RNA.[60]

This result indicates that much of the nucleolar RNA is present in the NRN and, further, that the additional RNA present is insufficient to dilute out the predominant differences that exist. Inasmuch as the RNA of the NRN is much more easily isolated, studies in this laboratory are in progress to isolate larger amounts of the RNA, using the NRN as a source. The evidence thus far is that this is a very useful procedure.

NUCLEOLAR ENZYMES

The enzymes of the nucleolus have been studied in isolated nucleolar preparations in recent months. The enzymes of RNA metabolism are of particular interest and somewhat puzzling results have been obtained with respect to the enzymatic ribonuclease. It came as no surprise that RNA polymerase activity was contained in the nucleolus,[34,42] although the results obtained with these enzymes with respect to activity do not correlate well with the biosynthetic activity determined from *in vivo* studies. In liver nucleoli *in vitro*, the rates of labeling of RNA are higher than those obtained in tumors, and this disagrees with the experimental findings obtained in the *in vivo* studies alluded to above.

Recent studies have been made on the RNase activities of nucleolar preparations, after it was demonstrated that RNase was present in nucleoli[63] and that the amount increased in nucleoli of thioacetamide-treated rats, in which the rate of nucleolar biosynthesis reactions is elevated. Further studies demonstrated that much of the nucleolar RNase activity was "latent", i.e. could be released by treatment with Triton X-100 or by freezing and thawing.

As shown in Table 8, the nucleolar ribonuclease activities differ markedly in tumors, normal, and regenerating liver. The activity of this enzyme in nucleoli was higher than that in the whole nuclear preparations in all tissues studied. However, much of the activity in the tumors was bound to inhibitors. The most surprising results with this enzyme have been obtained with regenerating liver. In this tissue, the total amount of enzyme and the amount of free enzyme were significantly greater than in normal liver and in the tumors. It may be suggested that this enzyme either serves as a control mechanism or as a part of the ribosomal apparatus, although neither of these conclusions can be supported by evidence at the present time. It would seem that in the tumors the high biosynthetic activity for nucleolar RNA is uncoupled from a destructive mechanism, or control mechanism, which would affect the total product released from the nucleolus. Alternatively, it is possible that the RNase subserves another function, such as a participating role in movement of messenger strands across the ribosomal surface.

TABLE 8. FREE RIBONUCLEASE ACTIVITY IN THE NUCLEI AND NUCLEAR SUBFRACTIONS FROM NORMAL AND NEOPLASTIC TISSUES

Fraction	Specific activity (ng RNase/mg protein)
Normal rat liver	
Nuclei	2.1
S_1	2.8
S_2	7.4
Nucleoli	18.0
Regenerating rat liver—18 hr	
Nuclei	9.0
S_1	11.2
S_2	17.6
Nucleoli	30.8
Walker 256 carcinosarcoma	
Nuclei	0
S_1	0.6
S_2	9.1
Nucleoli	10.6
Novikoff hepatoma	
Nuclei	0
S_1	0
S_2	5.7
Nucleoli	8.6
Ehrlich ascites tumor	
Nuclei	0
S_1	0
S_2	12.0
Nucleoli	14.2

CONCLUSIONS

Four major differences have been found between the nucleoli of neoplastic tissues and other tissues: (1) The nucleoli of neoplastic cells biosynthesize an RNA which is low in adenylic acid and uridylic acid by comparison to that of other tissues. (2) The nucleolar 6S RNA of the tumors studied differ significantly from that of normal liver in U.V. base composition. (3) The rate of synthesis of nucleolar RNA in the tumors studied is approximately twice that of regenerating liver and 7 times that of normal liver. (4) The RNase content of nucleoli of neoplastic cells is substantially lower than that of nontumor tissues. It is not certain whether any of these findings will lead to effective chemotherapeutic developments for neoplastic diseases, although it does seem that if such findings can be verified and extended, approaches to chemotherapy may well be delineated.

The role of the nucleolus is unequivocally that of 28S RNA synthesis, without which ribosomes cannot be formed. It is very unlikely that the nucleolus subserves only this role, and many claims have been made that the nucleolus synthesizes 18S RNA, m-RNA and other types of RNA species. In general, however, the worker in modern-day nuclear chemistry is confronted by the serious problems delineating the roles of the molecules already known and at the same time with an expanding horizon involving new information with respect to cell structure and cell function. This is a most exciting time, despite the frustrations which stem from inadequate methods and the many unknowns.

REFERENCES

1. T. H. MONTGOMERY, *J. Morph.*, **15**, 265, 1898.
2. T. O. CASPERSSON and J. SCHULTZ, *Proc. Nat. Acad. Sci. U.S.*, **26**, 507, 1950.
3. T. O. CASPERSSON, *Cell Growth and Cell Function*, New York, Norton, 1950.
4. J. BRACHET, *C.R. Soc. Biol., Paris*, **133**, 88, 1940.
5. J. BRACHET, *Chemical Embryology*, New York, Interscience, 1950.
6. J.-E. EDSTROM, *J. Biophys. Biochem. Cytol.*, **8**, 39, 1960a.
7. J.-E. EDSTROM, *Ibid.*, **8**, 47, 1960b.
8. J.-E. EDSTROM, W. GRAMPP and N. SCHOR, *J. Biophys. Biochem. Cytol.*, **11**, 549, 1961.
9. R. P. PERRY, A. HELL and M. ERRERA, *Biochim. Biophys. Acta*, **49**, 47, 1961.
10. R. P. PERRY, M. ERRERA, A. HELL and H. DURWALD, *J. Biophys. Biochem. Cytol.*, **11**, 1, 1961.
11. R. P. PERRY, *Proc. Nat. Acad. Sci., U.S.*, **48**, 2179, 1962.
12. R. M. FRANKLIN and D. BALTIMORE, *Cold Spring Harb. Symp. Quant. Biol.*, **27**, 175, 1962.
13. M. L. BIRNSTIEL, M. CHIPCHASE and J. BONNER, *Biochem. Biophys. Res. Commun.*, **6**, 161, 1961.
14. M. L. BIRNSTIEL, J. H. RHO and M. I. H. CHIPCHASE, *Biochim. Biophys. Acta*, **55**, 734, 1962.
15. M. L. BIRNSTIEL, M. I. H. CHIPCHASE and B. B. HYDE, *Biochim. Biophys. Acta*, **76**, 454, 1963.
16. M. L. BIRNSTIEL, M. I. H. CHIPCHASE and W. G. FLAMM, *Biochim. Biophys. Acta*, **87**, 111, 1964.
17. E. H. MCCONKEY and J. W. HOPKINS, *Proc. Nat. Acad. Sci., U.S.*, **51**, 1197, 1964.
18. R. DESJARDINS and H. BUSCH, *Texas Repts. Biol. Med.*, **22**, 444, 1964.
19. H. BUSCH, *Histones and Other Nuclear Proteins*, New York, Academic Press, 1965.
20. M. MURAMATSU, J. L. HODNETT, and H. BUSCH, *J. Biol. Chem.*, **241**, 1544, 1966.
21. M. MURAMATSU, J. L. HODNETT, W. J. STEELE and H. BUSCH, *Biochim. Biophys. Acta*, **123**, 116, 1966.
22. D. GROGAN, R. DESJARDINS and H. BUSCH, *Cancer Res.*, **26**, 775, 1966.
23. D. D. BROWN and J. B. GURDON, *Proc. Nat. Acad. Sci., U.S.*, **51**, 39, 1964.
24. F. M. RITOSSA and S. SPIEGELMAN, *Proc. Nat. Acad. Sci., U.S.*, **53**, 737, 1965.

25. A. FICQ, *Exptl. Cell Res.*, **9**, 286, 1955.
26. A. FICQ and J. BRACHET, *Exptl. Cell Res.*, **11**, 135, 1956.
27. C. H. WADDINGTON and J. L. SIRLIN, *Exptl. Cell Res.*, **17**, 582, 1959.
28. R. C. KING and R. G. BURNETT, *Science*, **129**, 1674, 1959.
29. M. J. OLSZWESKA, *Exptl. Cell Res.*, **16**, 193, 1959.
30. M. MURAMATSU, K. SMETANA and H. BUSCH, *Cancer Res.*, **23**, 510, 1963.
31. M. L. BIRNSTIEL and B. B. HYDE, *J. Cell Biol.*, **18**, 41, 1963.
32. W. J. STEELE and H. BUSCH, *Biochim. Biophys. Acta*, **119**, 501, 1966.
33. R. P. PERRY, *Proc. Nat. Acad. Sci., U.S.*, **51**, 2197, 1962.
34. T. S. RO and H. BUSCH, *Cancer Res.*, **24**, 1630, 1964.
35. G. SIEBERT, J. VILLALOBOS, JR., T. S. RO and H. BUSCH, *J. Biol. Chem.*, **241**, 71, 1966.
36. J. BRACHET, H. CHANTRENNE, and F. VANDERHAEGHE, *Biochim. Biophys. Acta*, **18**, 544, 1955.
37. H. BUSCH and R. DESJARDINS, *Exptl. Cell Res.*, **40**, 353, 1965.
38. R. MAGGIO, P. SIEKEVITZ and G. E. PALADE, *J. Cell Biol.*, **18**, 267, 1963a.
39. R. MAGGIO, P. SIEKEVITZ and G. E. PALADE, *Ibid.*, **18**, 293, 1963b.
40. K. J. MONTY, M. LITT, E. R. M. KAY and A. L. DOUNCE, *J. Biophys. Biochem. Cytol.*, **2**, 127, 1956.
41. H. BUSCH, M. MURAMATSU, H. R. ADAMS, K. SMETANA, W. J. STEELE and M. C. LIAU, *Exptl. Cell Res. Suppl.*, **9**, 150, 1963.
42. T. S. RO, M. MURAMATSU and H. BUSCH, *Biochem. Biophys. Res. Commun.*, **14**, 149, 1964.
43. K. TSUKADA and I. LIEBERMAN, *J. Biol. Chem.*, **239**, 1564, 1964.
44. A. SIBATANI, S. R. DE KLOET, V. G. ALLFREY and A. E. MIRSKY, *Proc. Nat. Acad. Sci., U.S.*, **48**, 471, 1962.
45. R. R. BENSLEY, *Science*, **96**, 389, 1942.
46. A. W. POLLISTER and H. RIS, *Cold Spring Harbor Symp. Quant. Biol.*, **12**, 147, 1947.
47. I. B. ZBARSKI and G. P. GEORGIEV, *Biochim. Biophys. Acta*, **32**, 301, 1959.
48. G. P. GEORGIEV and V. L. MANTIEVA, *Biokhimiya*, **25**, 143, 1960.
49. G. P. GEORGIEV, *Biokhimiya*, **26**, 1095, 1961.
50. G. P. GEORGIEV and J. S. CHENTSOV, *Exptl. Cell Res.*, **27**, 570, 1962.
51. G. P. GEORGIEV, O. P. SAMARINA, M. I. LERMAN, M. N. SMIRNOV and A. N. SERETZOV, *Nature, Lond.*, **200**, 1291, 1963.
52. K. SMETANA, W. J. STEELE and H. BUSCH, *Exptl. Cell Res.*, **31**, 198, 1963.
53. W. J. STEELE and H. BUSCH, *Biochim. Biophys. Acta*, **119**, 501, 1966.
54. W. J. STEELE, N. OKAMURA and H. BUSCH, *J. Biol. Chem.*, **240**, 1742, 1965.
55. K. SHANKAR NARAYAN, W. J. STEELE and H. BUSCH, *Exptl. Cell Res.*, 1966 (in press).
56. M. MURAMATSU and H. BUSCH. *J. Biol. Chem.*, **240**, 3960, 1965.
57. R. DESJARDINS, K. SMETANA, W. J. STEELE and H. BUSCH, *Cancer Res.*, **23**, 1819, 1963.
58. R. DESJARDINS, K. SMETANA and H. BUSCH, *Exptl. Cell Res.*, **40**, 127, 1965.
59. S. JACOB, W. STEELE and H. BUSCH, *Cancer Res.*, **27**, 52, 1967.
60. N. OKAMURA and H. BUSCH, *Cancer Res.*, **25**, 693, 1965.
61. R. DESJARDINS, K. SMETANA, D. GROGAN, K. HIGASHI and H. BUSCH, *Cancer Res.*, **26**, 97, 1966.
62. M. MURAMATSU, and H. BUSCH, *Cancer Res.*, **24**, 1028, 1964.
63. J. VILLALOBOS, JR., W. J. STEELE and H. BUSCH, *Biochem. Biophys. Res. Commun.*, **17**, 723, 1964.

DISCUSSION

T. TAKAHASHI: Dr. Busch, I would like to ask you about the mechanism of conversion of 45S RNA to 35S RNA.

H. BUSCH: The precise mechanisms are not known since the structure and possible cleavage products are not defined. Because the products are well defined by sedimentation, it seems that definitive enzymatic reactions are involved; we have referred to the enzyme as a "convertase". *In vitro* studies of its function have been complicated by the presence of RNase in the nucleolar preparations.

H. NAORA: Have you tested the possibility that 6S RNA is not capable of transferring amino acids?

H. BUSCH: Studies on the transfer activity of the 6S RNA are now being made but 6S nucleolar RNA has much less transfer activity. The base compositions differ from those of transfer RNA.

M. AMANO: Are there any differences in relative RNA contents of 6S nucleolar RNA between liver and kidney cells? Do you think that the different base ratio of kidney nucleolar RNA is due to the different amount of 6S RNA or other reasons?

H. BUSCH: The 6S nucleolar RNA of kidney has not yet been studied but if it is like that of liver and the tumors, it is likely that it is not as rapidly synthesized as 28S RNA, i.e. it is apparently not an export product. Whether 6S nucleolar RNA is a nucleolar product or not is also not clear, but from the close similarity of its base composition to the overall nuclear product, it seems more likely that the 6S RNA is an extranucleolar product.

S. SPIEGELMAN: There exists a 5S fraction in bacteria which is probably analogous to the "6S" shown in your data. It has been possible to demonstrate by competition molecular hybridization experiments that the "5S" component has a unique sequence, different from both the ribosomal and transfer RNA varieties.

Y. KAWADE: We found that the 42S RNA contains components corresponding to both the 18S and 30S ribosomal RNA. Dr. Busch, in your paper, the 18S RNA was missing or very small in amount, in the nucleolus. Do you have any idea about where the 18S RNA is?

H. BUSCH: Although we do not find 18S RNA in the nucleoli and do not find evidence for its synthesis by isotope experiments, we cannot rule out the possibilities that: (a) 18S RNA may be rapidly transferred out of the nucleoli, (b) 18S RNA may be formed on the nucleolar surface, or in the nuclear ribonucleoprotein network, or (c) 18S RNA may be lost during isolation. However, studies on the residue insoluble in 0.15 M NaCl and 2 M NaCl, show that there is much 18S RNA which is as rapidly labelled as 28S RNA. This residue, which contains the nuclear ribonucleoprotein network (NRN), contains both the nucleolus and the radiating network that is attached to the nuclear membrane. Segregation of functional activity of the nucleoli and extranucleolar elements of the NRN seems likely, although its functional value is not clear.

AN *IN VIVO* SYNTHESIS OF RNA
IN THE NUCLEOLI
OF MOUSE ASCITES TUMOR CELLS

MITSUO IZAWA

Biology Division, National Cancer Center, Research Institute,
Tsukiji, Chuo-ku, Tokyo, Japan

AUTORADIOGRAPHIC studies of RNA synthesis have clearly indicated that a nucleolus is an active intranuclear site of RNA synthesis.[1-3] Recent investigations of RNA synthesis in an anucleolate mutant of *Xenopus* developing embryos[4] and of the hybridization between DNA prepared from nucleolar and extranucleolar fractions of HeLa cell nuclei,[5] *Drosophila* adults[6] or *Xenopus* embryos[7] having different numbers of nucleolar organizer regions and ribosomal RNA extracted from the proper organism have proved more directly that the nucleolus is a special site of synthesis of ribosomal RNA. But it is still obscure whether the RNA synthesized can proceed to further maturation steps upon ribosomal particles of the nucleolus *in situ*.

On the other hand, the organization of the nucleolus has been studied extensively by electron microscopy[8,9] and differential centrifugation.[10,11] We have already reported methods yielding nearly complete solubilization of isolated nucleoli of a mouse ascites tumor cell using both a polyanion, sodium polyethylene sulfonate, and an anionic detergent, sodium deoxycholate.[11] It has also been found that the nucleolar lysate can be fractionated into subnucleolar components by means of sucrose zone sedimentation or differential centrifugation.[11]

In the present report, an *in vivo* RNA synthesis in the nucleoli will be described in terms of their organization. The findings strongly suggest that the nucleolus is a site of ribosomal particle formation in addition to that of ribosomal RNA synthesis.

MATERIALS AND METHODS

A mouse ascites tumor cell used was induced originally by a Rous sarcoma virus, Schmidt-Ruppin strain, kindly given by Prof. T. Yamamoto, Institute of Medical Sciences, University of Tokyo.

The procedures of isolation of nucleoli, solubilization of the isolated nucleoli and fractionation of the lysate into subnucleolar components have been described elsewhere.[11]

Chemicals. Uridine-2-[^{14}C] (specific activity, 23–30 mC/mmole) was purchased from the New England Nuclear Corporation, Boston, Mass., U.S.A. Inorganic ortho-[^{32}P] (specific activity, 85.5 C/mg P) was obtained from the Radiochemical Center, Amersham, England. Actinomycin D was a gift of Merck, Sharp and Dohme, Rahway, N.J., U.S.A. 5-fluorouracil was given by Hoffman-LaRoche, Basel, Switzerland, through the Kyowa Hakko Kogyo Co., Tokyo, Japan. Actidione, a cycloheximide, was a product of the Upjohn Company, Kalamazoo, Mich., U.S.A.

Incorporation experiments. In order to analyze *in vivo* RNA synthesis in the subnucleolar components the sucrose zone sedimentation[11] was employed exclusively as the fractionation procedure. RNA synthesis was examined by *in vivo* incorporation of uridine-2-[^{14}C] injected intraperitoneally at 2 μC per mouse for an appropriate period. After analysis of absorbancy at 260 mμ of the subcomponents fractionated by zone sedimentation the sample was treated twice with cold 5% trichloroacetic acid to remove acid-soluble materials. A bovine serum albumin (approximately 1 mg) was added as a carrier. The RNA in the acid-insoluble fraction was hydrolyzed by 1.5 ml of the acid at 90° for 20 min. The radioactivity in the hydrolysate (1 ml) was determined by a Packard scintillation spectrometer using 10 ml of Bray's scintillator.[12]

When the effect of actinomycin D, 5-fluorouracil and actidione on the pattern of incorporation among the subcomponents was tested, those agents dissolved in saline of an appropriate concentration were administrated in the same manner before or after injection of the radioactive precursors as described in the text.

RESULTS

Solubilization and Fractionation of the Isolated Nucleoli

Since the procedures of solubilization and fractionation of isolated nucleoli were described in detail previously,[11] only the results obtained are briefly summarized here. The nucleoli lyzed by the treatment with both 0.15% sodium polyethylene solfonate and 1% sodium deoxycholate in a buffer containing 0.02 M Tris-HCl (pH 7.5), 0.02 M KCl and 0.002 M MgCl$_2$ were fractionated by a differential centrifugation or sucrose zone sedimentation. By these treatments almost complete solubilization of the nucleoli could be accomplished, because only 7 and 12% of RNA and protein in the lysate were sedimented by centrifugation at 25,000 × g for 10 min. It was also observed that the nucleolar RNA was distributed in both the 105,000 × g sedimentable and non-sedimentable fractions in a ratio of 1/0.7.

After the lysate was centrifuged at 25,000 × g for 10 min; the supernatant layered on 5–30% sucrose gradient in the buffer was centrifuged at 35,000 rev/min for 160 min in a SW-39 bucket of a Spinco L ultracentrifuge. The results are seen in Fig. 1. There are two distinct peaks, i.e. a rapidly sedimented one occupying approximately 15% of the total 260 mμ absorbing materials and a slowly sedimented one holding a major part of the materials. Between these peaks there was an intermediate portion in which no significant peak was observed. As shown in Fig. 1, the former peak corresponded with the position of labeled ribosomes and the latter one with the position of labeled ribosomal RNA.

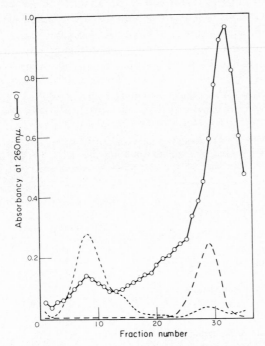

FIG. 1. Sucrose zone sedimentation pattern of the nucleolar lysate of mouse ascites tumor cells. The isolated nucleoli were solubilized by the action of both 0.15% sodium polyethylene sulfonate and 1% sodium deoxycholate in a buffer containing 0.02 M Tris-HCl (pH 7.5), 0.02 M KCl and 0.002 M MgCl₂. The lysate was centrifuged at 25,000 × g for 10 min. The supernatant fraction of the centrifugation was layered on 5–30% sucrose gradient in the buffer. The sedimentation was carried out by a centrifugation at 35,000 rev/min for 160 min using a SW-39 bucket in a Spinco L ultracentrifuge. The dotted and broken lines represent the position of labeled cytoplasmic ribosomes and ribosomal RNA respectively spun together with the nucleolar lysate.

In view of these findings it was concluded that there were ribosomal particles in the nucleoli and that the nucleolar RNA was distributed both in the ribosomal particles and in the non-sedimentable soluble fraction.

Some Properties of RNA and Ribosomal Particles in the Subnucleolar Components

In order to obtain more evidence on the similarity between the ribosomal particles in the nucleoli and ribosomes, attempts to compare the chemical nature of these fractions were made. The nucleotide composition of a bulk of RNA in both the $105,000 \times g$ sedimentable and non-sedimentable fractions analyzed by a method of Katz and Comb[13] proved very similar to each other and resembled that of ribosomal RNA of the tumor cells (Table 1). This result was confirmed by an analysis on the composition of newly synthesized RNA. The RNA in both the rapidly and slowly sedimented peaks on the zone sedimentation labeled with inorganic ortho-[^{32}P] (200 μC per mouse, 30 min) had nearly identical nucleotide compositions with that of the bulk of nucleolar RNA, and was of a typical ribosomal type (Table 1).

TABLE 1. NUCLEOTIDE COMPOSITION OF A BULK OF RNA (A_{260}) AND NEWLY SYN-
THESIZED RNA (^{32}P) IN THE NUCLEOLI AND THE SUBNUCLEOLAR COMPONENTS
OF MOUSE ASCITES TUMOR CELLS

The nucleotide composition on alkaline hydrolysate (0.3 N KOH, 37°C, 18 hr) of RNA in these fractions was analyzed by a method of Katz and Comb.[13] The composition of ribosomal RNA is also shown as comparison.

Fractions analyzed	Nucleotide composition (mol %)				$\dfrac{A + U}{G + C}$
	UMP	GMP	CMP	AMP	
Nucleolar RNA (A_{260})	22.2	35.3	28.6	13.9	0.56
RNA in the rapidly sedimenting peak (A_{260})	21.5	35.4	29.0	14.1	0.55
RNA in slowly sedimenting peak (A_{260})	23.7	36.2	26.0	13.9	0.60
Nucleolar RNA (^{32}P)	22.6	34.2	28.9	14.3	0.58
RNA in the rapidly sedimenting peak (^{32}P)	21.0	36.7	28.1	14.2	0.55
RNA in the slowly sedimenting peak (^{32}P)	21.4	35.3	28.4	14.9	0.57
Ribosomal RNA (A_{260})	21.0	34.3	29.1	15.6	0.58

The other way to characterize the RNA in the subnucleolar components was an analysis on molecular species of extracted RNA using the sucrose zone sedimentation. The RNA in the $105,000 \times g$ sedimentable fraction extracted by a sodium dodecyl sulfate-phenol method was layered over 5–40% sucrose gradient in a solution containing 0.01 M Tris-HCl (pH 7.5) and 0.14 M NaCl. The centrifugation was run at 39,000 rev/min for 4 hr in a SW-39 bucket of a Spinco L ultracentrifuge. It was found that the RNA consisted of both 28S and 18S without a 4S to 6S region, but the pattern was not so clear as was that of the ribosomal one.

When the nucleolar lysate was fractionated by zone sedimentation using the medium without magnesium ions, only the fast sedimenting peak was shifted significantly to the lighter portion and separated into two sub-components discretely. These components corresponded with the peaks of labeled subunits of the tumor cell ribosomes when they were spun together with the nucleolar lysate under the same conditions.

From these results together with the ones reported previously[11] a conclusion could be drawn that there were ribonucleoprotein particles in the nucleoli which have similar characters to the ribosomes.

Incorporation of Uridine-2-[¹⁴C] into the Subnucleolar Components

To clarify RNA synthesis in the nucleoli in connection with their structural organization described above, kinetics of the *in vivo* incorporation of uridine-2-[¹⁴C] into the RNA of subnucleolar components was studied. Results are shown in Figs. 2a and 2b. Newly synthesized RNA in the nucleoli during a very short labeling period, i.e. 1 to 2 min, was localized exclusively near the top of the sedimentation which corresponded to the position of ribosomal RNA. With a certain time lapse the radioactivity began to appear in the intermediate portion and then in the rapidly sedimenting peak (Fig. 2a). After 30 min the synthesized RNA was accumulated in the rapidly sedimenting one along the curve of absorbancy at 260 mμ with a distinct peak. The peak became more clear and on the contrary the radioactivity on the top area of the sedimentation began to fall significantly after 100 min (Fig. 2b). Finally after 12 hr significant radioactivity could be not detected from any subnucleolar component on sedimentation. These facts clearly demonstrated that newly synthesized RNA in the nucleoli was first ribosomal RNA, or its precursor, and gradually matured to ribosomal particles. The ribosomal particles completed in the nucleoli might be transferred to an extranucleolar part of the nuclei and then to the cytoplasm successively.

Effect of Actinomycin D, 5-fluorouracil and Actidione on the Incorporation of Uridine-2-[¹⁴C] into the Subnucleolar Components

The mechanism of synthesis of ribosomal RNA and/or the formation of ribosomal particles was studied through analyses on the effect of agents relating to RNA and protein syntheses on the pattern of incorporation into the subcomponents.

When actinomycin D, an inhibitor of DNA- dependent RNA synthesis,[14] was administered at a concentration of 1 mg per kg body weight; 2 min prior to injection of the precursor, the RNA synthesis was not observed in any subcomponent on the sedimentation. But it was noteworthy that this agent had no inhibitory effect on the formation of ribosomal particles from the nucleolar RNA synthesized before the administration of the inhibitor.

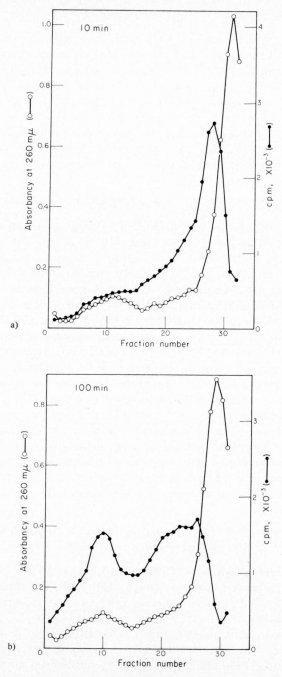

FIG. 2. Time course of the *in vivo* incorporation of uridine-2-[¹⁴C] into the sub-nucleolar components of mouse ascites tumor cells: (a) 10 min, (b) 100 min. Uridine-2-[¹⁴C] was administrated intraperitoneally at 2 μC per a mouse. The sub-nuclear components were obtained as described in Fig. 1. Open and closed circles represent absorbancy at 260 mμ and radioactivity of each fraction on the sedimentation respectively.

This result means that once formed, nucleolar RNA can be matured to the ribosomal particles without further concurrent RNA synthesis.

Next, the effect of 5-fluorouracil on the pattern of incorporation was tested. It has been demonstrated that 5-fluorouracil can be incorporated into RNA molecules in place of uracil[15] where it blocks the ribosome formation in bacterial cells.[16] It was checked separately that the agent given at a concentration of 25 mg per kg body weight after 2 min of addition of the precursors did not inhibit significantly both RNA and protein syntheses in the nucleolar fraction. Influence on the pattern of RNA synthesis in the presence of 5-fluorouracil supplied in the same manner is seen in Fig. 3.

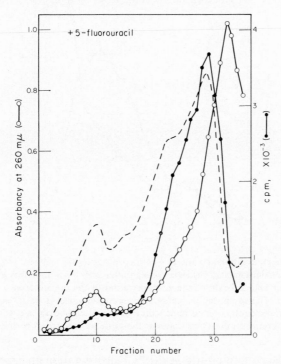

Fig. 3. Effect of 5-fluorouracil on the *in vivo* incorporation of uridine-2-[14C] into the subnucleolar components of mouse ascites tumor cells. 5-fluorouracil was given in the same manner before 2 min of the administration of the precursor at 25 mg per kg body weight. The subnucleolar components were obtained as described in Fig. 1. Open and closed circles are the same with Fig. 2. Dotted line represents the pattern of control (minus 5-fluorouracil) at the same incorporation period.

It was clearly shown that the synthesized RNA distributed only around the slowly sedimenting peak without moving it to the peak of ribosomal particles even after 30 min of incorporation. The RNA synthesized in the presence of 5-fluorouracil in the tumor cell nucleoli might be incapable of moving to further steps upon the ribosomal particle formation, that is to possibly protein assembling ones.

The effect of actidione, a potent inhibitor of protein synthesis,[17] on the location of synthesized RNA among the subnucleolar components was studied. When the antibiotic was provided at a concentration of 5 mg per kg body weight in 2 min prior to administration of the precursors, it did not prevent the RNA synthesis in the nucleoli, but did inhibit protein synthesis by

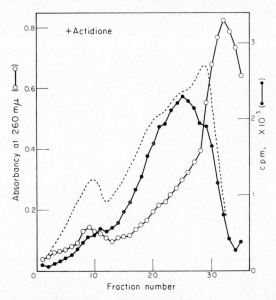

Fig. 4. Effect of actidione on the *in vivo* incorporation of uridine-2-[14C] into the subnucleolar components of mouse ascites tumor cells. Actidione was given in the same manner before 2 min of the administration of the precursor at 5 mg per kg body weight. The subnucleolar components were obtained as described in Fig. 1. Open and closed circles are the same with Fig. 2. Dotted line represents the pattern of control (minus actidione) at the same incorporation period.

over 90%. The pattern of uridine-2-[14C] incorporation on zone sedimentation in the presence of the agent was shown in Fig. 4. Actidione also blocked the appearance of synthesized RNA into the peak of ribosomal particles of the nucleoli. On this occasion, however, the pattern was different from that of 5-fluorouracil, so that the RNA was distributed in a wider range on the sedimentation including both the slowly sedimenting peak and the intermediate portion. This fact suggests a possibility that the maturation process of the ribosomal RNA in the nucleoli on the ribosomal particles can be advanced until a certain point without simultaneous protein synthesis. To finish the formation of ribosomal particles, however, the protein synthesis has to be accompanied at the same time.

DISCUSSION

It has been demonstrated by the sucrose zone sedimentation on the nucleolar lysate that the nucleolus consists of the following subcomponents; the rapidly sedimenting peak which has the same sedimentation profile with the ribosomes and the slowly sedimenting one located close to the position of ribosomal RNA.[11] The location of ribosomal particles in pea seedling nucleoli has been reported previously by Birnstiel et al.[10] It has also been considered that these two peaks correspond to the granular and the fibrous portions respectively observed in electron micrograms.[8,9] In this investigation, however, the fractionation of the intermediate portion between two peaks was not satisfactorily done and should be improved to distinguish each component on the way to maturation of the ribosomal particles in the nucleoli in situ.

Results of the molecular species and nucleotide composition of RNA extracted from the ribosomal particles of the nucleolus indicate that there is the ribosomal RNA consisting of both 28S and 18S components (Table 1). These findings agree with the results reported on the similarity of the molecular species[10,18] and the nucleotide composition[19,20] between nucleolar and cytoplasmic RNAs. These might be also sustained by recent findings on the localization of DNA cistron for both 28S and 18S ribosomal RNAs in the nucleolus or nucleolar organizer region.[5-7]

Time course experiments on the incorporation of uridine-2-[^{14}C] into the subnucleolar components fractionated by the sucrose zone sedimentation (Figs. 2a and 2b) have revealed that the nucleolus is not only the site of ribosomal RNA synthesis but also of ribosomal particle formation. This finding has been strongly supported by the recent electron autoradiographic studies of Granboulan and Granboulan[21] and Karasaki.[22] They have observed that pulse labeled RNA in the nucleoli of cultured monkey kidney cells and developing newt embryos has been found only in the fibrous portion of their ultrastructures and the grains can be observed subsequently on the granular one of them. As mentioned above, these fibrous and granular portions may be equivalent to the slowly sedimenting fraction and the rapidly sedimenting ones respectively in the present experiment.

While in the presence of 5-fluorouracil, the RNA and protein synthesized in the nucleoli had no quantitatively significant alteration, the synthesized RNA was not transferred to the ribosomal particles (Fig. 3). In a bacterial cell, E. coli, it has been found that 5-fluorouracil inhibits the ribosome formation at certain intermediate steps from ribosomal RNA, and that particles called "fluorouracil particles" accumulate in the cell.[26] The ribosomal RNA synthesized in the nucleoli of the tumor cells with this analogue may contain the fluorouracil molecule in it and be unable to assemble protein molecules on it. Thus the RNA stays nearly on the top of the zone sedimentation.

It also has to be noted here that the maturation process of the ribosomal RNA upon the ribosomal particles in the nucleolus is not interrupted by the treatment with actinomycin D, but ceases without simultaneous protein synthesis. This evidence indicates that the maturation process is essentially governed by reactions on protein synthesis and its assembly on the RNA. Similar observations on the effect of actinomycin D and puromycin on the formation of ribonucleoprotein particles in a nuclear fraction prepared from cultured HeLa and L cells have been reported by Tamaoki and Mueller[23] and Tamaoki.[24] Their findings may originate phenomena started in the nucleoli of those cells and be common with the results reported in the present investigation.

It is still uncertain whether the protein molecules which are assembled with the ribosomal RNA in the nucleolus can be synthesized properly in it. This question together with a precise analysis of the process of ribosome formation in nucleoli of higher organisms remains to be solved. Such study is now under progress in our laboratory and will be reported elsewhere.

SUMMARY

RNA synthesis in the nucleoli of mouse ascites tumor cells in terms of their structural organization were investigated. The organization of the nucleolus was analyzed by fractionation of the nucleolar lysate into sub-components using sucrose zone sedimentation. The lysate could consist of the following three fractions: the rapidly sedimenting peak having the same sedimentation profile with the ribosomes, the slowly sedimenting one which corresponded nearly to the position of ribosomal RNA and the intermediate portion between these peaks without any peak.

The pulse labeled RNA was localized exclusively in the slowly sedimenting peak as the ribosomal RNA or its precursor. Radioactivity was transferred gradually to the intermediate portion and then to the rapidly sedimenting peak. After 30 min of labeling the RNA in the rapidly sedimenting one was distributed along the curve of absorbancy at 260 mμ with a distinct peak. These findings clearly proved that the nucleolus was not only the site of ribosomal RNA synthesis but also the one of ribosomal particle formation. The maturation process of the RNA upon the particles in the nucleolus ceased at certain intermediate steps when protein synthesis was blocked, but was not essential for this to occur simultaneously with RNA synthesis.

ACKNOWLEDGEMENTS

The author would like to thank Dr. H. Naora for his advice and Miss K. Kawashima for her expert technical assistance. This work was supported in part by grants from the U.S. Public Health Service (CA-06986) to Dr. H. Naora, and a grant-in-aid from the Ministry of Education of Japan (No. 94044).

REFERENCES

1. P. WOODS, *Brookhaven Symp. Biol.*, **12**, 153, 1959.
2. R. P. PERRY, *Exptl. Cell Res.*, **20**, 216, 1960.
3. M. AMANO and C. P. LEBLOND, *Exptl. Cell Res.*, **20**, 250, 1960.
4. D. D. BROWN and J. B. GURDON, *Proc. Nat. Acad. Sci., U.S.*, **51**, 139, 1964.
5. E. H. MC CONKEY and J. W. HOPKINS, *Proc. Nat. Acad. Sci., U.S.*, **51**, 1197, 1964.
6. F. M. RITOSSA and S. SPIEGELMAN, *Proc. Nat. Acad. Sci., U.S.*, **53**, 737, 1965.
7. H. WALLACE and M. L. BIRNSTIEL, *Biochim. Biophys. Acta*, **115**, 296, 1966.
8. N. GRANBOULAN and P. GRANBOULAN, *Exptl. Cell Res.*, **34**, 71, 1964.
9. K. SMETANA, K. S. NARAYAN and H. BUSCH, *Cancer Res.*, **26**, 786, 1966.
10. M. L. BIRNSTIEL and M. I. H. CHIPCHASE and B. HYDE, *Biochim. Biophys. Acta*, **76**, 454, 1963.
11. M. IZAWA and K. KAWASHIMA, *Biochim. Biophys. Acta*, **155**, 51, 1968.
12. G. A. BRAY, *Anal. Biochem.*, **1**, 279, 1960.
13. S. KATZ and D. G. COMB, *J. Biol. Chem.*, **238**, 3065, 1963.
14. E. REICH and I. H. GOLDBERG, *Progr. Nucleic Acid Res. and Mol. Biol.*, **3**, 184, 1964.
15. C. HEIDERBERGER, *Ibid.*, **4**, 1, 1965.
16. S. OSAWA, *Ibid.*, **4**, 161, 1965.
17. C. W. YOUNG, P. F. ROBINSON and B. SACKTOR, *Biochem. Pharmacol.*, **12**, 855, 1963.
18. W. S. VINCENT, *Proc. 11th Intern. Congr. Genet., The Hague*, 1963, Vol. 2, p. 342, Ed. J. Geerts, Pergamon Press, London.
19. J. E. EDSTRÖM, W. GRAMP and N. SCHOR, *J. Biophys. Biochem. Cytol.*, **11**, 549, 1961.
20. J. E. EDSTRÖM and J. G. GALL, *J. Cell Biol.*, **19**, 279, 1963.
21. N. GRANBOULAN and P. GRANBOULAN, *Exptl. Cell Res.*, **38**, 604, 1965.
22. S. KARASAKI, *J. Cell Biol.*, **26**, 937, 1965.
23. T. TAMAOKI and G. C. MUELLER, *Biochim. Biophys. Acta*, **108**, 73, 1965.
24. T. TAMAOKI, *J. Mol. Biol.*, **15**, 624, 1966.

DISCUSSION

H. BUSCH: (1) What types of RNA are present in the 40S and 60S nucleolar particles?

(2) Was it possible to demonstrate nucleolar 18S RNA?

M. IZAWA: (1) We have not yet tested RNA molecular species in 60S and 40S subnucleolar components appearing in magnesium ion free medium.

(2) When total nucleolar RNA was extracted by an SDS-phenol method, it was found that there is an RNA component having a similar sedimentation pattern to 18S ribosomal RNA, but the pattern was not clear like ribosomal 18S.

RNA SYNTHESIS
IN THE CHROMATIN FRACTION
OF RAT LIVER AND ASCITES
HEPATOMA AH-130 CELLS

M. AMANO and T. FUKUDA

Biology Division, National Cancer Center Research Institute,
Tsukiji, Chuo-ku, Tokyo, Japan

RNA metabolism in different kinds of normal and cancer cells has been studied extensively by means of histochemical and biochemical techniques.[1] Nuclear structures such as chromatin[2-5] and the nucleolus[4-11] are playing important roles in RNA synthesis. It is necessary to isolate these nuclear structures in sufficient amounts for biochemical studies. The methods of isolation have been developed by Chauveau et al.[12] for normal liver nuclei, by Takahashi et al.[13] for Novikoff's ascites tumor cell nuclei and by Muramatsu et al.[14] and Desjardins et al.[15] for the nucleoli of normal rat liver and Walker tumor cells.

To compare the RNA metabolism of normal and cancer cells originating from same organ, rat liver and ascites hepatoma AH-130 cells are used in this experiment. An attempt was made to isolate nuclear components such as the nucleus, nucleolus and chromatin. The molecular species of RNA which are present and newly synthesized in the chromatin fraction were studied by the analysis of nucleotide compositions, sedimentation patterns in sucrose density gradient centrifugation and fractionation in gel filtration Sephadex G-200.

MATERIALS AND METHODS

Materials. Male albino rats of the Wister strain, weighing 250–300 g and fed *ad libitum*, were killed by decapitation for the normal liver. The livers were perfused with cold saline solution to remove blood cells. Female albino rats of the Donryu strain, weighing 150 g were injected intraperitoneally with 0.5 ml of ascites fluid of the tumor bearing rat 7 days after transplantation. Ascites hepatoma cells were harvested from rats 5 days after transplantation. Ascites fluid was centrifuged for 10 min at 250 rpm to collect hepatoma cells without erythrocytes. Normal rat was anesthetized with ether and injected with 1 mC of radioactive orthophosphate intravenously *via* the jugular vein. Tumor bearing rat was injected with 0.5 mC intraperitoneally.

Reagent grade chemicals were used for all analysis; deoxyribonuclease (DNase) from Worthington was electrophoretically pure and crystallized; polyvinyl potassium sulfate (PVS) from Wako Chemical Co.; Dowex ion exchange resin from Dow Chemical Co. and Sephadex G-200 from Pharmacia. Radioactive phosphate was purchased from the Isotope Society of Japan.

Fractionation of chromatin. The nuclei of rat liver were isolated by the method of Muramatsu *et al.*[14] From ascites hepatoma cells, the nuclei were isolated by the combination of the modified method of Takahashi *et al.*[13] and the hypertonic sucrose solution of Chauveau *et al.*[12] Hepatoma cells were suspended in a 3 mM $CaCl_2$ solution and homogenized with a Emanuel-Chaikoff homogenizer. Homogenate was mixed with an equal volume of 0.5 M sucrose solution containing 3 mM $CaCl_2$. The crude nuclear fraction obtained by centrifugation at 300 g for 10 min was suspended with 2.3 M sucrose solution and centrifuged again at 40,000 g for 50 min. The isolated nuclei from normal and hepatoma cells suspended in 0.25 M sucrose containing 10 μg PVS per ml were sonicated with a 150 W, 20 KC sonic oscillator (Umeda Electric Co.) for 2 min. The nucleolar fraction was isolated by the method of Muramatsu *et al.*[14] He attempted to obtain the chromatin fraction from the extranucleolar portion of the nucleus by addition of magnesium.[16] All supernatants of the nucleolar purification procedures of rat liver nuclei were mixed with equal volumes of water and made up at the final concentration of 0.75 mM of $MgCl_2$ and centrifuged at 18,000 g for 10 min. The clear supernatant and white precipitate were referred to as nuclear sap and chromatin fraction, respectively. To obtain the chromatin fraction from hepatoma cell nuclei, $MgCl_2$ was added to the extranucleolar fraction at the concentration of 1 mM and centrifuged at 10,000 g for 15 min.

RNA extraction. The chromatin fraction was suspended in 0.01 M Tris-HCl buffer solution pH 7.6 containing 0.14 M NaCl and 0.001 M $MgCl_2$ in a Tefron homogenizer. The suspension was vigorously shaken for 10 min and an equal volume of 80% phenol containing 0.1% 8-hydroxyquinoline was added and shaken again for 10 min. Nucleic acids in aqueous phase were precipitated by addition of two volumes of chilled ethanol and washed twice with cold 70% ethanol. Extracted nucleic acids dissolved in 2 ml of 0.01 M Tris-HCl buffer solution pH 7.6 containing 0.001 M $MgCl_2$ were incubated with 40 μg of DNase at 37°C for 20 min. The resulting deoxyribonucleotides were separated twice from the RNA fraction by solubilization and precipitation.

Analysis of RNA in sucrose density gradient centrifugation and fractionation on Sephadex G-200 column. About 400 μg RNA was layered on linear gradients of 5 to 40% sucrose solution and centrifuged at 39,000 rpm for 4 hr with swinging bucket rotor RPS-40 in a Hitachi 40P ultracentrifuge.

About 35 fractions were collected after puncturing the bottom of the tube. The optical density at 260 mμ of each fraction was determined. For the distribution of radioactive RNA, radioactivities were counted on the precipitate of each fraction after washing twice with cold 0.5 N perchloric acid, with 1 mg of bovine serum albumin as a carrier.

Sephadex G-200 column was used to separate RNA molecules by the method of Wood and Zubay.[17] Two ml of RNA, containing 20 units of OD_{260}, dissolved in the solvent, 0.15 M NaCl, 0.15 M sodium acetate pH 5.0, were applied to the top of the Sephadex G-200 column and fractions of 2 ml were collected.

Chemical determinations. Nucleic acids were precipitated with cold 5% trichloroacetic acid (TCA) and centrifuged. The precipitate was washed twice by suspending in cold 5% TCA solution and centrifuging. Nucleic acids were extracted twice with 5% TCA solution at 90°C for 15 min. DNA and RNA were determined by the Burton's[18] and Webb's[19] methods using calf thymus DNA and *E. coli* ribosomal RNA as standards.

Determination of nucleotide components in RNA hydrolysates. All nucleic acids and protein fractions of whole nuclei precipitated with cold 5% perchloric acid were hydrolyzed in 0.3 N KOH at 37°C for 18 hr. DNA and protein were separated by acidification after RNA hydrolysis. The extracted and fractionated RNA by gel filtration was hydrolyzed in the same way. Isolation of four ribonucleotides of RNA hydrolysates were performed by the method of Katz and Comb.[20]

For the analysis of radioactive base compositions, radioactive inorganic phosphates in separated nucleotide fractions were removed by Yanagita's method[21] before counting.

RESULTS

RNA content in chromatin fraction. To fractionate the chromatin from the extranucleolar portion of the nucleus, magnesium concentration and the condition of centrifugation were selected from the standpoint of the complete recovery of DNA and least amount of RNA in the precipitate. Nucleic acids contents varied depending on the magnesium concentration and the centrifugal force. The conditions for normal liver chromatin separation were reported previously.[22] The conditions for hepatoma cells were slightly different but under such conditions mentioned above, almost all DNA and about 65% of extranucleolar RNA were obtained in chromatin fraction.

Analysis of chromatin RNA by sucrose density gradient centrifugation and gel filtration. The sedimentation pattern of RNA extracted from the chromatin fraction of normal liver cells (Fig. 1) was different from that of ascites hepatoma cells (Fig. 2). Certain amounts of RNA having heterogeneous molecular sizes between 18S and 4S (center and right peaks in Fig. 1) were found in the chromatin RNA of normal liver cells. These RNA molecules were polydispersed but separated clearly by gel filtration from ribosomal and soluble RNA (Fig. 3). The relative amount of this intermediate molecular-sized RNA was 26.9% of the total chromatin RNA. On the other hand, the amount of intermediate molecular-sized RNA in chromatin fraction from hepatoma cells was less than 10% of the total chromatin RNA.

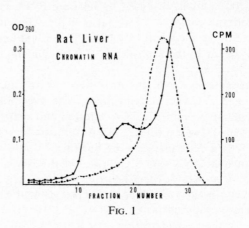

FIG. 1

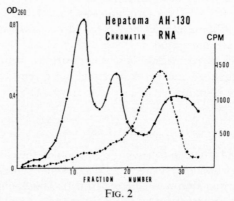

FIG. 2

FIGS. 1, 2. Sedimentation pattern of chromatin RNA by sucrose density gradient (5% to 40%) centrifugation (39,000 rpm, 4 hr). Optical density at 260 mµ and radio-activities are indicated by the solid line and broken line respectively.

Chromatin RNA of rat liver cells (Fig. 1) showed three peaks but the middle and right peaks were not separated well. The right peak was shifted slightly to the heavier position than that of 4S soluble RNA under these conditions of sucrose density gradient centrifugation. Radioactivities were almost all distributed in the region between 18S ribosomal and 4S soluble RNA 30 min after injection of radio-active phosphate.

Chromatin RNA of hepatoma cells (Fig. 2) were separated clearly into three different sizes of RNA. The 18S ribosomal and 4S soluble RNA, however, were not separated as well as in the case of cytoplasmic RNA. The distribution of radioactive RNA 30 min after injection was similar to that of rat liver chromatin.

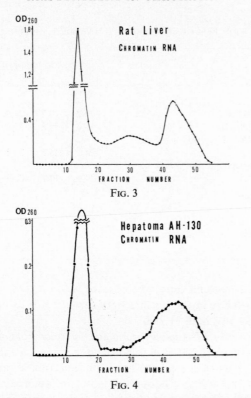

FIG. 3

FIG. 4

FIGS. 3, 4. Gel filtration elution patterns (Sephadex G-200) of chromatin RNA from rat liver and hepatoma cells. Three peaks can be identified in the solution profiles of normal rat chromatin RNA from left to right corresponding to ribosomal, intermediate molecular-sized and soluble RNAs (Fig. 3). But two peaks only can be seen in the RNA from the hepatoma chromatin fraction corresponding to ribosomal and soluble RNAs with a small amount of intermediate molecular-sized RNA (Fig. 4).

RNA synthesized in chromatin during short labelling with radioactive phosphate. Radioactive RNA was localized almost exclusively in the intermediate molecular-sized RNA species after 30 min labelling of chromatin RNA of normal liver (Fig. 1) and ascites hepatoma cells (Fig. 2).

The base compositions of newly synthesized RNA in chromatin fraction are shown in Table 1. The ratio of radioactive A + U/G + C after 15 min in normal liver chromatin is 1.06 and after 30 min in hepatoma chromatin 0.95. These values are quite different from that of ribosomal RNA but not exactly same as A + T/G + C of rat liver DNA.[25]

Base compositions of RNA hydrolysates from nuclear and chromatin fractions. The total nuclear RNA of normal liver cells had a higher ratio of A + U/G + C 0.87 than that of hepatoma cells 0.58 as shown in Table 1. The A + U/G + C ratio of chromatin RNA was 0.95 from normal liver but

TABLE 1. BASE COMPOSITIONS OF RNA (MOL.% OR COUNT%)

	U	G	C	A	$\dfrac{A+U}{G+C}$
Rat liver					
Nuclear RNA (OD)	29.2	29.2	24.3	17.3	0.87
Chromatin RNA (OD)	30.0	30.8	20.6	18.6	0.95
Chromatin RNA ^{32}P, 15 min labelling (count)	24.7	23.5	25.2	26.7	1.06
Intermediate molecular-sized RNA from chromatin (OD)	29.6	30.1	20.3	20.0	1.00
Hepatoma AH-130					
Nuclear RNA (OD)	22.9	34.5	28.9	13.7	0.58
Chromatin RNA (OD)	22.6	35.3	26.7	15.5	0.62
Chromatin RNA ^{32}P, 30 min labelling (count)	28.9	27.9	23.5	19.7	0.95

only 0.62 from hepatoma cells. The ratio of A + U/G + C of intermediate molecular-sized RNA in the normal liver chromatin fraction separated by gel filtration was also high at 1.00.

Specific activities of RNA in chromatin and nucleolar fractions. The specific activity of chromatin RNA is not much different from that of nucleolar RNA in normal liver cells until one hour after injection of radioactive phosphate as shown in Table 2. The specific activity of nucleolar RNA in hepatoma cells, on the other hand, is about four times higher than that of chromatin RNA.

TABLE 2. SPECIFIC ACTIVITIES OF RNA EXPRESSED AS COUNTS PER MIN PER MG RNA

Time intervals after injection	Normal rat liver		Hepatoma AH-130	
	Chromatin	Nucleolus	Chromatin	Nucleolus
30 min	12,320	16,680	497	2014
1 hr	20,300	37,200	1346	7492

DISCUSSION

The base compositions of the total nuclear RNA of rat liver cells are quite different from that of ascites hepatoma AH-130 cells, as shown in Table 1. The ratio of A + U/G + C of hepatoma cell nuclear RNA is lower than that of liver cell. The nucleolar RNA is characterized by the GC rich ribosomal RNA in rat and guinea pig liver,[24,8] regenerating liver[10] and also in tumor cell.[24] If the relative amount of nucleolar RNA in the total nuclear RNA is much higher in tumor cells than normal cells, the A + U/G + C ratio of total nuclear RNA of hepatoma cells becomes lower than that of liver cells. The percentage of nucleolar RNA to total nuclear RNA, however, is 25% in liver and 30% in hepatoma cells.[22,23] According to the results of chromatin RNA analysis by sucrose density gradient centrifugation and gel filtration, a relatively large amount of intermediate molecular-sized RNA was found in liver chromatin RNA but very little in that of hepatoma (Figs. 1 and 2). The base compositions of intermediate molecular-sized RNA are characterized by high $A + U/G + C$ ratios (Table 1). Thus, the high $A + U/G + C$ ratio of liver nuclear RNA is due to the large amount of this intermediate molecular-sized RNA.

It was concluded from these results that the RNA synthesized in chromatin of both normal rat liver and hepatoma cells is actually AU rich and intermediate molecular-sized RNA.

SUMMARY

Chromatin fraction was isolated from the extranucleolar portion of the nucleus by adding $MgCl_2$ at a concentration of 0.75 mM for normal liver and 1 mM for hepatoma. The chromatin fraction contained all the DNA and 40% of the RNA of the extranucleolar portion of the nucleus for normal liver and all DNA and 65% RNA for hepatoma.

About 27% of chromatin RNA in normal liver was intermediate molecular-sized RNA between ribosomal and soluble RNAs. A similar RNA molecular species was found in the chromatin RNA of hepatoma but the amount of this RNA species was less than 10%.

The synthesizing RNA in the chromatin was this intermediate molecular-sized RNA in both normal and hepatoma cell nuclei. This newly synthesized RNA in chromatin has high $A + U/G + C$ ratio of about 1.00.

ACKNOWLEDGEMENTS

The authors wish to thank Dr. H. Naora for encouragement through this work and Miss T. Akino for her technical assistance.

This work was in part supported by grants to Dr. H. Naora from the U.S. Public Health Service (CA-06986) and the Rockefeller Foundation.

REFERENCES

1. H. BUSCH, *An Introduction to the Biochemistry of the Cancer Cell*, p. 95. Academic Press, New York, 1962.
2. R. M. S. SMELLIE, In *Progress in Nucleic Acid Research*, **1**, 46. Ed. by J. N. Davidson and W. E. Cohn. Academic Press, New York, 1963.
3. R. P. PERRY, *Proc. Nat. Acad. Sci.*, **48**, 2179, 1962.
4. D. D. BROWN and J. B. GURDON, *Proc. Nat. Acad. Sci.*, **51**, 139, 1964.
5. M. AMANO, C. P. LEBLOND and N. J. NADLER, *Exptl. Cell Res.*, **38**, 314, 1965.

6. J.E.EDSTRÖM, *J. Biophys. Biochem. Cytol.*, **8**, 47, 1960.
7. M.I.H.CHIPCHASE and M.L.BIRNSTIEL, *Proc. Nat. Acad. Sci.*, **50**, 1101, 1963.
8. R.MAGGIO, P.SIEKEVITZ and G.E.PALADE, *J. Cell Biol.*, **18**, 293, 1963.
9. E.H.McCONKEY and J.W.HOPKINS, *Proc. Nat. Acad. Sci.*, **51**, 1197, 1964.
10. M.MURAMATSU and H.BUSCH, *J. Biol. Chem.*, **240**, 3960, 1965.
11. F.M.RITOSSA and S.SPIEGELMAN, *Proc. Nat. Acad. Sci.*, **53**, 737, 1965.
12. J.CHAUVEAU, Y.MOULE and C.ROUILLER, *Exptl. Cell Res.*, **11**, 317, 1956.
13. T.TAKAHASHI, R.B.SWINT and R.B.HURLBERT, *Exptl. Cell Res.* Suppl. **9**, 330, 1963.
14. M.MURAMATSU, K.SMETANA and H.BUSCH, *Canc. Res.*, **23**, 510, 1963.
15. R.DESJARDINS, K.SMETANA, W.J.STEELE and H.BUSCH, *Canc. Res.*, **23**, 1819, 1963.
16. D.G.COMB, R.BROWN and S.KATZ, *J. Mol. Biol.*, **8**, 781, 1964.
17. P.S.WOODS, and G.ZUBAY, *Proc. Nat. Acad. Sci.*, **54**, 1705, 1965.
18. K.BURTON, *Biochem. J.*, **62**, 315, 1956.
19. J.M.WEBB, *J. Biol. Chem.*, **221**, 635, 1956.
20. S.KATZ and D.G.COMB, *J. Biol. Chem.*, **228**, 3065, 1963.
21. T.YANAGITA, *J. Biochem.*, **55**, 260, 1964.
22. M.AMANO, *Exptl. Cell Res.* **46**, 169, 1967.
23. T.FUKUDA, M.IZAWA and M.AMANO, Unpublished.
24. M.MURAMATSU and H.BUSCH, *Canc. Res.*, **24**, 1028, 1964.
25. E.CHARGAFF, In *The Nucleic Acids* vol. 1, p. 356. Ed. by E.Chargaff and J.N.Davidson, Academic Press, New York, 1955.

THE MAINTENANCE OF RNA SYNTHESIS IN ISOLATED RAT LIVER NUCLEI

MASAOKI YAMADA and SUNAO IWATA

Laboratory of Cytochemistry, Department of Anatomy, School of Medicine
Tokushima University, Tokushima, Japan

IT IS known that a considerable loss of nuclear material and activity occurs during the isolation of nuclei (Stern et al.[1], Kay et al.[2], Logan et al.[3]). In preliminary reports of biochemical activities of the isolated nuclei many limits were due to the successful preservation of activities (Allfrey et al.[4-7]). In most of the studies on roughly isolated nuclei it is rather difficult to show nuclear activities as distinct from the activities of subcellular components. In particular for the study of nuclei, intranuclear RNA synthesis has to be separated from that in the cytoplasm. Few tissues are suitable for nuclear isolation and analysis by reliable methods. Evidence has been clearly obtained of RNA synthesis in the nuclei of thymocytes (or lymphocytes) isolated in a sucrose-calcium medium (Stern et al.[1], Allfrey et al.[4]). Their medium has proved fully satisfactory for nuclear isolation from other kinds of tissue. Liver tissue is usually employed for cell fractionation studies. However, little is known about nuclear activity itself as distinct from that of the cytoplasm. An attempt as been made to discover how to maintain the activity of RNA synthesis after enucleation from the liver. The results obtained concern this problem and may indicate the occurrence of nuclear RNA synthesis independent of cytoplasmic components.

MATERIALS AND METHODS

The livers of albino rats (Wistar) were employed for the isolation of nuclei. Prior to the isolation, the blood was removed from the heart by perfusion with 0.9% saline solution. About 10 g of liver tissue were obtained per rat, weighing about 150 g. The tissue was minced to remove connective tissue, then homogenized with 200 ml of a medium containing 0.25 M sucrose, 5 mM magnesium chloride and 0.54 mM EDTA, buffered with 25 mM Tris at pH 7.2. The homogenate was filtered through nylon cloth and the filtrate was centrifuged down at $1000 \times g$ for 15 min. The sediment was resuspended and the centrifugation was repeated. Next, the sediment was resuspended in a sucrose

111

medium of higher concentration (0.88 M), then was recentrifuged at $450 \times g$ for 15 min. The pellet then consisted of nuclei still contaminated with a few subcellular particles. For further purification, the pellet was resuspended in a medium containing 2.4 mM citric acid and 0.88 M sucrose, adjusted to pH 7.2, and was centrifuged at $450 \times g$ for 15 min. This final sediment contained only nuclei. All procedures were carried out at 1–2 °C. The nuclei thus obtained were used for chemical analyses, radioactive incorporations *in vitro* and cytofluorophotometry of nucleic acids.

The chemical analyses were performed for phosphorus of nucleic acids (Fiske *et al.*[8]), for DNA with the diphenylamine reaction (Dische[9]), for RNA with the orcinol reaction (Ceriotti[10]) and for protein with the biuret (Robinson *et al.*[11]) and the micro-biuret methods (Zamenhof[12]). For the purpose of RNA synthesis in the nuclei, a radioactive precursor for RNA was introduced into the nuclear preparation. The nuclei were suspended in an incubation medium containing 0.25 M sucrose, 5 mM magnesium chloride and 12 mM sodium chloride buffered with 25 mM Tris at pH 7.2, where 2.0 ml of the medium contained 2×10^7 nuclei in one tube containing 20 µg of DNA-phosphorus. For the incorporation into the nuclear RNA, 0.5 µC of ^{32}P-phosphate (S.A. 85 C/mg P) or 1.0 µC of uridine-2-^{14}C (S.A. 30 mC/mM) with non-labelled carrier was added to every tube containing the nuclear suspension, and incubated for various periods at 38 °C. The labelled nuclei were washed twice with the incubation medium containing 1 mM phosphate buffer (pH 7.2) and then washed successively three times with the incubation medium only, for the phosphorus assay. The nuclear sediments were collected carefully and were treated in the routine way and their radioactivities were measured with a gas-flow counter (Hitachi). For the subfractionation of RNAs from the labelled nuclei, the thermal SDS-phenol method (Georgiev *et al.*[13]) was used, and the radioactivity of each RNA fraction was measured.

To ascertain the purity of the isolated nuclei, the cytofluorophotometry of nucleic acids (Caspersson and Rigler[14–15]) was used. A microfluorophotometer constructed by the authors' group (Yamada *et al.*[16]) was employed for the evaluation of the nuclear preparation. According to this method, by which the ratio of nuclear RNA to DNA was measurable in an individual nucleus, the possibility of a nuclear preparation contaminated with cytoplasmic particles was easily excluded.

RESULTS

Chemical analyses of the isolated nuclei: Since the number of nuclei can be calculated in the nuclear fraction by using a hemocytometer, the mean amounts of DNA, RNA and protein are estimated in a single nucleus. The amount of DNA per nucleus is about 12–14 pg and that of RNA is around 2 pg, which is about 15 % of the DNA content. The mean amount of protein per nucleus is approximately 25 pg. Comparing the amount of DNA in the

nuclear fraction with that in the total homogenate, the recovery of fractionated nuclei seems to be 25 % of the total nuclei of the homogenate.

Evaluation of the purity of nuclei: The purity of the nuclear fraction was
judged directly from a smear of the nuclear fraction stained by the acridine
orange technique for nucleic acids (Rigler[15]). In a good preparation (Fig. 1)
there is only nuclear fluorescence (yellowish green for DNA and orange for
nuclear RNA) without any extranuclear fluorescence, while ribosomal orange
fluorescence is easily detectable when cytoplasmic contaminants are present.

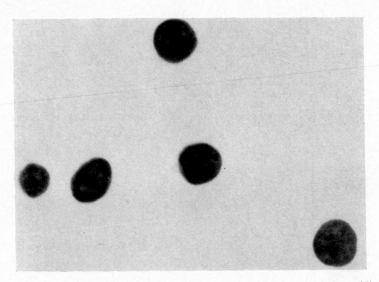

Fig. 1. A negative photograph of the fluorescence of nuclei stained with the acridine
orange method.

From the fluorescence spectrum of a single nucleus (Fig. 2), the maximum of
the peak is at 530 mμ which is identical with the peak of DNA. However, the
peak of RNA at 590 mμ is less apparent in the nucleus. The per cent ratio
of the amount of RNA to DNA is around 13 % which corresponds closely
to the value obtained from chemical assays.

The uptake of phosphoric acid and uridine into the nuclei: The purified
nuclei were incubated with ^{32}P-phosphate, and the uptake of radioactivity
was followed for 3 hr of incubation. The radioactivity of the nuclei increases
rapidly within the first hour and then slows down (^{32}P in Fig. 3). An
increase of radioactivity in the nuclei is also observed when uridine-2-^{14}C
is added (^{14}C in Fig. 3). This uptake of uridine requires phosphate at the
same level as in the uptake of ^{32}P-phosphate (0.05 mM). Accordingly, the
nuclear fraction seems to be lacking in inorganic phosphate. In the uptake
of ^{32}P-phosphate, the radioactivity extractable with cold TCA is apparently
seen during the first hour, when it is compared with the activity remaining

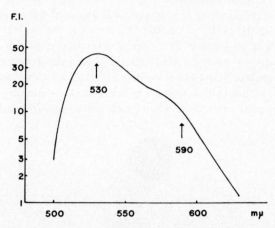

FIG. 2. A fluorescence spectrum of the nucleus stained with Rigler–Caspersson's method.

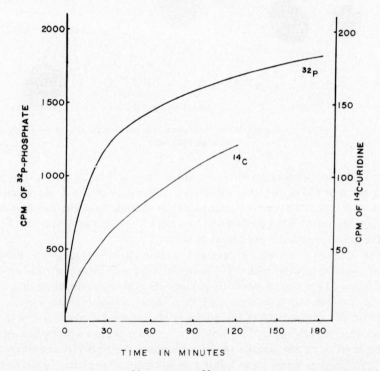

FIG. 3. The incorporation of ^{14}C-uridine or ^{32}P-phosphate into the isolated nuclei.

after cold TCA extraction (Fig. 4). This indicates that the ratio of the TCA-extractable radioactivity to the TCA-non-extractable one is initially higher and later becomes lower.

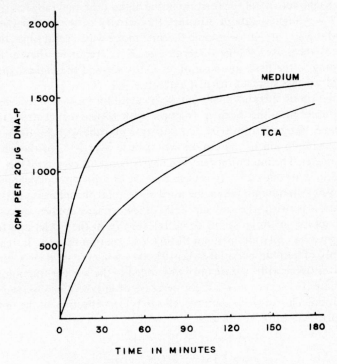

FIG. 4. CPM after washing with the medium or TCA.

The incorporation of ^{32}P-phosphate into RNAs: The subfractionation of RNAs was carried out with the thermal SDS-phenol method after 60 min of labelling with ^{32}P-phosphate. The RNAs of the nuclei were fractionated into three thermal ranges; 0–40°C, 40–55°C and 55–65°C (Table 1). The amount of phosphorus in the lowest thermal range was around 80% of the

TABLE 1. THE SUBFRACTIONATION OF RNAs LABELLED WITH ^{32}P-PHOSPHATE

No.	Total P (μg)			Total ePM			Specific activity		
	0–40 °C	40–55 °C	55–65 °C	0–40 °C	40–55 °C	55–65 °C	0–40 °C	40–55 °C	55–65 °C
1	161.0	37.0	3.2	1527	1285	888	9.5	34.7	281.5
2	137.6	21.0	2.7	1073	732	777	7.8	33.6	292.1
3	49.8	4.3	<1.0	558	215	212	11.2	50.0	—
4	125.0	3.3	<1.0	160	65	160	1.3	20.3	—

total RNA phosphorus, the amount in the middle range was less than 20%
and the amount in the highest range was less than 2%. So, the specific activity of the phosphorus in experiments 3 and 4 was too low to estimate. However, the specific activity of the fraction of the highest thermal range as shown
in 1 and 2 was highest (about 30 times the activity of the lowest thermal
range), and the activity of the middle thermal range was several times higher
than that of the lowest range. Georgiev et al.[13] reported that m-RNA
corresponding to the base composition of DNA showed the highest specific
activity and was resistant to thermal extraction.

The inhibition of ^{32}P-phosphate uptake by actinomycin-S: ^{32}P-phosphate
was introduced into the nuclear fraction with varied concentrations of
actinomycin-S, ranging from 0.04 µg to 10 µg/2 ml/tube with 2×10^7 nuclei.
The per cent inhibition of the uptake was a little over 30% of the activity
of control nuclei. The inhibition rate was nearly constant even at higher concentrations of actinomycin-S. It is clear that the phosphorus uptake relating
to actinomycin-S is not more than one third of the total phosphorus uptake.

The relation between the cell sap and ^{32}P-phosphate uptake: A certain
amount of soluble material seems to be released from the nuclei and to be
transfered into the isolation medium during the isolation process. In respect
of the supply of soluble materials, the 105,000 × g supernatant of a homogenate of liver tissue with sucrose only was added to the nuclear fraction with
^{32}P-phosphate. However, only a slight decrease of uptake was observed and
no acceleration. This suggests that the cell sap is least effective on the uptake
of phosphoric acid into the nucleus.

DISCUSSION

For this study of nuclear activities, 0.25 M (isotonic) sucrose medium has
been empirically used by adding a very small amount of cation. Bivalent
cations, especially calcium ions, protect nuclei against injury and
clumping during isolation (Schneider et al.[17], Hogeboom et al.[18], Yamada
et al.[19]). So that 0.25 M sucrose containing 1.8–6 mM calcium chloride has
been effectively used for the study of isolated nuclei (Allfrey et al.[4], Logan
et al.[3], Wilczok et al.[20], Maggio et al.[21], Fisher et al.[22], Henshaw et
al.[23]). The most successful isolation has been performed on thymocyte
nuclei with this sucrose medium by Allfrey, Mirsky and Osawa[4]. They
showed that 0.25 M sucrose was optimal for preventing loss of the activities of
thymocyte nuclei. In spite of the success of this medium with thymocyte
nuclei, it is not entirely suitable for purifying liver nuclei. It seems, at least,
to be necessary to keep deoxyribonucleoprotein (DNP) in a native state and
also to keep enzymes related to RNA synthesis in an active state for the study
of nuclear activities. Because of these necessities, 0.25 M sucrose medium in
combination with small amounts of EDTA and magnesium chloride was
used here. The use of this combined medium is supported by the preservation

of DNP (Zubay et al.[24]) and RNA-polymerase (Fox et al.[25]). Moreover, EDTA was so used to prevent nuclear aggregation resulting from the presence of cation. To obtain highly purified nuclei, two successive steps were used; one was the step of centrifugation in 0.88 M sucrose and the other was the step of washing with very dilute citric acid. In this way a suitable purity was attained. Data show a good agreement between chemical assay and cytofluorophotometric assay of the ratio of RNA to DNA. The amount of RNA of an individual nucleus, being around 15% of the DNA, is lower than that of previous reports (Hogeboom et al.[18], Chauveau et al.[26], Sporn et al.[27], Maggio et al.[21], Fisher et al.[28]). This low value may depend upon a removal of cytoplasmic contaminants and also upon a release of RNA from nuclei during the isolation process. Moreover, a lack of phosphoric acid may be indicated by the fact that ^{14}C-uridine uptake into the nuclei increases on adding phosphate. Even with lack of these materials the uptake of phosphoric acid or uridine into the nuclear fraction apparently occurs, and radioactive phosphorus incorporated into the TCA soluble fraction of nuclei is higher during the first hour than that later. This indicates that a rapid uptake into the acid soluble fraction is initiated before the incorporation into RNA. In respect to the distribution of the radioactivities in the RNAs fractionated with the thermal SDS-phenol method, the highest specific activity was shown in the fraction of the highest thermal range, in which the base analysis had been shown to be the DNA-type (m-RNA) by Georgiev et al.[13] This result is also supported here by the fact that partial blocking of ^{32}P-uptake occurs in the presence of actinomycin-S. This blocking of m-RNA is said to be similar to that of actinomycin-D (Kawamata et al.[29]). From the basis of present experiments, it is evident that RNA synthesis is maintained in liver nuclei even after their purification.

By analogy with its effect on thymocyte nuclei, sodium chloride has been tested and found to have effects on liver nuclei similar to those on thymocyte nuclei (Allfrey et al.[7]). However, it seems unsuitable to use calcium ions in the medium for isolation of liver nuclei because it seems to inhibit the activity of RNA-polymerase of the liver as reported by Fox et al.[25] In the present study of liver nuclei, calcium chloride can be effectively replaced by magnesium chloride. The use of magnesium is also demonstrated by supplying a lack of magnesium, which has been removed during the isolation process as suggested by Naora et al.[30] Moreover, the difference in the maintenance of the nuclear activities between thymocytes and liver cells may be dependent on different functions of possible organ properties of RNA-polymerase, its related enzymatic system and priming activities of the synthesis. Some of these may reflect on the lack of phosphoric acid in the nucleus. Even with excess cell sap, which contains sufficient phosphoric acid, there was no acceleration of RNA synthesis, which suggests the presence of a slight depressing action of the cell sap. However, this has still to be proved.

SUMMARY

To obtain nuclear RNA synthesis isolated from the liver cell, an effective method was introduced for the isolation of nuclei, the purity of which was evaluated by chemical and cytofluorophotometric assays. Such isolated nuclei were active in synthesizing RNAs independent of cytoplasmic components, in which m-RNA synthesis was initiated. A slight difference was observed between the maintenance of activities in liver nuclei and thymocyte nuclei.

ACKNOWLEDGEMENTS

The authors wish to express their thanks to Professor T. Caspersson and Dr. R. Rigler, Jr., Karolinska Institute, Sweden, for performing cytofluorophotometry, and to Prof. J. Kawamata, Research Institute for Microbial Diseases, Osaka, for a gift of actinomycin-S. This work was also supported by a grant (No. 71002, 1966) from the Ministry of Education of Japan.

REFERENCES

1. H. STERN and A. E. MIRSKY, J. Gen. Physiol., 37, 177, 1953.
2. E. R. M. KAY, R. M. S. SMELLIE, G. H. HUMPHREY and J. N. DAVIDSON, Biochem. J., 62, 160, 1956.
3. R. LOGAN, A. FICQ and M. ERRERA, Biochim. Biophys. Acta, 31, 402, 1959.
4. V. G. ALLFREY, A. E. MIRSKY and S. OSAWA, J. Gen. Physiol., 40, 451, 1957.
5. V. G. ALLFREY, In The Cell, p. 279. Ed. by J. Brachet and A. E. Mirsky, Academic Press, New York, 1959.
6. V. G. ALLFREY, J. H. FRENSTER, J. W. HOPKINS and A. E. MIRSKY, Ann. N. Y. Acad. Sci., 88, 722, 1960.
7. V. G. ALLFREY, R. MEUDT, J. W. HOPKINS and A. E. MIRSKY, Proc. Nat. Acad. Sci., 47, 907, 1961.
8. C. H. FISKE and Y. SUBBAROW, J. Biol. Chem., 66, 375, 1925.
9. Z. DISCHE, Mikrochemie, 8, 4, 1930.
10. G. CERIOTTI, J. Biol. Chem., 214, 59, 1955.
11. H. W. ROBINSON and C. G. HOGDEN, J. Biol. Chem., 135, 707, 1940.
12. S. ZAMENHOF, In Methods in Enzymology, vol. III, p. 696. Ed. by S. P. Colowick and N. Kaplan, Academic Press, New York, 1957.
13. M. I. LERMAN, V. L. MAUT'EVA and G. P. GEORGIEV, Biokhimiya, 29, 518, 1964.
14. T. CASPERSSON, G. LOMAKKA and R. RIGLER, JR., In IInd Int. Congr. Histo- and Cytochem. p. 88. Ed. by T. H. Schiebler et al., Springer, 1964.
15. R. RIGLER, JR., Acta Physiol. Scand., 67, Suppl., 267, 1966.
16. M. YAMADA, A. TAKAKUSU, K. YAMAMOTO and S. IWATA, Arch. Histol. Jap., 27, 389, 1966.
17. W. C. SCHNEIDER and M. L. PERTERMAN, Cancer Res., 10, 751, 1950.
18. G. H. HOGEBOOM, W. C. SCHNEIDER and M. J. STRIEBICH, J. Biol. Chem., 196, 111, 1952.
19. M. YAMADA, A. TAKAKUSU, S. IWATA and F. FUJISAWA, J. Nara Med. Assoc., 17, 153, 1966.
20. T. WILCZOK and K. CHORAZY, Nature, 188, 517, 1960.
21. R. MAGGIO, P. SICKEVITZ and G. E. PALADE, J. Cell Biol., 18, 267, 1963.
22. W. D. FISHER and G. B. CLEINE, Biochim. Biophys. Acta, 68, 640, 1963.

23. E. C. HENSHAW and H. H. HIATT, *J. Mol. Biol.*, **8**, 479, 1964.
24. G. ZUBAY and P. DOTY, *J. Mol. Biol.*, **1**, 1, 1959.
25. C. F. FOX, W. S. ROBINSON, R. HASELKORN and S. W. WEISS, *J. Biol. Chem.*, **239**, 186, 1964.
26. J. CHAUVEAU, Y. MOULE and CH. ROUILLER, *Exptl. Cell Res.*, **11**, 317, 1956.
27. M. B. SPORN, T. WANKO and W. DINGMANN, *J. Cell Biol.*, **15**, 109, 1962.
28. R. F. FISHER, D. J. HOLBROOK and J. L. IRVIN, *J. Cell Biol.*, **17**, 231, 1963.
29. J. KAWAMATA and M. IMANISHI, *Nature*, **187**, 112, 1960.
30. H. NAORA, A. E. MIRSKY and V. G. ALLFREY, *J. Gen. Physiol.*, **44**, 713, 1961.

DISCUSSION

H. BUSCH: Why is the activities *in vitro* so low when compared to those *in vivo*?

M. YAMADA: There is (1) a possible release of RNA and its relating enzymatic system during the nuclear isolation or (2) a removal of cytoplasmic contaminants which contain RNA.

A. DI MARCO: Did you measure the P/O ratio on these preparations? It is in fact possible that the RNA synthesis is limited by the available energy.

M. YAMADA: Yes, I think so, but I have not yet measured it in my experiments.

II. RNA–PROTEIN SYNTHESES AND CELL DIFFERENTIATION

MESSENGER RNA
FOR FIBROIN SYNTHESIS
IN POSTERIOR SILKGLANDS

YOSHIAKI MIURA and HIROWO ITOH

Department of Biochemistry, Chiba University School of Medicine,
Chiba City, Japan

SILK was once one of Japan's most important exports. Therefore, biological research on silkworms has been extensively carried out in this country.

Studies on the synthetic mechanism of silk protein are now being carried out by three main groups of Japanese biochemists: Professor K. Shimura of Tohoku University, Doctor H. Shigematsu of Sericultural Research Institute and myself at Chiba University. Professor Shimura's group has succeeded in isolating messenger RNA-like nucleic acid from polysome fractions of silkgland's homogenate;[1] while Dr. Shigematsu recently reported on RNA extracted from silkglands by alkaline phenol.[2] This has a potent enhancing effect of ^{14}C-glycine incorporation into fibroin.

This report will concentrate on describing our efforts to isolate the messenger RNA for the synthesis of fibroin.

Fibroin, the main component of silkprotein, produced by the common silkworm, *Bombyx mori*, is a fibrous protein, whose molecular weight is around 84,000. It contains about 43% of glycine, 32% of alanine and 15% of serine. Since a molecule of fibroin contains a high percentage of glycine, we can use ^{14}C-labeled glycine as a favorable marker of fibroin synthesis in silkglands.

In 1958, we published a paper entitled Fibroin Synthesis and Ribonucleic Acid Metabolism in the Silkgland,[3] in which we emphasized that the active incorporation of ^{14}C-orotate into RNA is not always parallel with extensive incorporation of ^{14}C-glycine into fibroin, although the existence of intact RNA is essential.

The introduction of a newer concept concerning messenger RNA into the mechanism of protein biosynthesis led us again to consider this matter in all its aspects. If the active labeling of RNA occurs before the fibroin synthesis, then a fairly large amount of messenger RNA should be synthesized before the beginning of fibroin synthesis and kept intact until it will be used as a template. In other words, we would like to postulate a relatively stable messenger RNA for fibroin synthesis.

123

Figure 1 shows the result of a comparative study of the *in vitro* incorporation of ^{14}C-orotate into RNA of the various subcellular fractions on the
third and sixth days of the fifth instar.

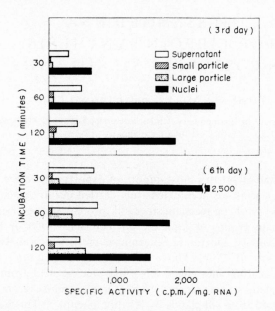

FIG. 1. *In vitro* uptake of ^{14}C-orotate by RNA of various subcellular fractions isolated from posterior silkgland cells on the 3rd and 6th days of the 5th instar.

On the third day, about 2 days before the actual synthesis of fibroin,
practically no incorporation of ^{14}C-orotate into large or small particle
fractions was observed, even after 120 min of incubation. Whereas, on the
sixth day, during the mostactive fibroin-synthesizing-stage, a fairly large
amount of radioactivity was detected in the large particles where fibroin
synthesis is actually carried out. Furthermore, it was demonstrated that the
specific activity of RNA in the large particles continued to increase for
120 min, in contrast with a gradual decrease of the specific activity of the
nuclear RNA.

From these results, it seems likely that rapidly labeled RNA is observed in
the nuclear fraction on the third day just before the fibroin synthesis and it
will emigrate from the nuclei into large particles in the cytoplasm when the
fibroin synthesis occurs on the 6th day.

In order to obtain this rapidly labeled RNA, we have extracted nucleic
acid from posterior silkglands without homogenation, since the nuclei of
silkglands are easily destroyed by the Waring blender.

Silkglands were soaked in a solution containing 5 % sodium dodecyl sulfate for 1–2 hr. Then, 5 volumes of 0.1 M tris-buffer (pH 7.5) and 5 volumes

of 90% phenol saturated with 0.1 M tris-buffer were mixed with one volume of tissue and vigorously shaken for 10 min at room temperature. The aqueous layer was separated by centrifugation. This procedure was repeated twice. Fibrous nucleic acids were precipitated from the aqueous layer, after an addition of 3 volumes of ethanol and $\frac{1}{10}$ volume of 20% potassium acetate.

The ratio of RNA to DNA in fibrous nucleic acid varies between 10 and 40 according to the date of the fifth instar. The DNA content remains relatively high in the early period of the fifth instar, whereas it begins to decrease from the fifth day. The fibrous precipitations were not obtained when it was previously treated by DNase. Besides the nucleic acid, fibrous precipitates contained traces of protein and polysaccharides. Treatment with lysozyme or proteinase did not have any effect on the formation of fibrous precipitation of the nucleic acid. Thus, the necessary condition for the formation of fibrous precipitation may be attributed to the presence of DNA.

The eluation pattern of the crude fibrous nucleic acid from methylated albumin column chromatography is shown in Fig. 2. There are, at least, two kinds of nucleic acids besides low-molecular nucleotides. The low-molecular nucleotides increase remarkably after treatment with RNase while a sharp peak of DNA began to appear in the remaining part of the elution pattern of the high-molecular nucleotides.

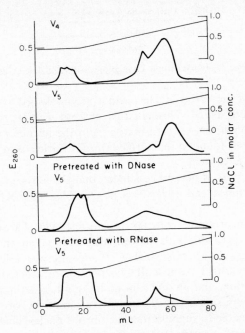

FIG. 2. Elution pattern of fibrous nucleic acid from methylated albumin column chromatography. V_4 means silkglands dissected on the fourth day of the fifth instar. V_5 means silkglands dissected on the fifth day of the fifth instar.

Figure 3 shows the data of hyperchromic change of the fibrous nucleic acid during the course of a gradual rise of temperature. There are at least two kinds of nucleic acids. Sharp transitions were observed around 47°C and 70°C. Treatment with RNase did not change this pattern but these two transition points faded out through treatment with DNase. From the data obtained with calf-thymus DNA, the hyperchromic change at 70°C indicates the dissolution of the double-stranded DNA, whereas the hyperchromic change around 47°C might depend upon the dissolution of the DNA–RNA complex, because the latter type of the complex has a lower melting point.[4]

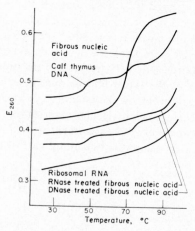

FIG. 3. Hyperchromic change of various nucleic acids.

Purification of the fibrous nucleic acid is accomplished by the salting-out-process with 10% NaCl. Through this process, a high molecular RNA, probably ribosomal RNA, was isolated from the fibrous nucleic acid. The base ratio of RNA found in the purified fibrous nucleic acid is characterized by its high G + C content.

From these results, it may be concluded that the fibrous nucleic acid is an RNA–DNA complex and its RNA is neither ribosomal RNA nor transfer RNA.

[14]C-orotate incorporates into fibrous nucleic acid. Through the *in vitro* incorporation experiment with minced silkglands taken on the fourth day of the fifth instar, it was revealed that ^{14}C-orotate was able to incorporate into fibrous nucleic acid. Almost all incorporated radioactivity was found in the nuclear fraction, and an addition of 10 μg/ml of actinomycin D to the incubation medium suppressed the incorporation of ^{14}C-orotate into RNA. Thus, it is reasonable to suppose that the fibrous nucleic acid obtained from silkglands contains messenger RNA for the fibroin synthesis. In order to verify this assumption, the following experiments on the incorporation of ^{14}C-glycine into protein were carried out.

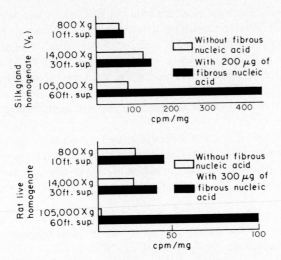

FIG. 4. Stimulating effect of silkglands' nucleic acid on the incorporation of ^{14}C-glycine into protein.

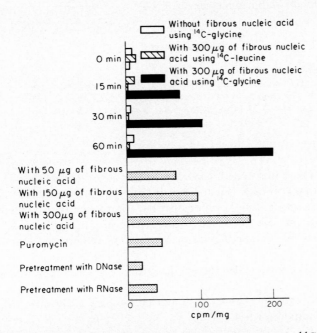

Fig. 5. Stimulating effect of silkgland fibrous nucleic acid on the ^{14}C-amino acid's incorporation system obtained from supernatant fraction of rat liver homogenate after centrifugation at $105,000 \times g$ for 60 minutes.

Figure 4 depicts the biological activities of fibrous nucleic acid. *In vitro* incorporation of ^{14}C-glycine into protein was assayed with each supernatant fraction of silkgland's homogenate after centrifugations. The stimulating effect of the fibrous nucleic acid on the incorporation of ^{14}C-glycine into protein was observed only in the case when the supernatant fraction of a $105,000 \times g$ centrifugation for 60 min was used. This supernatant fraction contains a sufficient amount of the necessary components except free messenger RNA.

Figure 5 gives the experimental results using rat liver homogenate instead of silkgland's homogenate. The addition of the fibrous nucleic acid obtained from silkglands enhanced the incorporation of ^{14}C-glycine into liver protein but no stimulating activity was observed when ^{14}C-glycine was replaced by ^{14}C-leucine.

Similar fibrous nucleic acid may be obtained from rat liver. The enhancing effect of rat liver fibrous nucleic acid on the amino acid incorporation into protein was more remarkable in the case of ^{14}C-leucine.

These results may be interpreted that fibrous nucleic acid secured from silkglands contains messenger RNA for the synthesis of glycine-rich and leucine-poor fibroin. The small amount of DNA that accompanies the messenger RNA might stabilize messenger RNA.

REFERENCES

1. K. SHIMURA, *Seikagaku*, **37**, 145, 1965.
2. H. SHIGEMATSU, H. TAKESHITA and S. ONODERA, *J. Biochem.*, **60**, 140, 1966.
3. S. TAKEYAMA, H. ITOH and Y. MIURA, *Biochim. Biophys. Acta*, **30**, 233, 1958.
4. H. M. SCHULUMAN and D. M. BONNER, *Proc. Nat. Acad. Sci., U.S.*, **48**, 53, 1962.

DISCUSSION

S. SENO: You have mentioned that DNA will act as to stabilize the m-RNA in the system that you have isolated, but is there any possibility that DNA might act as messenger as in the model that Dr. Naora mentioned yesterday?

Y. MIURA: In the late 5th instar, where active fibroin synthesis occurs, DNA is going to fade out. So, I guess there is little possibility that DNA acts as messenger.

S. SPIEGELMAN: Have you determined the structure of the complex between the RNA and DNA?

Y. MIURA: No, not yet. However, by our preliminary experiment, Mg^{++} would play an important role.

S. SPIEGELMAN: I might note that we have found a very useful procedure to avoid nuclease degradation. It consists of extracting in the presence of sodium dodecyl sulphate (SDS) and pronase.

Y. MIURA: I would like to use pronase in addition to SDS.

H. BUSCH: We have studied DNA–RNA complexes of liver but found low-molecular weight RNA present—probably due to the presence of ribonucleases. In similar complexes from Novikoff hepatomas, there is RNA with a variety of sedimentation constants. Some rapidly labeled RNA, high in A or U, sediments in the 60S region. Thus, this chromatin fraction contains RNA which is heterogenous with respect to sedimentation and probably there are a number of types of RNA present in these classes.

Y. MIURA: We have succeeded in isolating a similar fibrous RNA from the nucleoli of rat liver. However, the chemical analysis is not yet finished.

ON THE GLUCOSE-6-PHOSPHATE
DEHYDROGENASE ACTIVITY
AND ISOZYME OF RATS*

SAMUEL H. HORI and SAJIRO MAKINO

Zoological Institute, Faculty of Science, Hokkaido University, Sapporo, Japan

THE glucose-6-phosphate dehydrogenase (G6PD), the first step in the pentose phosphate shunt, is regarded as an important enzyme which provides reduced nicotinamide adenine dinucleotide phosphate for various synthetic reactions.

Recent genetical and biochemical studies on the G6PD of human red blood cells (RBC) have revealed that this enzyme is under the control of a gene located in the X-chromosomes, the dose effect is not observed in females, and that in heterozygous female carriers of the gene for G6PD deficiency, RBC consists of two populations, one normal and one deficient for G6PD (Beutler et al., 1962; Childs et al., 1958; Gross et al., 1958; Motulsky et al., 1959; Porter et al., 1962; Tarlov et al., 1962; Tönz and Rossi, 1964).

At about that time, it was reported by Lyon (1961) that one out of two X-chromosomes in female mice appeared to be genetically inert, and that either paternal or maternal X can be inert by chance. Similar findings had been made by Russell and her colleagues (Russell, 1961, 1963, 1964). In accordance with these genetical findings, some cytologists reported that sex chromatin observed in mammalian female cells (Barr, 1966) is of an X-chromosomal origin, and either paternal or maternal X can be heteropycnotic (Ohno and Hauschka, 1960; Ohno and Makino, 1961; Ohno and Weiler, 1961; Ohno et al., 1959). Such a condensed status of an X-chromosome does not seem to favor the synthesis and release of messenger RNA, so that the possibility was suggested of its metabolic inertness (Taylor, 1960; Hsu, 1962).

These genetic and cytologic findings provided a probable explanation for the aforementioned biochemical findings on human RBC G6PD; that is, no difference in the RBC G6PD activity between male and female, and mosaicism of RBC populations in females heterozygous for G6PD deficiency are both easily explained if one of the two X's in females is genetically inert. Likewise, the findings of Grumbach et al. (1962), and Harris et al. (1963) that

* This work is supported by a grant from Ministry of Education for the Co-operative Research (Cancer), No. 94002, 1966.

a quantitative relationship did not exist between the number of X-chromosomes and the RBC G6PD activity can also be attributed to dosage compensation by heteropycnosis of X-chromosomes in excess of one.

Although much of evidence had been accumulated in favor of the single-active-X hypothesis, several recent findings indicated that the hypothesis should not be accepted in its simplest form (Russell, 1964). It is beyond the scope of the present study to review indications unfavorable to it. It is worthwhile to point out here, in relation to RBC G6PD deficiency in humans, that the deficiency does not result from the direct action of a gene, but is a more remote consequence of the presence of a genetic defect (Marks *et al.*, 1959a, b). As a matter of fact, subjects with G6PD deficiency of RBC have normal levels of this enzyme in their white blood cells (WBC) and livers (Marks *et al.*, 1959a, b).

Boyer and Porter (1962) have concluded from the electrophoretic studies of WBC G6PD from subjects with RBC G6PD deficiency that the deficiency controlling locus is closely linked with the structural locus. However, the question is still unanswered as to why the deficiency controlling locus is functional only in RBC and skins (Davidson *et al.*, 1963; Gartler *et al.*, 1962; DeMar and Nance, 1964), but not in WBC and livers.

In addition to man, sex linkage of G6PD gene has been reported in Drosophila (Young *et al.*, 1964) in horse and donkey (Trujillo *et al.*, 1965;

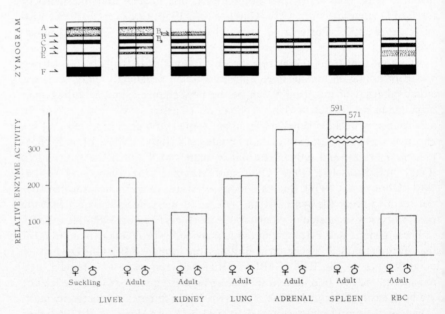

FIG. 1. Relative G6PD activities in the livers of suckling and adult rats, and in five other organs of adult rats, in addition to schematic representations of their zymograms.

Mathai *et al.*, 1966), and in wild hares (Ohno and Poole, 1965). In contrast, Shaw and Barto (1965) reported its autosomal linkage in *Peromyscus*.

Although this sex linkage has not been proven in rats, it seemed interesting to collect information on the phenotype of G6PD controlling gene and on factors affecting the phenotype in this experimental animal. This is dealt with in the present study.

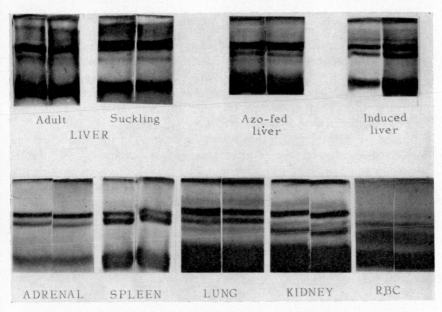

FIG. 2. Photographs of zymograms obtained from various organs of rats. Absence of bands A and B in zymograms from the livers of azo-dye fed male rats cannot be taken as an actual change induced by azo-dye feeding, in view of the fact that a clear-cut demonstration of bands A and B is often impossible even with samples from normal rats.

Our first findings on the G6PD of rats were (1) that hepatic enzyme activity was about twice as high in adult females as in males, and (2) that such a difference occurred in company by a difference in electrophoretically distinguishable molecular forms of the enzymes (Hori *et al.*, 1967) (Figs. 1 and 2). Zymograms obtained from water extracts of rat livers showed 6 bands, designated as A to F. Among these, bands C and F are major ones, and the rest equally minor. It was speculated that the sum of activities of minor bands was less than 10% of the total activity. The enzyme activities of bands D and E were roughly at the same level in males, while in females band D showed apparently more activities than band E. Although it was very difficult to demonstrate clearly bands A and B, the above-mentioned sex difference in relation to bands D and E was a consistent finding.

In order to determine whether sex difference was similarly present in other organs of adult rats, the lung, kidney, adrenal, spleen, and RBC were subjected to the tests for total activities and zymograms. As a result, sex differences were not detected in these organs. In this way the peculiarity of the hepatic enzymes was indicated. (Figs. 1 and 2).

Two questions arise from these findings: (1) do the observed sex differences exist in sexually immature animals?, and (2) if so, what causes them?

Examination of the livers of suckling rats (1–2 weeks old) has revealed that the enzyme activity is at about the same level in both sexes, though the values are 75% and 36% of those for adult males and females, respectively, and that zymograms of both sexes exhibit only band E, but not band D (Figs. 1 and 2). It was thus considered that band D enzyme appeared during maturation more abundantly in females than in males. Zymograms obtained from the other organs of suckling rats were similar to those from adults, and the total activities were already at about the adult level.

Based on the assumption that sexual dimorphism of the hepatic enzyme described above might be brought about by sex hormones, the enzyme activities and zymograms of gonadectomized, adult rats were investigated in the next step of the present study.

The results showed that neither orchiectomy nor ovariectomy induced changes in the total activity (Figs. 3A, B), and zymogram (Figs. $4A_1$, B_1).

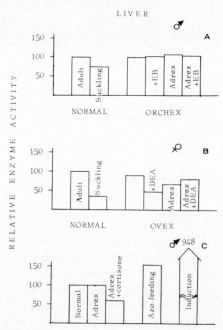

FIG. 3. Relative G6PD activities in the livers of male and female rats under several experimental conditions. EB: estradiol benzoate; DEA: dehydroepiandrosterone.

On the other hand, subcutaneous injections of dehydroepiandrosterone (DEA; 2 mg per day for 5 days) were found to strikingly suppress the hepatic enzyme activity of ovariectomized rats (Fig. 3), and the zymograms obtained from these animals exhibited bands D and E of equal activity (Fig. $4B_2$), as

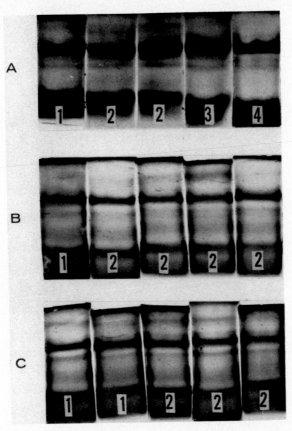

FIG. 4. Photographs of zymograms from gonadectomized rats. A: orchiectomized rats without further treatment (1); those received EB injections (2); orchiectomized and adrenalectomized rats without further treatment (3); those received DEA injections.

in the case of normal males. In contrast, estradiol benzoate (0.2 mg per day for 5 days) had no effect on the enzyme in orchiectomized rats (Fig. $4A_2$). In addition, no inhibitory action of DEA was found in the lung and kidney of ovariectomized rats.

These observations suggest that the low enzyme activity in adult males may be due to inhibition of the enzyme activity, or of enzyme synthesis by androgenic hormones, and that the inhibition is specific for hepatic enzymes. It was uncertain, however, that DEA exerted its influence directly upon hepatic G6PDs. In order to test this, the in vitro effect of DEA was first examined in crude hepatic enzyme preparations.

As a result, DEA exhibited 84.5% inhibition of enyzme activity at a concentration of 10^{-4} M, the data being comparable to those of Marks and Banks (1960). Their findings, as well as those of Hershey et al. (1963), have also indicated that the inhibitory effect of DEA in vitro is not limited to hepatic G6PD, but also holds for G6PDs from various sources. Accordingly, our present finding that the in vivo effect of DEA is restricted only to the

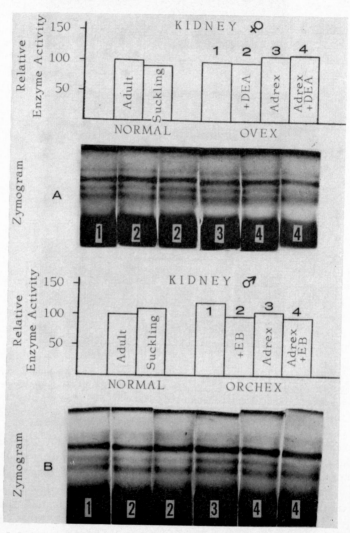

FIG. 5. Relative G6PD activities and zymograms of the kidney under various conditions.

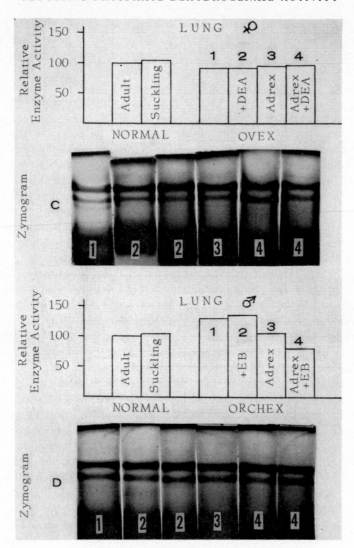

FIG. 6. Relative G6PD activities and zymograms of the lung under various conditions.

liver suggests the possibility that DEA may exert its effect through complex interactions with other endocrine systems.

In this connection, our preliminary data on the effect of bilateral adrenalectomy on G6PD seem interesting. Adrenalectomy of both normal and orchiectomized rats had no effect upon G6PD of the liver, lung and kidney (Figs. 4–6). On the contrary, a considerable decrease in G6PDs of the liver, but not those of the lung and kidney, was observed following adrenalectomy of ovariectomized rats (Fig. 3B). In addition, no further decrease of the activity of these enzymes was produced by injection of DEA into adrenal-ectomized and ovariectomized rats (Fig. 3B). In spite of such a decrease in the total enzyme activity, the zymograms showed a similar pattern to those of normal females (Figs. 4C 1, 2).

The above findings imply that DEA might exert its influence upon hepatic G6PD through these adrenal glands. Actually cortisone acetate was found to be a potent inhibitor of hepatic G6PDs (Fig. 3C).

The next question as to the sex difference of G6PD was whether it is possible to raise the enzyme activity of males up to the level of females. This was accomplished by two methods. One was the dietary induction of hepatic G6PD (Tepperman and Tepperman, 1958, Hori, 1966). In this case, the activity increased up to ten-fold (Fig. 3C). Zymograms of the livers of in-duced activity showed an intensified band D, a feature resembling those of females (Fig. 2). It was in this case that bands A and B were constantly and clearly demonstrated. The other is to feed rats on azo dye; feeding of male rats on diet containing 0.06% 3′-methyl-4-dimethyl-aminoazobenzene for 20 days or more resulted in about 50% or more increase of the total activity (Fig. 3C). Again, the zymogram obtained from azo dye-fed rats showed an intensified band D (Fig. 2). These findings strongly favor the view that band D enzyme increases to a greater extent than the others whenever the total activity is required to increase. In addition, speculation is possible that the band D enzyme in males is in a repressed condition. Stimulation of band D enzyme was also observed in lactating females.

In contrast to the stimulation of enzyme activity in these experiments, it was discovered that males castrated 7 weeks previously did not show appre-ciable changes in enzyme activity and in zymograms. The males were 4 weeks old at the time of castration. This finding indicates that the male phenotype in regard to activity and molecular forms of the G6PD had already been determined at 4 weeks after birth, and when once determined, had not been affected by removal of the testes. Evidently, the effects of androgenic hor-mones no hepatic G6PD during the first 4 weeks of post-natal life seem to be maintained for several weeks in the absence of hormones.

Evidence has been accumulating on hormonal control of specific enzyme synthesis (Ahren and Hamberger, 1962; Civen and Knox, 1959; Henning et al., 1963; Kenney and Kull, 1963; Lee and Lardy, 1965; Lee et al., 1959; Liao and Williams-Ashman, 1962; Mueller et al., 1958; Paik and Cohen, 1960; Sols et al., 1965; Tata, 1963).

Sex hormone dependence on certain enzymes, detectable by means of electrophoresis, has also been reported (Allen and Hunter, 1960; Shaw and Koen, 1963). The data of Shaw and Koen (1963) demonstrated the presence of a highly active kidney esterase in adult male mice, and its absence in im-mature and female mice. This male enzyme occurred in the kidneys of several strains of mice, but not in other organs of mice, or in any organ of rats. They also found that injection of testosterone, but not of estrogen, into female mice induced the male enzyme activity, though testosterone had no *in vitro* effect on the enzyme.

A similar testosterone induction of a sex-associated protein was reported by Bond (1962). This protein separated by DEAE-cellulose column chromatography exists in the livers of adult male rats, and can be induced in females by the injection of testosterone. Estradiol showed, in this instance, an inhibitory effect on the production of this enzyme in male protein.

Interesting was his finding that a considerable amount of this protein was already present in 5-week-old males, and that it did not decrease markedly following long-term castration. This is in conformity with the present data in that the effects of androgen appear to last for several weeks.

Evidence already given suggests, together with that obtained in the present study, that androgenic hormones are inhibitory to certain sex-associated proteins, but stimulative to other sex-associated proteins in the rat liver, and that estrogenic hormones have reverse effects in some cases, but do not in others.

The cause or causes remain to be explored why the G6PD activity increases markedly in females during maturation. Estrogenic hormones do not appear to be potent stimulators of the hepatic enzyme. Although growth hormone might be responsible for the activity increase, Glock and McLean (1955) reported that growth hormone has no *in vitro* effect on G6PD. However, ineffectiveness *in vitro* does not necessarily mean ineffectiveness *in vivo*, so that effect of growth hormone might be an indirect one, if any.

Although our knowledge at this time is too meager to permit enunciating the mechanism involved in the activity increase of G6PD in females, there is one thing which merits consideration in regard to this problem. It is the fact that polyploidization takes place so actively in growing rats that more than two-thirds of hepatic cells become polyploid, mostly tetraploid, when rats are grown up (Alfert and Geschwind, 1958; Beams and King, 1942; Clara, 1930; Fautrez *et al.*, 1955; Frazer and Davidson, 1953; Jacobj, 1925; Leuchtenberger *et al.*, 1951; McKellar, 1949; Naora, 1957; Pasteels and Lison, 1950; Swartz, 1956; Thomson and Frazer, 1954).

Such high percentages of polyploid cells is a peculiar phenomenon in the liver, and the polyploidization is known to be controlled by growth hormone (DiStefano *et al.*, 1955; Helweg-Larsen, 1949; Leuchtenberger *et al.*, 1954). Although the metabolic significance of polyploidization in the liver has not been clarified, it appears unlikely that doubling of genome would not exhibit any dose effect with regard to the protein synthesis. This speculation does not necessarily mean, however, that doubling of genome is a prerequisite to the doubling of protein synthesis.

Details of the method employed in the present study will appear in the coming issue of *J. Fac. Sci. Hokkaido Univ.* Ser. 6.

Added in proof: The data suggesting the induction of hepatic G6PD by estradiol were obtained after this manuscript had been written and published in *J. Histochem. Cytochem.* (1967) **15**, 530.

REFERENCES

AHREN, K. and HAMBERGER, L. (1962) *Acta Endocrinol.*, **40,** 265.
ALFERT, M. and GESCHWIND, I.I. (1958) *Exp. Cell Res.*, **15,** 230.
ALLEN, J.M. and HUNTER, R.L. (1960) *J. Histochem. Cytochem.*, **8,** 50.
BARR, M.L. (1966) In *Int. Rev. Cytol.* **19,** Ed. by Bourne, G.H. and Danielli, J.F., Academic Press, New York.
BEAMS, H.W. and KING, R.L. (1942) *Anat. Rec.*, **83,** 281.
BEUTLER, E., YEH, M. and FAIRBANKS, V.F. (1962) *Proc. N.A.S.*, **48,** 9.
BOND, H.E. (1962) *Nature*, **196,** 242.
BOYER, S.H. and PORTER, I.H. (1962) *Proc. N.A.S.*, **48,** 1868.
CHILDS, B., ZINKHAM, W., BROWNE, E.A., KIMBRO, E.L. and TORBERT, J.V. (1958) *Bull. Johns Hopkins Hosp.*, **120,** 21.
CIVEN, M. and KNOX, W.E. (1959) *J. Biol. Chem.*, **234,** 1787.
CLARA, M. (1930) *Z. mik. anat. Forsch.*, **22,** 145.
DAVIDSON, R.G., NITOWSKY, H.M. and CHILDS, B. (1963) *Proc. N.A.S.*, **50,** 481.
DEMARS, R. and NANCE, W.E. (1964) *Wister Inst. Symp. Monograph* No. **1,** 35.
DISTEFANO, H.S., DIERMEIER, H.F. and TEPPERMAN, J. (1955) *Endocrinol.* **57,** 158.
FAUTREZ, J., PISI, E. and CAVALLI, G. (1955) *Nature*, **176,** 311.
FRAZER, S.C. and DAVIDSON, J.N. (1953) *Exp. Cell Res.*, **4,** 316.
GARTLER, S.M., GANDINI, E. and CEPPELLINI, R. (1962) *Nature*, **193,** 602.
GLOCK, G.E. and MCLEAN, P. (1955) *Biochem. J.*, **61,** 390.
GROSS, R.T., HURWITZ, R.E. and MARKS, P.A. (1958) *J. Clin. Incest.*, **37,** 1176.
GRUMBACK, M.M., MARKS, P.A. and MORISHIMA, A. (1962) *Lancet*, **I,** 1330.
HARRIS, H., HOPKINSON, D.A., SPENCER, N., COURT-BROWN, W.M., and MANTLE, D. (1963) *Ann. Human Genet.*, **27,** 59.
HELWEG-LARSEN, H.FR. (1949) *Acta Pathol. Microbiol. Scand.* **26,** 609.
HENNING, H.V., SEIFFERT, I. and SEUBERT, W. (1963) *Biochem. Biophys. Acta*, **77,** 345.
HERSHEY, F.D., LEWIS, C., JOHNSTON, G. and MASON, S. (1963) *J. Histochem. Cytochem.*, **11,** 218.
HORI, S.H. (1966) *Gann*, **57,** 85.
HORI, S.H., KAMADA, T. and MATSUI, S. (1967) *J. Fac. Sci. Hokkaido Univ.* Ser. 6, in press.
HSU, T.C. (1962) *Exp. Cell Res.*, **27,** 332.
JACOBJ, W. (1925) *Arch. Ent. Organismen*, **106,** 124.
KENNEY, F.T. and KULL, F.J. (1963) *Proc. N.A.S.*, **50,** 493.
LEE, Y.P. and LARDY, H.A. (1965) *J. Biol. Chem.* **240,** 1427.
LEE, Y.P., TAKEMORI, A.E. and LARDY, H. (1959) *J. Biol. Chem.*, **234,** 3051.
LEUCHTENBERGER, C., VENDRELY, R. and VENDRELY, C. (1951) *Proc. N.A.S.*. **37,** 33.
LEUCHTENBERGER, C., HELWEG-LARSEN, H.FR. and MURMANIS, L. (1954) *Lab. Invest.*, **3,** 245.
LIAO, S. and WILLIAMS-ASHMAN, H.G. (1962) *Proc. N.A.S.*, **48,** 1956.
LYON, M.F. (1961) *Nature*, **190,** 372.
MARKS, P.A., GROSS, R.T. and HURWITZ, R.E. (1959a) *Nature*, **183,** 1266.
MARKS, P.A. and GROSS, R.T. (1959b) *J. Clin. Invest.*, **38,** 2253.
MARKS, P.A. and BANKS, J. (1960) *Proc. N.A.S.*, **46,** 447.
MATHAI, C.K., OHNO, S. and BEUTLER, E. (1966) *Nature*, **210,** 115.
MCKELLAR, M. (1949) *Am. J. Anat.*, **88,** 263.
MOTULSKY, A.G., KRAUT, J.M., THIEME, W.T. and MUSTO, D.F. (1959) *Clin. Res.*, **7,** 89.
MUELLER, G.C., HERRANEN, A.M. and JERRELL, K.F. (1958) *Recent Progr. Hormone Res.*, **14,** 95.
NAORA, H. (1957) *J. Biophys. Biochem. Cytol.*, **3,** 449.
OHNO, S., KAPLAN, W.D. and KINOSITA, R. (1959) *Exp. Cell Res.*, **18,** 415.

OHNO, S. and HAUSCHKA, T.S. (1960) *Cancer Res.*, **20**, 541.
OHNO, S. and MAKINO, S. (1961) *Lancet*, **1**, 78.
OHNO, S. and WEILER, C. (1961) *Chromosoma*, **12**, 362.
OHNO, S. and POOLE, J. (1965) *Science*, **150**, 1737.
PAIK, W.K. and COHEN, P.P. (1960) *J. Gen. Physiol.*, **43**, 683.
PASTEELS, J. and LISON, L. (1950) *Compt. Rend. Acad. Sci.*, **230**, 780.
PORTER, I.H., SCHULZE, J. and MCKUSICK, V.A. (1962) *Nature*, **193**, 506.
RUSSELL, L.B. (1961) *Science*, **133**, 1795.
RUSSELL, L.B. (1963) *Science*, **140**, 976.
RUSSELL, L.B. (1964) *Trans. N.Y. Acad. Sci.* Ser. II, **26**, 726.
SHAW, C.R. and KOEN, A.L. (1963) *Science*, **140**, 70.
SHAW, C.R. and BARTO, E. (1965) *Science*, **148**, 1099.
SOLS, A., SILLERO, A. and SALAS, J. (1965) *J. Comp. Cell. Physiol.*, **66**, suppl. 1, 23.
SWARTZ, F.J. (1956) *Chromosoma*, **8**, 53.
TARLOV, A.R., BREWER, G.J., LARSON, P.E. and ALVING, A.S. (1962) *Arch. Intern. Med.*,
    109, 209.
TATA, J.R. (1963) *Nature*, **197**, 1167.
TAYLOR, J.H. (1960) *J. Biophys. Biochem. Cytol.* **7**, 455.
TEPPERMAN, H.M. and TEPPERMAN, J. (1958) *Diabetes* **7**, 478.
THOMSON, R.Y. and FRAZER, S.C. (1954) *Exp. Cell Res.*, **6**, 367.
TÖNZ, O. and ROSSI, E. (1964) *Nature*, **202**, 606.
TRUJILLO, J.M., WALDEN, B., O'NEIL, P. and ANSTALL, H.B. (1965) *Science*, **148**, 1603.
YOUNG, W.J., PORTER, J.E. and CHILDS, B. (1964) *Science*, **143**, 140.

DISCUSSION

F. H. KASTEN: I do not have anything personally to add to this nice contribution but would like to mention that as I recall, there is a sexual dimorphism in the histological structure of one of the salivary glands of the adult rat. Since polysaccharide synthesis is involved in this organ, it might be interesting to study the isozymes of glucose-6-phosphate dehydrogenase in this instance.

PRIMARY INDUCTION IN AMPHIBIAN DEVELOPMENT, ACTIVE GROUPS OF THE EFFECTIVE AGENTS AND HETEROGENEOUS CELL COMPOSITION OF THE REACTOR TISSUES

Izumi Kawakami, Kiichiro Ave, Naoi Sasaki and
Munefumi Sameshima

Department of Biology, Faculty of Science, Kyushu University,
Fukuoka, Japan

The mechanisms of cell differentiation certainly are a major subject of modern biology, and attempts have been made by many investigators to approach this subject by employing the techniques of modern molecular biology. Cell differentiation, however, remains a mystery of life, and even under the present situation extensive study will be required to clarify this subject, because some interesting and important phenomena have been left without any conceivable explanation. One of these phenomena is primary induction in embryonic development.

The present paper is concerned with the induction for the development of amphibian embryos; the search for the active groups of the neuro- and mesoderm-inductive proteins obtained from guinea-pig liver and bone marrow, and the separation of the different kinds of cells destined to differentiate into definite tissues from presumptive epidermis of amphibian embryos.

INDUCTIVE SUBSTANCES IN PRIMARY INDUCTION AND BIOLOGICAL TESTS OF THEIR ACTIVITIES

Heterologous tissues, such as liver, kidney, and bone marrow of guinea-pig and rat have been proved to have the capacity of primary induction, and the active substances have been isolated and purified to some extent (Yamada, 1961; Kawakami and Iyeiri, 1964; Tiedemann and Tiedemann, 1964). The essential substance should be proteins themselves and not RNAs as they lose their inducing activity when treated with proteolytic enzymes (Toivonen and Kuusi, 1948; Hayashi, 1958, 1959; Tiedemann, Tiedemann and Kesselring, 1960).

For the analysis of the active groups of the inductor proteins, a crude ribo-nucleoprotein fraction precipitated with dihydrostreptomycin sulfate from the supernatant of centrifuged guinea-pig liver homogenate was used as the original neuro-inductive sample, and the precipitate obtained at pH 4.7 from the supernatant of the homogenate of guinea-pig bone marrow as the mesoderm-inductive sample. With these test protein samples the experimental system was prepared by wrapping them with the presumptive epidermis from *Triturus* gastrula, and the inductive effects of the samples were checked by observing the morphologic configuration of tissues differentiated from the epidermis after 12 days of incubation at 22 °C.

In the test with the liver ribonucleoprotein fraction, the inductions of the solely neural tissue were evoked in 29 cases (81 %) out of 36 observed, while with the extract of bone marrow, mesodermal tissues were mainly formed in all the 33 cases observed; solely mesodermal tissues in 27 cases and mesodermal tissues with some neural tissues in 6 cases.

CHEMICAL MODIFICATION OF THE NEURO-INDUCTIVE RIBONUCLEOPROTEIN FRACTION

Modification of guanidyl and amide groups on the inductor protein by treating with formaldehyde and alanine resulted in a complete suppression of the overall inducing activity of the samples. A remarkable but not complete suppression of the activity was produced by modifications of threonine- and serine-hydroxyl groups with formic acid (positive cases/total cases: 6/25, 24 %), and amino groups with acetic anhydride (9/24, 38 %), formaldehyde and acetamide (9/28, 32 %), or phenylisocyanate (13/35, 37 %).

Modifications of phenol and imidazol groups with iodine or diazobenzensulfonic acid did not change the character of the neural inductivity of the samples, but increased the frequency of induction, evoking neural tissue differentiation in 23 out of 24 test cases in iodine treatment and in all the 26 test cases in the diazobenzensulfonic acid treatment. An increase in the induction frequency was also obtained by reducing disulfide bonds with thioglycollic acid and iodoacetamide (27/29, 93 %), or thioglycollic acid and p-chloromercuribenzoic acid (23/23, 100 %).

Modification of serine-hydroxyl groups with phosphorous oxychloride, of carboxyl groups with methanol in the presence of hydrochloric acid, and of sulfhydryl groups with p-chloromercuribenzoic acid or iodoacetamide produced no effect on the neuro-inductive activity of the protein.

The effect of modification of free amino groups on the inductive activity has been studied by several workers (Lallier, 1950; Kuusi, 1953; Tiedemann and Tiedemann, 1959). According to these authors, formaldehyde, ketene, and nitrous acid almost completely destroyed the overall capacities of inductive tissues. The agents and conditions which were employed in their studies, however, were relatively nonspecific in their reactions with amino groups. The reagents and reaction conditions employed in the present study were those acting more specifically on the target amino acid residues, and in some instances the inducing effects were observed by using different kinds of reagents, each having a similar specific activity to the protein. Therefore, the data of the present study will give a reliable basement to assume that guanidyl and/or amide groups are the most important for the neuro-inductive activity of the protein. Among these two groups we draw particular attention to the guanidyl groups, because protamine, which has arginines only as the basic amino acid and no side-chain amide groups, becomes neuro-inductive when conjugated with RNA (Hayashi, Kawakami and Sasaki, unpublished).

The increase of neuro-inductive capacity following the cleavage of disulfide bonds seems to suggest an unfolding of the protein structure exposing the amino acid residues having neuro-inductive activity. The increase in neural induction activity of mesoderm-inductive substance after treating with some reagents, to which we will refer later, may be due to a similar mechanism.

CHEMICAL MODIFICATION OF MESODERM-INDUCTIVE PROTEIN

In order to determine the active groups of mesoderm-inductive substance, the modifications of the amino acid residues of the extract, of guinea-pig bone marrow were carried out by employing treatments similar to those used in the case of neuro-inductive substance. Through these observations it was found that the effects of modification of phenol and imidazole groups by treating with iodine and of guanidyl and amide groups with formaldehyde in the presence of a high concentration of alanine resulted in remarkable changes in the inducing capacity of the substance. The modification with iodine resulted in complete suppression of the mesoderm-inductive capacity of bone marrow extract with the appearance of a strong neuro-inductive capacity, in contrast to the case of the neuro-inductive substance, in which these modifications increased the frequency of induction. The blockage of guanidyl and amide groups yielded an identical result as in the case of the neuro-inductive substance; i.e. mesoderm-inductive capacities of the bone marrow extract were completely lost.

Modifications other than those mentioned above brought no remarkable change in the mesoderm-inductive capacity of the extract, though the modification of amino groups with acetic anhydride, and of sulfhydryl and disulfide bonds with sodium thioglycollate and iodoacetamide increased the frequency of the neural induction by the bone marrow extract.

It may be concluded from these data that the guanidyl and/or amide groups and the phenol and/or imidazol groups play important roles in mesoderm-inductive activity.

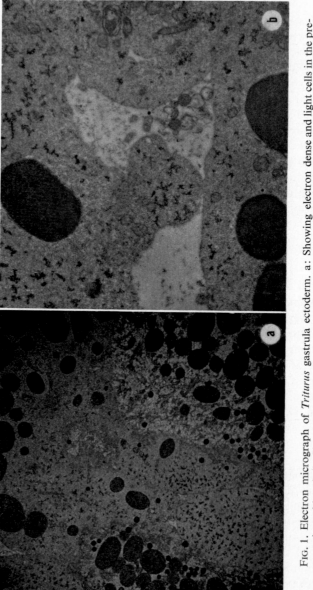

Fig. 1. Electron micrograph of *Triturus* gastrula ectoderm. a: Showing electron dense and light cells in the presumptive epidermis. b: Showing a destroyed electron dense cell in the epidermis, which had been brought into contact with a piece of guinea-pig bone marrow.

DIFFERENT CELL TYPES IN THE AMPHIBIAN
PRESUMPTIVE EPIDERMIS

To understand the mechanism of primary induction, more profound knowledge of the reactor tissue, i.e. the presumptive epidermis of the gastrula embryo, must be accumulated. On account of the complexity of the differentiation mechanism, however, little is known about the physico-chemical processes underlying the differentiation of the reactor cells, though several explanation concepts have been proposed. Among these concepts, multipotency of embryonic tissues has been long accepted without any question of the capacity of presumptive epidermis to form both ectodermal and mesodermal tissues. In other words, it has been taken for granted that the presumptive epidermis consists of developmentally homogeneous cells, and that neural and mesodermal differentiations from the epidermis are simply induced by the action of an inductive substance which is responsible for the cellular specialization. In the field of primary induction most investigators have conducted their experiments with this idea; Nieuwkoop and Igtevecht (1954) assumed two factors, one for activation and the other for transformation, and Toivonen and Saxén (1955) assumed the actions of neuralizing and mesodermalizing factors in primary induction. Their ideas were based on the premise that the presumptive epidermis is homogeneous in its cell composition and that each cell has dual potencies for differentiating neural and mesodermal cells. The same view was held by us (Kawakami and Yamana, 1959).

Eakin and Lehmann (1957), observed electronmicroscopically two kinds of cells, electron dense and light ones in *Xenopus* gastrula ectoderm. They attributed this difference in electron density to a difference in phase of differentiation of the cells. We detected these two kinds of cells in *Triturus* gastrula ectoderm (Fig. 1a), and saw cytolysis of the electron dense cells when mesoderm-inductive bone marrow tissue was brought into contact with the presumptive epidermis (Fig. 1b), suggesting that the two kinds of cells might be different in their differentiation destination. To approach this problem, an electrophoretic separation of these two cell types was attempted in order to determine whether or not the presumptive epidermis is truely heterogeneous in cellular composition.

Cells from the presumptive epidermis of *Triturus* gastrula, the dorsal blastopore lip from middle gastrula, and the fore-brain from an embryo in early tailbud stage were dissociated with Ca- and Mg-free Niu-Twitty solution containing EDTA (0.038 %). Then the dissociated presumptive epidermal cells were treated for 130 min with Holtfreter solution (M/11) acidified with lactic acid (pH 3.0) (Okano and Kawakami, 1959). By this treatment the dissociated cells were artificially neuralized. On the other hand, the dissociated cells were mesodermalized by treating for 4 hours with the extract of guinea-pig bone marrow (Yamada, 1958; Katoh, 1962). All these dissociated cells were suspended in glycerine of 1.09 specific gravity for the following experiment. A 2-cm column of the glycerine-cell sample was introduced into the cathodal arm of a U-glass tube between Na acetate-Na Veronal buffers of different pHs as shown in Fig. 3.

The electrophoresis apparatus employed was nearly the same as that described by Kolin (1955). Electrophoresis was carried out at 4°C at 120 V for 30 min. On starting the current, a potential gradient developed in the glycerine column containing the isolated cells, and separation of cell types was effected. As shown in Fig. 2 and Fig. 3A, the dissociated cells from the presumptive epidermis were distributed into three zones, labelled N, M_1, and M_2. The populations of cells of the N, M_1, and M_2 zones were 35, 41, and 24% of the original cell number, respectively.

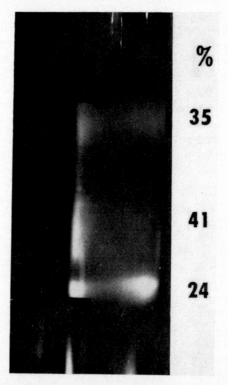

FIG. 2. Photograph showing electrophoretic spectra of the cells isolated from presumptive epidermis. The percentage of cell numbers shown in the figure refers to the cells in the N, M_1, and M_2 zones identified in Fig. 3.

The treatment of the dissociated cells with acidified solution resulted in a loss of 68% of the original number of cells. Under electrophoresis the remaining cells (32%) migrated to the position of the N zone, and no cells were distributed at the levels of M_1 and M_2 (Fig. 3B).

An almost identical electrophoretic spectrum of cells was obtained from cells dissociated from fore-brain. The small number of cells distributed at the level corresponding to M_2 may be due to mesodermal cells contaminating the brain (Fig. 3C).

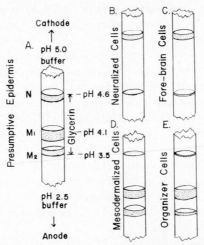

FIG. 3. Schematic drawings of the electrophoretic spectra of isolated cells from presumptive epidermis (A), of neuralized cells (B), of isolated cells from fore-brain (C), of mesodermalized cells (D), and of isolated cells from the organizer (E). The pH value at each level of N, M_1, and M_2 is indicated in (A).

After the treatment with the bone marrow extract the dissociated presumptive epidermal cells decreased to 66% of the original cell number, and were distributed upon electrophoresis at the levels of M_1 and M_2 (Fig. 3D).

The dorsal blastopore lip differentiates itself into mesodermal tissues accompanying a small amount of neural tissue, so that, for the present, it was chosen as a typical mesodermal tissue of embryos in an early developmental stage. The cells isolated from the tissue distributed in two zones corresponding to M_1 and M_2 upon electrophoresis, except for a small number of cells at the level of N (Fig. 3E).

DISCUSSION AND SUMMARY

In the present study, we referred to two sets of experimental data. In the first study we showed that both neuro- and mesoderm-inductive substances have the specific amino acid residues which are responsible for the active inductive capacities of the substances. From this we imply the protein nature of the essential inductor contained in the extract of bone marrow and liver.

The inductive activities of the extracts of guinea-pig liver and bone marrow were lost by treating with the medicaments that induce the modification of guanidyl and/or amide groups, and the mesoderm-inductive samples became neuro-inductive after being exposed to the agents which result in the modification of phenol and/or imidazole groups. These data led us to believe that the participation of guanidyl and/or amide groups induces the activation of neural differentiation, and when acting in conjugation with phenol and/or imidazole groups the activation of mesodermal differentiation.

In the second study we demonstrated the heterogeneity of the cell composition of the presumptive epidermis of *Triturus* gastrula. This has forced us to modify our view of the induction mechanism. The present observation certainly shows that the presumptive epidermis consists of at least three cell types, and each of these cell types responds differentially to the environmental stimulants, such as those of neuralizing and mesodermalizing agents. Moreover, one of these cell types showed the same electrophoretic behavior as that shown by the cells of specifically neuralized epidermis and of the differentiated neural tissue, and the other two cell types behaved as did the mesodermalized epidermis and normal mesodermal tissues. These results positively refute the concept of the multipotentiality of presumptive epidermis, at least as far as the ectodermal and mesodermal differentiations are concerned.

Although further studies are required for the establishment of the concept, the fact that the presumptive epidermis is made up of heterogeneous cell components and that the number of cells which disappeared after the neuralizing treatment is comparable to the cell number in the M_1 and M_2 zones in electrophoresis and likewise the number that disappeared following mesodermalizing treatment to the cell number in the N zone suggest strongly the possibility that the neuro-inductive agent selectively destroys the M_1 and M_2 cell types that are destined to become mesodermal cells, and conversely, that the mesoderm-inductive agent destroys the N type cells that are destined to become ectoderm. Electron microscopic observations of presumptive epidermis mesodermalized by guinea-pig bone marrow regularly showed cytolysis of the electron dense cells as already described. Evidently these cells correspond to the N type cells.

By way of summary it may be reasonable to conclude that the active groups specific to each of the neuro- and mesoderm-inductive factors function by killing the cells belonging to another clone though this does not exclude the possibility that they may activate cells belonging to their own clone.

REFERENCES

EAKIN, R. M. and LEHMANN, F. E. (1951) *Roux' Arch.*, **150**, 177.
HAYASHI, Y. (1958) *Embryologia*, **4**, 33.
HAYASHI, Y. (1959) *Develop. Biol.*, **1**, 247.
KATOH, A. K. (1962) *Exp. Cell Res.*, **27**, 427.
KAWAKAMI, I. and YAMANA, K. (1959) *Mem. Fac. Sci. Kyushu Univ.*, Ser. E, **2**, 171.
KAWAKAMI, I. and IYEIRI, S. (1964) *Exp. Cell Res.*, **33**, 516.

KOLIN, A. (1955) *Proc. Nat. Acad. Sci.*, **41**, 101.
KUUSI, T. (1953) *Arch. Biol.*, **64**, 189.
LALLIER, R. (1950) *Experientia*, **6**, 92.
NIEUWKOOP, P.D. and NIGTEVECHT, G.V. (1954) *J. Embryol. Exp. Morph.*, **2**, 175.
OKANO, H. and KAWAKAMI, I. (1959) *Mem. Fac. Sci. Kyushu Univ.*, Ser. E, **2**, 184.
TIEDEMANN, H., TIEDEMANN, H. and KESSELRING, K. (1960) *Z. Naturf.*, **15**, 312.
TIEDEMANN, H. and TIEDEMANN, H. (1964) *Reve. Suisse Zool.*, **57**, 41.
TOIVONEN, S. and KUUSI, T. (1948) *Ann. Soc. Zool.-Bot. Fenn. Vanamo*, **13**, No. 3, 1.
TOIVONEN, S. and SAXÉN, L. (1955) *Exp. Cell Res.* (Suppl.), **3**, 346.
YAMADA, T. (1958) *Experientia*, **14**, 81.
YAMADA, T. (1961) In *Advances in Morphogenesis*. Ed. by M. Abercrombie and J.Brachet. Academic Press, New York.

DISCUSSION

J.D.EBERT: (1) If your argument is correct, one would expect to find a high incidence of cell death in induced tissues during normal embryogenesis. Is there evidence of cell death? For example in the neural tube?

(2) Please relate your current hypothesis to Tiedemann's recent arguments relating to the change in cell affinity resulting from action of "mesodermalizing factor".

I.KAWAKAMI: (1) We have not observed cell death in normal embryo, but under electron microscopy a destruction of cells was always observed in the electron dense cells when bone marrow tissue, which is mesoderm-inductive, was brought in contact with a presumptive epidermis.

(2) I have no idea about that point, as we have not succeeded in dividing these cell species alive. We must study as to whether or not these cell species have the ability to self-differentiate into neural and mesodermal tissues before I can give an answer to your question.

MODE OF ACTION OF INSECT BRAIN HORMONE— AN ELECTRON MICROSCOPIC STUDY

Hironori Ishizaki

Department of Developmental Biology, Zoological Institute,
College of Science, Kyoto University, Kyoto, Japan

The postembryonic development of the insect is characterized by periodic growth and molting with the terminal event of metamorphosis. The initiation and direction of development at each stage are primarily regulated by the hormones. Thus the cells of most tissues require stimulation by the prothoracic gland hormone, ecdysone, to initiate the development at a critical time in each intermolting stage. The brain hormone, a hormone originating from the neurosecretory cells of the brain, turns on the prothoracic gland so that the gland secretes ecdysone. The brain hormone is regulated, to be secreted periodically to insure the regular advancement of development.

Because of the essential role of the brain hormone in the insect's development, the biology and chemistry of the brain hormone have long been the main project in our laboratory and we now know that this hormone is a protein.[1] It has been highly purified, 0.03 μg of the most purified preparation effectively evoking adult development in an assay pupa of the Eri-silkworm, *Samia cynthia ricini*. In parallel with the purification work the cytological, mainly electron microscopic, study of the prothoracic gland cell has been initiated to obtain knowledge on the mode of action of the brain hormone. The present paper deals with the finding that a distinct change in the distribution pattern of the chromatin is the earliest response, as far as revealed by electron microscopy, of the cell to the brain hormone.

The prothoracic gland of the Eri-silkworm was used as material. The pupa of this species debrained shortly after pupation cannot initiate adult development, remaining as the pupa for a considerable time, since its prothoracic gland fails to secrete ecdysone in the absence of the brain hormone. Whenever the brain hormone is injected into such debrained pupae they initiate the adult development and moths emerge 3 weeks later. The prothoracic gland cells of debrained pupae, various periods of time after the brain hormone injection, were examined by electron microscopy.

The prothoracic gland cell of the debrained resting pupae, which is inactive with respect to the ecdysone secretion is characterized by the presence of attenuated cytoplasmic organelles, mitochondria being the only frequently encountered organelle. One day after the brain hormone injection the endoplasmic reticulum, ribosomes, a new type of the cytoplasmic inclusion which is most probably the secretory granule, and still other types of cytoplasmic organelles appear in abundance. The possible secretory granule is characterized by a well-developed myeline structure in which several fat droplets are included (Fig. 1). This granule resembles a lysosome in its morphology. However, it cannot be regarded as a lysosome since it gives a negative acid phosphatase reaction and the clear-cut profile of its secretion outwards the cell is observed in the later stage of the granule formation. Three days after the injection the population of the granules reaches its height. They then migrate towards the cell periphery, assemble to form larger aggregates in each of which many granules are included, and finally become wrapped in the plasma membrane at the cell surface to be thrown out into the hemocoel (Fig. 2). There is no way at present to identify the chemical nature of the substances contained in this granule, but it is highly possible that it contains ecdysone: this granule formation is the only sign of secretory activity seen in the cell and the timing of its secretion well coincides with the critical period when the body has just been activated by ecdysone as determined by the experiment of removal of the body part containing the prothoracic gland. If it is true that the granule represents the profile of ecdysone elaboration and secretion, its morphology is of special interest from the endocrinological view point because it has been the general experience that almost no clear-cut profile of either formation or secretion of the secretory product is observed in the steroid-producing cells in vertebrates as well as in invertebrates.[2,3]

However, far more interesting in relation to the mechanism of hormone action is the finding that the earliest response of the cell to the brain hormone injection is a change in the distribution pattern of the chromatin in the nucleus which precedes the cytoplasmic changes. Figures 2 and 3 are the nuclei of the resting and the brain hormone-activated prothoracic gland cells, respectively. A sharp contrast is readily noticed, i.e. the heterochromatin, the densely aggregated mass of chromatin, has a tendency to attach to the nuclear

FIG. 1. A representative field of the cytoplasm of a prothoracic gland cell 3 days after the brain hormone injection. Arrows point to the cytoplasmic inclusion which is presumably the ecdysone-containing secretory granule. Os-fixed and stained with uranyl before dehydration and with lead after sectioning. N: A part of the nucleus. ×24,000.

FIG. 2. A part of a prothoracic gland cell near the cell surface. Five days after the injection. A large aggregate of secretory granules is about to be released into the hemocoel. Stained with uranyl and lead after sectioning. ×19,000.

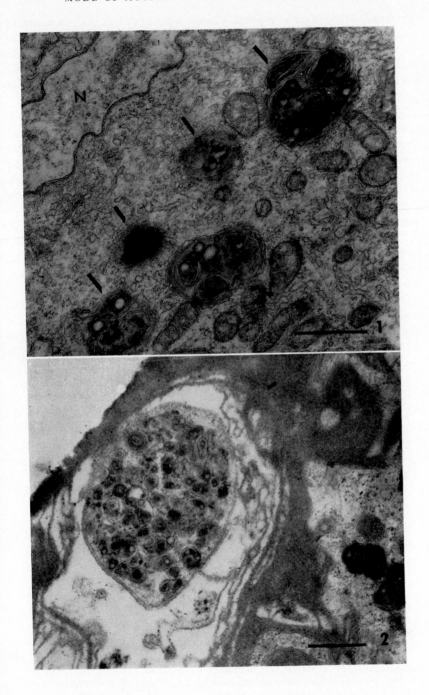

membrane in the inactive nucleus, while it is detached from the nuclear membrane in the activated nucleus. This change is already apparent 1 hr after the injection, the earliest time so far examined, when no difference is observed in the cytoplasm. Once the chromatin has been detached from the nuclear membrane, it remains so thereafter at least up to 5 days after the injection, the latest time so far examined.

This simple finding itself does not allow much discussion about the mechanism of hormone action. But in view of the current trends that much attention is directed to the nucleus as the site of the hormone action the present finding has some bearing on the matter. The change in the chromatin distribution observed here may, as a tentative explanation, suggest an alteration in the adhesive force between the chromatin and the nuclear membrane and may possibly reflect a change in the electric charge of the chromatin. It may be possible that these changes result in a change in the configuration, or the coiling state, of the chromosomal material. The puffing studies have demonstrated a quick response of certain specific gene loci to ecdysone.[4,5] An interesting fact is that the similar puffings are produced with certain degrees of regularity by such simple and non-specific factors as a change of K^+/Na^+ ratio in the cell[6] and various unphysiological conditions of the medium to which the cells are exposed.[7] Though it may be still premature to correlate directly the present finding with the puffing studies, it may be speculated that the brain hormone, and possibly other hormones, first alters some unknown physiological conditions of the cell through which the physico-chemical, possibly the coiling, property of the DNA molecule as a whole is altered, and then certain specific gene loci which have been prepared, according to the tissue and stage specificity, ready to react when appropriate conditions are introduced are activated. What the author conceives is, in short, that the change in configuration of the entire DNA molecule might be an agency that leads to the differential gene activation and its study may be a promising subject to analyze the mode of hormone action.

ACKNOWLEDGEMENT

The author wishes to thank Professor M. Ichikawa for his encouragement throughout this work.

FIG. 3. A part of the nucleus of an inactive prothoracic gland cell of a debrained resting pupa. The nucleus of this cell is highly polyploid, very large and exhibits many complicated processes. This photograph shows only a small portion of the nucleus where long processes run parallel. Note the adherence of the chromatin to the nuclear membrane. Stained with lead after sectioning. ×18,000.

FIG. 4. A part of the nucleus of a prothoracic gland cell activated by brain hormone 2 days after brain hormone injection. Note the detachment of the chromatin from the nuclear membrane. Treatment same as in Fig. 3. ×18,000.

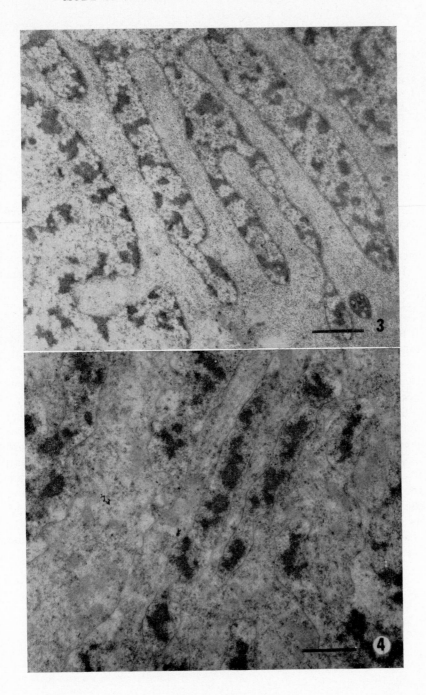

REFERENCES

1. M. ICHIKAWA and H. ISHIZAKI, *Nature*, **198**, 308, 1963.
2. A. K. CHRISTENSEN and D. W. FAWCETT, *J. Biophys. Biochem. Cytol.*, **9**, 653, 1961.
3. B. SCHARRER, *Z. Zellforsch.*, **64**, 301, 1964.
4. U. CLEVER and P. KARLSON, *Exp. Cell Res.*, **20**, 623, 1960.
5. U. CLEVER, *Chromosoma*, **12**, 607, 1961.
6. H. KROEGER, *Nature*, **200**, 1234, 1963.
7. U. CLEVER, In *Genetic Control of Differentiation, Brookhaven Symposia in Biology*, No. 18. Brookhaven National Laboratory, Upton, New York, 1966.

DISCUSSION

H. BUSCH: What is the nature of chromatin not normally seen after osmium fixation in mammalian cells?

H. ISHIZAKI: What I can say is only that different tissues react with heavy metals in different ways.

E. ZEUTHEN: Have I understood rightly that you think the chromosome might spiral down the projecting tube made of nuclear membrane?

H. ISHIZAKI: Probably right. Reconstruction by means of serial sections may answer this for sure. This is now in progress.

F. H. KASTEN: I believe that what Dr. Busch is getting at is that with osmium fixation of mammalian cells very little condensed chromatin is observed compared with aldehyde fixation. The fact that you see so much nuclear chromatin in this insect material after osmium fixation raises some questions as to the nature of this particular chromatin.

S. SENO: Concerning the question of Dr. Kasten, I should like to point out one thing. Recently, Dr. Mirskey and coworkers reported that in the presence of arginine-rich histone, chromatin turns to a diffuse structure which is active in gene expression. On losing arginine-rich histone the chromatin fibers form an aggregate which is inactive in gene expression. So I think that the dense chromatin structure may mean an inactive state of the gene, and the diffuse structure of chromatin, seen after the injection of the hormone, an active state of the gene and this may be somehow concerned with the changes of the chromatin structure, apart from the problem of fixation artefacts.

THE EFFECT OF 5-BROMO-DEOXYURIDINE
ON THE SYNTHESIS
OF 7S AND 19S ANTIBODIES

YUJIRO NAMBA and MASAO HANAOKA

Department of Pathology, Institute for Virus Research,
Kyoto University, Kyoto, Japan

AT PRESENT, it is well known that antigens induce the formation of several classes of antibodies;[1-6] the antibodies synthesized early belong to 19S globulin (IgM), while 7S (IgG) antibodies appear later. The efforts, yielding two different conclusions, have been made by many investigators to reveal the cellular origin of these two antibodies.[2,6-11] Some investigators contend that these two classes of antibodies are synthesized by the same cell series, while others, including ourselves, are of the opinion that the two classes of antibodies are formed by different cell series. The contention of the former is that immunologically competent cells synthesize 19S antibodies in an early stage of their development, and that during the maturation process the information is switched over to the synthesis of 7S antibodies; the latter are of the opinion that the cells forming 7S antibodies are never of the mature type of the 19S forming cells, but are independent of each other cytogenically.[16,17] Opinions are unanimous, however, on the point that most of 19S-forming cells belong to reticulum cells and large basophilic lymphocytes including lymphogonia[12] which may be called homocytoblast transitional cells[13] or large lymphocytes,[14] i.e. the cells of the reticulo-lymphocytic system (RLS),[15] while the 7S antibodies are synthesized by plasmablasts and plasma cells, which may be distinguished from the cells of RLS morphologically. If 19S antibody formation could be selectively suppressed by damaging the proliferation of 19S antibody-forming cells without the suppression of 7S antibody formation, it would indicate that both classes of antibodies are produced by independent cell groups.

We have approached the problem by using 5-bromo-deoxyuridine (BUdR) which is known to be incorporated into DNA in place of thymidine in its keto form and cytidine in its enol form, damaging the proliferating cells selectively,[18-22] and the relationship between the cell proliferation and the synthesis of antibodies has been analysed by using tissue culture cells, which produce antibodies *in vitro*.

In this paper it is reported that the two classes of antibodies will be synthesized by two different cell series respectively, that the proliferation of the immunologically competent cells will not necessarily be accompanied by antibody formation, and that the cells are stimulated to produce antibodies even after the cells have ceased to proliferate.

MATERIALS AND METHODS

A/Jax mice weighing between 18 and 20 g and about 12 weeks old were used for the *in vivo* experiments. The lymph node cells from albino rabbits weighing about 2.5 kg served as the material in the *in vitro* experiments. As the antigen, wild type coli phage T-2 was used. For the immunization of mice, T-2 phage (10^9 or 10^{11} PFU per animal) was injected into the tail vein. For the *in vitro* observation, the popliteal lymph node cells from the rabbits immunized with the antigen dose of 10^{11} PFU in both hind-foot pads were used. The cells were harvested about two months after the challenge with the antigen. For the antibody titration, the blood of mice obtained by cutting femoral arteries and the blood from 4 mice were pooled. The blood samples were left standing overnight at 4°C, and inactivated at 56°C for 30 min. They were stored at −20°C. Assays of the virus-neutralizing activities were carried out by the agar-overlay method. The neutralization of phages by excess of antibodies followed single hit kinetics. Antibody activities were expressed by K-values which were calculated by measuring the reduction rate of plaque numbers. The assays of 2-mercaptoethanol (2-ME) sensitivity were carried out as follows: The antisera diluted to 1:4 (v/v) in the final concentration of 0.1 M 2-ME were incubated for 1 hr at 37°C. Subsequently, the samples were diluted as needed and the titration was performed immediately without removing 2-ME. By the 2-ME treatment, T-2 phage was neutralized non specifically. Therefore, correction was made in each case. In the dilution of 1:10 (v/v), 0.22 to 0.28 K was subtracted from the measured antibody activity, and in the dilution of 1:100 (v/v), 3.5 to 5.0 K was subtracted. The nonspecific activities were found to be not correlated with the real activities.

Sucrose gradient ultracentrifugation was carried out with the method described by Edelman and Kunkel.[23] A serum sample was diluted with PBS and layered over a gradient formed from 37% and 10% sucrose. The centrifugation was performed in the Hitachi ultracentrifuge with a swing rotor SW 40 at 105,000 g for 16 hr. Twelve to fifteen fractions were collected by boring through the bottom of the tube.

BUdR (Sigma) was injected into mice at a concentration of 3 mg per ml in PBS in a dose of 35 μg/mouse gram body weight, intraperitoneally, which proved to have no harmful effect on mice. For the *in vitro* experiments, (BUdR) at a concentration of 50 μg/ml was added to the tissue culture of lymph node cells from the immunized rabbits.

Bacto-hemagglutinin P (PHA) from the Difco Laboratories, Detroit, was added in a concentration of 2 μg/ml of the culture medium.

C^{14}-amino acids from algal hydrolysate, obtained from the New England Nuclear Corporation, were used and about 10^7 cells/ml were exposed to 5 μg of C^{14}-labeled amino acids. Gamma-globulin synthesis of rabbit lymph node cells in culture was checked by the coprecipitation reaction of the culture medium with unlabeled rabbit gamma-globulin and guinea pig anti-rabbit serum in the region of a slight antibody excess.

The tissue culture media used in the experiments were a modified Eagle's medium containing 20% autologous rabbit serum obtained before immunization and Medium 199 of Morgan, using each medium within two weeks of preparation. The rabbit lymph node cells for tissue culture were obtained from the popliteal lymph nodes taken two months after the challenge with antigen. The lymph node cells were teased out of the diced fragments, filtered through a stainless screen mesh number 250, washed and suspended in a fresh medium at the concentration of 10^7 cells/ml. T-2 was added to the final concentration of 5×10^9 PFU/ml. After 24 hr the medium was replaced with antigen-free medium. The cells were cultured in Leighton tubes in the final concentration of 6×10^6 to 1.5×10^7 cells/ml. The culture medium was replaced every two or three days. The culture medium was removed and was centrifuged at 15,000 g for 30 min for the purpose of removing the antigen and cell debris. The samples were stored at $-20\,^\circ$C until the titration of antibody activities.

RESULTS

The In Vitro Study of Antibody Formation

By the method described above the antibody producing capacity of the lymph node cells of rabbits was observed in culture. By culturing the cells after exposing them to the antigen for 24 hr, the antibody activities could be detected after 60 hr reaching the maximum titer on the 8th day and a high level of titer was maintained for 4 days (Fig. 1). In these experiments the culture medium was changed every 48 hr. Antibody activity was not detected in the absence of antigen, though the cells obtained one month after the antigenic challenge of the animal retained their antibody producing activity *in vitro*. When the cells were cultured first in the absence of the antigen for some period, antibody response was reduced precipitously. It is evident that the elevation of the antibody level after challenging with the antigen was due to the synthesis of antibodies and not the release of the antibodies which had been formed *in vivo* by the incorporation of C^{14}-amino acids into the gamma-globulin of medium in the above culture system (Fig. 2).

To see the relationship between antibody formation and cell proliferation using the above cultured cells, BUdR, in the concentration of 50 µg/ml was added to the culture on the $\frac{1}{2}$, 1, 2, 3, 4, 5 and 6th day. The data clearly de-

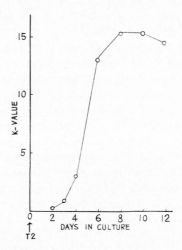

FIG. 1. Secondary antibody response *in vitro*. Rabbit popliteal lymph node cells obtained 60 days after the injection of T-2 phages were incubated with the antigens for 24 hours and cultured in antigen-free medium, changing this every 48 hours. Method: See text.

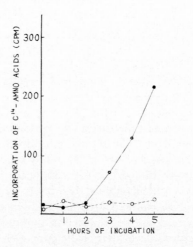

FIG. 2. Incorporation of C^{14}-amino acids into globulin in cultured medium using a similar experimental system to that in Fig. 1. The cells were cultured in the media containing 5 µc per ml. Method: See text. —●— Antigen added *in vitro*. ···○··· No antigen added *in vitro*.

monstrated that the formation of antibodies was severely suppressed in the presence of BUdR. The effect was more marked in those which were exposed to BUdR in earlier stages of culture, and there was actually no effect when the cells were exposed to BUdR on the 6th day of culture (Fig. 3). The experimental data given above indicate that the immunologically competent cells begin to proliferate in response to the antigenic stimulus but it is still a problem whether the proliferation of the immunologically competent cells is inevitably accompanied by antibody formation. To approach this problem,

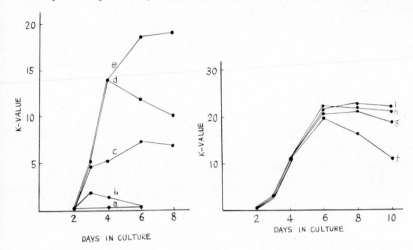

FIG. 3. Effect of BUdR on the secondary response *in vitro*. By using the system as in Fig. 1, 50 μg per ml of BUdR was added at various times after the antigenic challenge *in vitro*; a: on day 0 with antigen, b: 1 day after, c: 2 days after, d: 3 days after, e: control, no BUdR, f: 4 days after, g: 5 days after, h: 6 days after, and h: control, no BUdR.

phyto-hemagglutinin (PHA) was used. PHA induces the transformation of small lymphocytes to blastic lymphocytes followed by their proliferation.[24-27] However, it is still not decided whether such transformed cells synthesize gamma-globulin.[28-31,33]

In this tissue culture system, PHA also did not induce the synthesis of T-2 neutralizing antibodies. But the preincubation of the cells with PHA enhanced antibody formation but only in the presence of the antigen in the postincubation medium. These facts seem to suggest that PHA induces the proliferation of the immunologically competent cells and the antigen stimulated them to produce antibodies. This has been clearly evidenced through the experiment on the immunologically competent cells preincubated with both PHA and BUdR, where the cells were destroyed and sufficient antibody titer could not be detected (Table 1).

It has been shown above that PHA and the antigen stimulate the immunologically competent cells and it might be expected that they work

TABLE 1. EFFECT OF PHA AND BUdR DURING PREINCUBATION ON THE ANTIBODY FORMA-
TION OF CULTURED CELLS

Preincubate with	Preincubation period (day)			
	0	1	2	3
	Antibody titers on the 8th day of culture (K-value)			
–	15.0	1.60	0.50	0.35
BUdR (1 ml)	15.0	1.40	0.40	0.30
PHA (1 ml)		2.30	1.05	1.56
BUdR (1 ml) + PHA (1 ml)		0.80	0	0

The lymph node cells used were those from the rabbits challenged with T-2 phages 2 months prior to the harvest. After the preincubation the cells were treated with the antigen and cultured as in Fig. 1. Note the complete suppression of antibody formation by the preincubation with PHA and BUdR for 2 to 3 days. Method: See text.

synergistically in the antibody formation. But the experiment showed that antibody formation was suppressed when both of them were added simultaneously. The effects became manifest within 48 hr (Fig. 4a and Table 1).

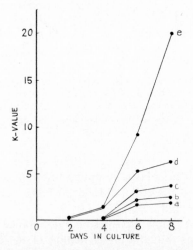

FIG. 4. Effect of PHA added simultaneously with or after the antigenic challenge on the antibody formation of lymph node cells *in vitro* in same culture system as in Fig. 1. Note that the addition of PHA suppresses the antibody formation when it is present in culture medium. a: PHA was added on day 0 with antigen, b: 1 day after the addition of antigen, c: 2 days after, d: 3 days after, e: control, no addition of PHA.

This suppressing effect was observed even when PHA was added 4 days after the antigenic challenge *in vitro* (Fig. 4b). This fact may be the result of the competition of antigen and PHA with the proliferation of antibody competent cells. However, it is uncertain whether the amount of antibodies produced per cell is reduced or the cells keep on producing antibodies in the same level even reduced in number.

From these findings the following conclusions are drawn: (a) Antibody competent cells can proliferate by the antigenic stimulation and produce the antibody, (b) BUdR can inhibit the proliferation of antibody competent cells, and (c) antibody competent cells have the ability to produce the antibody in contact with the antigen even after proliferation stimulated by the action of a factor other than the antigen, though the antibody formation is suppressed or inhibited by the co-existence of the antigen and the other stimulating factor of the cellular proliferation.

On the basis of these *in vitro* experiments, 19S and 7S antibody formation were observed *in vivo* under the effect of BUdR.

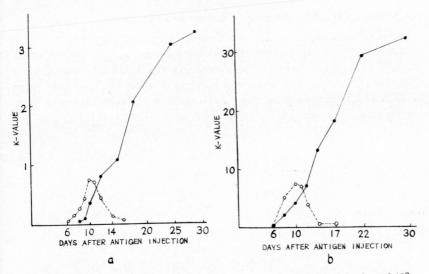

FIG. 5. Primary neutralizing antibody response of mice after the injection of 10^9 PFU (a) and 10^{11} PFU (b) of T-2 phages. ···○··· 2-mercaptoethanol (2-ME) sensitive antibody titer; —●— 2-mercaptoethanol (2-ME) resistant antibody titer.

Antibody Responses During Primary Challenge In Vivo

The patterns of antibody formation in the mice immunized with T-2 phage are shown in Fig. 5. Irrespective of antigen dose, the earliest detectable antibodies were 2-ME sensitive and were in the bottom fractions of the sucrose density gradient. Antibodies with these characteristics will be referred to as 19S antibodies. 2-ME resistant antibodies appeared a little later and the antibodies detectable on the 16th day were completely resistant to 2-ME treatment. As is shown in Fig. 6, these antibodies equilibrated in a band corresponding to 7S and will be referred to as 7S antibodies.

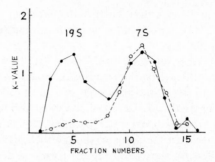

FIG. 6. Antibody activities of sucrose density gradient ultracentrifugal fractions of mouse sera, 10 days after the injection of 10^9 PFU of T-2 phages. —●— Total antibody titer; ···○··· antibody titer after 2-ME treatment.

The Effects of BUdR on Antibody Response In Vivo

The administration of BUdR in a dose of 35 μg/mouse gram body weight had almost no effect upon antibody formation when it was injected before the antigenic challenge. However, the administration of BUdR after the injection of the antigen resulted in abnormal patterns of antibody response. When BUdR was injected every day for 7 days after the antigenic challenge, anti-

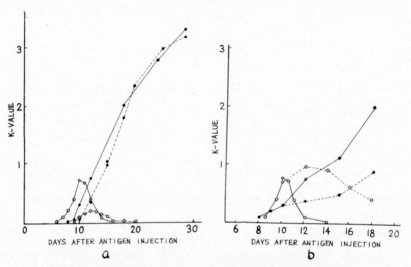

FIG. 7. Primary antibody response of BUdR-treated mice. BUdR (700 μg per mouse) was injected 24 hours (a) and 8 days (b) after the injection of 10^9 PFU of T-2 phages. Note the suppressive synthesis of 2-ME sensitive antibody (a). The prolonged duration of the synthesis of 2-ME sensitive antibody is found after the suppressive synthesis of 2-ME resistant antibody (b). —○— 2-ME sensitive antibody titer, no injection of BUdR; ···○··· 2-ME sensitive antibody titer, after injection of BUdR; —●— 2-ME resistant antibody titer, no injection of BUdR; ● 2-ME resistant antibody titer, after injection of BUdR.

body formation was almost completely blocked. However, when it was given once within 24 hr, only 19S antibody titer was markedly suppressed while 7S antibody titer rose to the level equivalent to that of the control, though there was some delay in the formation of both 19S and 7S antibodies (Fig. 7a). Similar delay in 7S antibody formation was also observed when BUdR was injected once during the third to the fifth day. In this instance, however, larger amounts of 19S antibodies were produced until 7S antibody titer rose to a considerable level.

When BUdR was given once on the 8th day of antigenic challenge, a new pattern of antibody response resulted, i.e. 7S antibody formation was suppressed on the 12th day and thereafter, while 19S formation was not only unaffected but rather enhanced, just as to compensate the suppressed 7S antibody formation (Fig. 7b). In these cases, considerable antibodies were found in the 19S fractions even on the 14th day of antigenic challenge (Fig. 8).

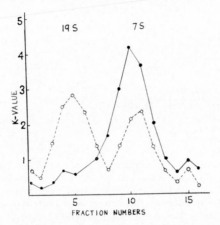

FIG. 8. Antibody activities of sucrose density gradient ultracentrifugal fractions of mouse sera 14 days after the injection of 10^9 PFU of T-2 phages. BUdR (700 µg per mouse) was injected 8 days after the antigenic challenge. Note the remaining of 19S antibody titer in sera of BUdR-treated mice. —●— Control; ⋯○⋯ BUdR treated.

This pattern of 19S antibody production appeared after the administration of BUdR on the 8th day was not altered by the administration of the same agent on the 10th day and thereafter.

Therefore, the findings support the view that these 19S antibodies were synthesized by the cells that cease to proliferate but not by the cells newly developed from the immunologically competent cells, because the cells in proliferation during their early stages of development could be damaged.

Feed-back Phenomena by the Infusion of 7S Antibodies

The modified pattern of antibody formation induced by the administration of BUdR on the 8th day was restored to the normal one by infusing isologous anti-T-2 globulin. When the mice were immunized with a larger amount of the antigen, a higher titer of the antibodies had to be administered in order to bring the modified pattern to normal. The anti-T-2 7S globulin was injected on the 10th or the 16th day after the antigenic challenge. Each treatment effected the suppression of 19S antibody formation (Fig. 9). Anti-BSA antibodies showed no noticeable effect.

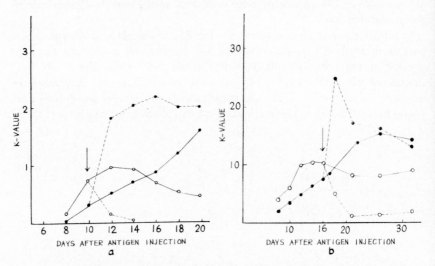

FIG. 9. Inhibition of 2-ME sensitive antibody synthesis by the injection of homologous anti-T-2 7S antibodies into mice which were treated with 700 µg per mouse of BUdR 8 days after the injection of 10^9 (a) and 10^{11} (b) of T-2 phages. Arrows indicate the infusion time of 7S antibodies at the 10th (a) and 16th (b) day in culture. —O— Control 2-ME sensitive antibody titer, no infusion of antibodies; ···O··· 2-ME sensitive antibody titer after the infusion of antibodies; —●— Control 2-ME resistant antibody titer, no infusion of antibodies; ···●··· 2-ME resistant antibody titer after the infusion of antibodies.

These findings indicate that antibody producing cells must be stimulated to produce antibodies even after the cells have ceased to proliferate in the case of 19S forming cells. Furthermore, these findings indicate that these cells keep on producing 19S antibodies for more than two weeks without cell division when 7S formation is blocked. This means that 19S antibodies will also be produced in the mature cells, and suggests that 19S producing cells are of a different series from 7S forming cells.

DISCUSSION AND CONCLUSIONS

Using BUdR as an inhibitor and PHA as a stimulator of the cellular proliferation, antibody formation was observed in the secondary response *in vitro* in lymph node cells of rabbits. These results indicate: (a) That antibody synthesis is suppressed by BUdR *in vitro* when it is added to the medium in early culture days. (b) That antibody synthesis is remarkably suppressed in PHA-stimulated and BUdR-treated cells even when they are followed by the antigenic challenge. (c) That PHA-stimulated cells enhance the antibody formation only in the presence of antigen. (d) That PHA does not act synergistically with the antigen when PHA and antigen are added to the medium simultaneously, and it may not be unreasonable to consider that PHA and antigen stimulate the cell in different directions and each direction of the proliferation competes with each other.

The above findings show that antigens have the ability to stimulate antibody competent cells both in the proliferation and the antibody formation. Therefore, by comparing the antibody response in the presence of BUdR with that in the absence of BUdR the proliferation and the antibody formation of antibody competent cells can be also followed *in vivo*.

As has been reported by some workers,[4] T-2 phage induces the formation of several classes of antibodies, and the major part of antibody activities belongs to either 19S or 7S antibodies. In the *in vivo* experiments, BUdR which affects specifically the activity of cell division, suppresses 19S formation when it is injected in the early stage of the primary immune response, while 7S antibody formation is scarcely affected. On the other hand, 7S antibody formation is considerably suppressed when BUdR is injected in the later stage of antibody formation while in such instances 19S antibody formation is not suppressed but it compensates the suppressed 7S antibody formation.

In the lymphoid matrix and medullary cords of lymph nodes or in other lymphoid tissues of animals immunized by T-2 or other antigens, reticulum cells begin to proliferate at the first stage of immune response followed by the transformation towards lymphogonia with a giant nucleolus, numerous cytoplasmic ribosomes and high mitotic activity. It has been confirmed that these cells contain IgM antibody in their cytoplasm by immunofluorescent techniques,[11,15] and humoral 19S (LgM) antibody titers rise in parallel with proliferations of these cells.[12,15] Therefore, when BUdR is injected into animals at the early stage of the immune response, it is reasonable that these 19S antibody forming reticulum cells or lymphogonia are affected by BUdR. Plasmablasts and plasma cells containing 7S (IgG) antibodies proliferate slowly at the later stage of the immune response in the medullary cord of lymph nodes and the periarteriolar region of mesenchymal tissues in parallel with the elevation of IgG antibody titers.[15,16,17] In the present experiments it is assumed that plasmablasts were affected by BUdR injected into immunized mice at a later stage of the immune response and the 7S antibody formation was suppressed.

The stimulated 19S formation seen after treating with BUdR in the later stage decreased to the control level by the injection of anti-T-2 7S antibodies. In such instances, 7S antibody titer rose instantaneously due to the infused

7S antibodies. But it decreased gradually to the suppressed 7S antibody level. Therefore, it is unreasonable to consider that in such processes 19S antibody-forming cells became 7S forming cells, and we are inclined to conclude as some other investigators[7,32] that there exists a regulatory mechanism between 7S and 19S antibody formation. The stimulated 19S formation was completely resistant to BUdR when it was injected on the 10th day and thereafter. Therefore, in this process 19S forming cells will not be proliferating. These facts will support the view that the cells ceased to divide must still be stimulated to produce antibodies, and agree with the immunocytological results that the mature large basophilic lymphocytes containing IgM increase in number in lymphatic tissues following the appearance of lymphogonia.[15]

The above results indicate that reticulum cells, lymphogonia and large basophilic lymphocytes, i.e. cells of the reticulo-lymphocytic system, are cytogenically different from the plasma cell series, or lymphogonia are not the precursors of plasma cells.

REFERENCES

1. D. C. BAUER and A. B. STAVITSKY: *Proc. Nat. Acad. Sci.*, **41**, 1667, 1961.
2. J. W. UHR, S. FINKELSTAIN and J. B. BAUMANN: *J. Exp. Med.*, **115**, 655, 1962.
3. J. W. UHR, S. FINKELSTAIN and E. C. FRANKLIN: *Proc. Soc. Exp. Biol. Med.*, **111**, 13, 1962.
4. J. W. UHR and S. FINKELSTAIN: *J. Exp. Med.*, **117**, 457, 1963.
5. A. A. BENEDICT, R. J. BROWN and R. AYENGAR: *J. Exp. Med.*, **115**, 195, 1962.
6. S. C. SVEHAG and B. MANDEL: *J. Exp. Med.*, **119**, 1, 1964.
7. K. SAHIER and R. S. SCHWARTZ: *J. Immunol.*, **95**, 345, 1965.
8. S. C. SVEHAG and B. MANDEL: *J. Exp. Med.*, **119**, 21, 1964.
9. M. D. SCHOENBERG, A. B. STAVITSKY, R. D. MOORE and M. J. FREEMAN: *J. Exp. Med.*, **121**, 577, 1965.
10. R. C. MELLORS and L. KORNGOLD: *J. Exp. Med.*, **118**, 387, 1963.
11. T. MASUDA: *Ann. Rep. Inst. Virus Res., Kyoto Univ.*, **6**, 132, 1963.
12. G. UNNO, M. HANAOKA, S. HASHIMOTO, K. IWAI and S. MORITA: *Acta Path. Jap.*, **4**, 75, 1954.
13. A. FAGRAEUS: *J. Immunol.*, **58**, 1, 1948.
14. G. J. V. NOSSAL, A. SCENBERG, G. L. ADA and C. M. AUSTIN: *J. Exp. Med.*, **119**, 485. 1964.
15. M. HANAOKA: *Ann. Rep. Inst. Virus Res., Kyoto Univ.*, **2**, 30, 1966.
16. S. AMANO: *Fundamentals of Hematology, Development and Function of Blood Cells.* Maruzen, Tokyo, 1948.
17. S. AMANO: *Ann. Rep. Inst. Virus Res., Kyoto Univ.*, **1**, 1, 1958.
18. J. W. LITTLEFIELD and E. A. COULD: *J. Biol. Chem.*, **235**, 1129, 1960.
19. L. CHEONG and M. A. RICH: *J. Biol. Chem.*, **235**, 1441, 1960.
20. P. D. LAWLEY and P. BROOKS: *J. Mol. Biol.*, **4**, 216, 1962.
21. H. S. SHAPIRO and E. CHERGAFF: *Nature*, **188**, 62, 1960.
22. T. C. HSU and C. E. SOMERS: *Proc. Nat. Acad. Sci.*, **17**, 396, 1961.
23. G. M. EDELMAN, H. G. KUNKEL and E. C. FRANKLIN: *J. Exp. Med.*, **108**, 105, 1958.
24. P. C. NOWELL: *Cancer Res.*, **20**, 462, 1960.
25. A. A. MACKINNEY, F. STOHLMANN and G. BRECHER: *Blood*, **19**, 349, 1962.

26. Y. TANAKA, L. B. EPSTEIN and F. STOHLMANN: *Blood*, **22**, 614, 1963.
27. M. W. ELVES, J. GOUGH, J. F. CHAPMAN and M. C. G. ISRAELS: *Lancet*, **I**, 306, 1964.
28. S. SELL, D. S. REWE and P. G. H. GELL: *J. Exp. Med.*, **122**, 823, 1965.
29. F. BACH and K. HIRSCHORN: *Exp. Cell Res.*, **32**, 592, 1963.
30. M. W. ELVES, S. ROATH, G. TAYLOR and M. C. G. ISRAELS: *Lancet*, **I**, 1292, 1963.
31. T. F. OBRIEN and A. H. COONS: *J. Exp. Med.*, **117**, 1062, 1963.
32. G. MOELLER and H. WIGZELL: *J. Exp. Med.*, **121**, 969, 1965.
33. F. PORENTI and R. CEPELLINI: *Biophys. Biochem. Acta*, **123**, 181, 1966.

DISCUSSION

B. THORELL: Did you actually measure the rate of proliferation, i.e. increase in the number of cells during the various *in vitro* stimulations, and how does this eventually relate to antibody formation? It is important for your conclusion about antibody formation and increase or not of the cell population.

M. HANAOKA: That is true, but in this condition total lymph node cells tend to decrease in number, therefore it is difficult to measure the number of antibody forming cells. Now we are trying to find out the relationship between antibody titer and the proliferation and the decrease of antibody forming cells as *in vivo* experiments.

J. D. EBERT: Do you base your conclusion that the induction of antibody formation is more than the increase in proliferation of cells synthesizing antibody (hypothesis of Jerne, Nossal and others) only on the failure of cells stimulated by phyto-hemagglutinin to show antibodies to T-2?

M. HANAOKA: We concluded that the antigen has two abilities, (1) to enhance the proliferation of the antibody competent cells and (2) to promote the synthesis of antibody, but PHA can only enhance the proliferation of the antibody competent cells.

ANTIBODY FORMATION
IN CULTURED MOUSE LYMPHOID CELLS

ATSUYOSHI HAGIWARA

Laboratory of Developmental Biology, Zoological Institute, College of Science.
Kyoto University

IN VITRO stimulation of cultured cells from a normal, or non-immunized, animal, by an antigen to induce a specific antibody is a dream of all immunologists.

Attempts in this direction have been made by many workers, but their results are more or less subliminal and contradictory in a sense that the antibody thus formed was too small in quantity to examine the exact quality.

Now, two technical devices provide us a much exact approach for investigation of this traditional problem. One is the Jerne's hemolytic plaque method[1] for the study of a single antibody-forming cell and another is the refinement of the cell culture technique to maintain lymphatic cells *in vitro* in good physiological conditions.

The first trial has been made to stimulate the lymphoid cells from normal mouse by incubating the cells with *E. coli* polysaccharide antigen. In this type of experiment, phytohemagglutinin (PHA), a well-known mitotic stimulant and potent blastogenic agent for lymphoid cells, was used to accelerate active blastogenesis and subsequent antibody formation. And in this experiment, the antibody formation was determined by a specific agglutinin reaction in the culture medium. In the second part of the experiment, sheep red cells (sRBC) were employed to stimulate the lymphoid cells from normal mice and antibody formation was determined by the hemolytic plaque method.

The culture condition I employed was rather simple, and some improvement may be needed in the composition of the culture medium in the future; however, polyvinylpyrrollidons (PVP) at the concentration of 1% in the culture medium was very effective in maintaining the cells viable up to 2 weeks as judged by the vital pigment extrusion method.

In the first experimental system, PHA culture, the antibody formation was determined by the antigen-coated latex particle agglutination reaction.[2] It was shown that the stimulation of the cells by antigen alone did not induce any detectable amount of antibody formation after 6-day incubation. In contrast, the cells did form antibody when they were incubated with PHA after antigenic stimulation. The control experiment showed that PHA alone does not induce antibody formation. These results are summarized in Table 1.

TABLE 1. ANTIBODY FORMATION IN PHA CULTURE

Cells incubated with	Antibody formation*
none	−
antigen	−
PHA	−
antigen + PHA	+

* Antigen-coated latex particle agglutinin reaction, positive (+) and negative (−), in 6-day culture medium.

Next, the daily titration of antibody formed in the culture medium was made. It is shown in Fig. 1 that the antibody began to appear at day 1 and reached the maximum between days 4 and 5.

In this system, actinomycin was inhibitory for the antibody formation only when the inhibitor was added to the culture within 24 hr after antigenic stimulation, and in the later phase of the culture, actinomycin was non-inhibitory for the antibody formation. In contrast, puromycin was inhibitory throughout the experimental period.

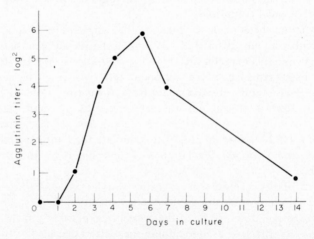

FIG. 1. Time course of antibody formation in PHA culture.

The next trial was made to detect antibody-forming cells by the plaque method. The lymphoid cells from the culture were plated by mixing with sRBC coated with antigen. Unfortunately, the cells recovered from the PHA culture contained large cell aggregates and these aggregates disturbed the smooth plating of the indicator cells, always producing non-specific plaque-like appearances around the aggregates. Thus the detection of the antibody-forming cells in this system was unsuccessful.

The second experiment was then set up by omitting PHA from the culture. I employed the mixed cell culture between the isologous spleen and thymic cells to accelerate blastogenesis. As shown in Fig. 2, the viable cell countings of the spleen-thymus mixed cell culture showed some increase as compared to that of the spleen or thymus cell culture. And microscopic observation showed that most of the cells underwent blastogenesis. The most favorable trait of this culture system was the prevention of cellular aggregate formation which was the most critical prerequisite for the hemolytic plaque method.

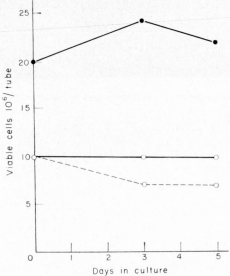

FIG. 2. Viable cell countings in spleen (O——O), thymus (O······O) and spleen-thymus (●——●) cell culture.

The plaque method of Jerne was modified so as to develop the plaque on microscopic slide glass in order to facilitate a more accurate determination of the plaque. Onto a clean slide glass, I placed 2 pieces of polyethylene gaskets having a diameter of 17 mm and a thickness of 0.1 mm. Into the gasket chamber, I dropped a mixture of lymphoid cells in Eagle's medium, sRBC and fresh guinea pig serum in modified barbital buffer. The lymphoid cells were appropriately diluted so as to develop less than 50 plaques in one gasket. Guinea pig serum was completely absorbed with sRBC before use, and the final concentrations of guinea pig serum and sRBC were 10% and 5% respectively. Three percent of gelatin was used to semi-solidify the mixture in the gasket chamber. The gaskets were then sealed with cover glass to

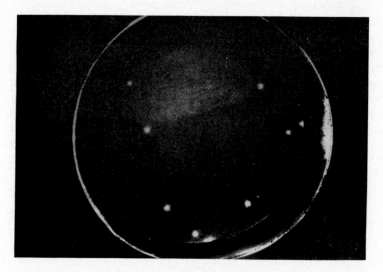

Fig. 3. Hemolytic plaques developed by gasket method.

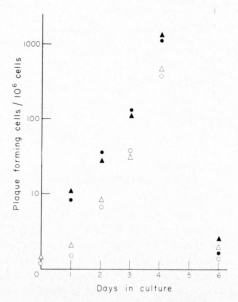

Fig. 4. Development of antibody-forming cells in culture of spleen and spleen-thymus with or without antigenic stimulation (△ spleen cell culture without antigen, ▲ spleen cell culture with antigen, ○ spleen-thymus cell culture without antigen, ● spleen-thymus cell culture with antigen).

keep the cells in complete monolayer condition. The composite slide glass was then incubated at 37° for 6 hr and the plaque counting was carried out by screening the slide glass under a low-power microscope. The hemolytic plaque developed by this method is shown in Fig. 3.

The daily determination of antibody-forming cells in lymphoid cell culture after antigenic stimulation was carried out. The results are summarized in Fig. 4. Four points must be noticed from this experiment.

Firstly, the number of the antibody-forming cells increased logarithmically after antigenic stimulation, namely, a 1000-fold increase was recorded at day 4 with the day 0 sample. Secondly, the number of antibody-forming cells decreased abruptly between days 5 and 6 after primary stimulation. At the present time, it remains unknown whether this decrease reflects actual repression of 19S antibody formation by 7S antibody formation or whether this simply reflects the declination of general synthetic activity of the cells. The former assumption is more probable since no abrupt changes in cell morphology and cell viability were noticed at this period of the culture. Thirdly, it was quite puzzling and unexpected that the antibody-forming cells appeared even in the non-stimulated cell culture. The same phenomenon was reported by Mishell and Dutton[3] who speculated that this increase was due to the presence of cross-reacting antigen in the serum of the culture. Fourthly, the preliminary work showed that spleen-thymus cell culture formed the antibody-forming cells at almost the same level as compared to that of spleen cell culture. The denominator of each value of the ordinate of this figure was expressed as 10^6 nucleated cells so that it is premature to conclude that blastogenesis in the spleen-thymus mixed cell population actually did not increase the number of the competent cells for the antibody formation.

ACKNOWLEDGEMENT

This investigation was supported (in part) by a PHS research grant RG-9469 from the Division of General Medical Science, U.S. Public Health Service.

REFERENCES

1. N. K. JERNE and A. A. NORDIN: *Science*, **140**, 405, 1963.
2. A. HAGIWARA: Symp. of Soc. for Cell Biol., in press.
3. R. I. MISHELL and R. W. DUTTON: *Science*, **153**, 1004, 1966.

DISCUSSION

B. THORELL: I think you might not be able to conclude very much about the effect of thymic cells, etc., because you seem to have a fairly high "base line", that is the antigen content in the medium. Considering the common S-shaped curve of response, it seems that your experiments are made on the "upper horizontal part", that is, response is almost maximal owing to the medium itself. You might, therefore, not expect very much effect from additional antigens, thymic cells, etc.

IN VITRO TRANSFER
OF CELLULAR IMMUNITY
BY TRANSFER AGENT OF RNA NATURE

S. Mitsuhashi, K. Saito and S. Kurashige

Department of Microbiology, School of Medicine, Gunma University, Maebashi, Japan

Infection of mice with a virulent strain 116-54 of *Salmonella enteritidis* causes severe septicaemia and general reticuloendotheriosis (Plate 1), and finally, all mice die 5 to 7 days after infection.[1] As has been true for experimental infections with other group salmonellae, e.g., *S. dublin*,[2] and *S. typhimurium*,[3] killed vaccines have increased the survival time of mice after challenge but have been largely ineffective in preventing death from the infection.[4-6]

In contrast, mice immunized with live vaccine of *S. enteritidis* or mice convalesced from the infection with virulent strain by the aid of antibiotics acquire antilethal resistance and survive the infection with 10^3 mouse MLD of virulent culture of the same organism. The serum of such mice does not show antilethal activity against infection with the same organism when administered passively into normal mice.[6,7] However, the mononuclear phagocytes (termed monocytes), obtained from the peritoneal cavity, subcutaneous tissue, or spleen of immunized mice, inhibit intracellular multiplication of virulent strain and resist cell degeneration caused by phagocytosis in the absence of antibody in cell culture medium.[8,9,10] This resistance was referred to as cellular immunity of monocytes[8] (Plates 2–4).

It was found that cellular immunity is transferable from immune to non-immune monocytes, through the supernatant fluid of cell culture and the microsomal (or ribosomal) fraction of immune monocytes. This agent responsible for the transfer of cellular immunity was referred to as transfer agent (TA), which is inactivated by treatment with RNase but not with DNase.[11-14] Simultaneously, it was found that cellular antibody was detectable in peritoneal monocytes, spleen and lymphnodes of immunized mice. This antibody was not detectable in the sera of mice immunized with live or killed vaccine or in the monocytes of mice immunized with killed vaccine.[15-17] This antibody was extracted from the abdominal monocytes or spleen of immunized mice, and is of macroglobulin nature in ultracentrifugal analysis.[17]

179

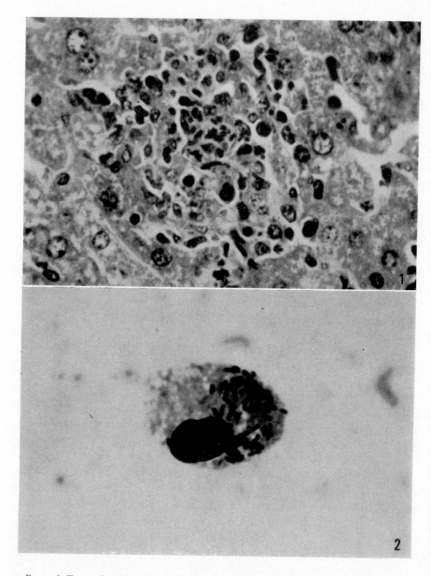

PLATE 1. Formation of typhom in the hepatic sinusoid of mouse by the infection with
S. enteritidis 116-54.

PLATE 2. Intracellular multiplication of a virulent strain 116-54 of *S. enteritidis* in
the monocyte of normal mouse. Three days after infection.

In the present article, the detection of cellular antibody in the peritoneal monocytes of mice treated with TA is presented and the role of monocytes in antibody formation will be discussed.

METHODS

Experimental animals, bacterial strains, and methods of immunization have been described previously.[8] The convalescent mice were obtained by an intravenous infection of 10^3 mouse MLD (10^{-5} mg, dry weight) of a virulent strain 116-54. The mice were administered subcutaneously with 5 mg of kanamycin per day, 24 hr prior to infection, and 1, 2, 4, 7, 10, 14, and 18 days after infection. They were used as immunized mice 24 days after the last injection of kanamycin.

Preparation of transfer agent (TA). Starting material for the preparation was collected from the abdominal cavity of immunized mice 5 days after intraperitoneal injection of 1 ml of glycogen solution per mouse. The collected cells were washed twice by centrifugation with Tris buffer. After homogenization for 5 min in a Teflon homogenizer, the homogenized cells were shaken for 15 min with an equal volume of water-saturated phenol in an ice bath. The aqueous phase was re-extracted twice with phenol. Phenol was removed by 5 extractions with washed ether. RNA was similarly extracted from the peritoneal monocytes of normal mice instead of immune mice. For the treatment of mice with TA, recipient mice received intraperitoneal injection of the phenol extracted RNA equivalent to 2×10^7 cells.

Assay of cellular antibody. The abdominal monocytes were collected from the mice which had been injected intraperitoneally with the phenol extracted RNA 5 days before the collection. Immune transfer and immune adherence hemagglutination[18,19] were used for the titration of cellular antibody, and details of these methods have been described previously.[17]

Cellular immunity. This was examined by infecting monocytes *in vitro* with a virulent strain 116-54 and was reported previously.[8]

Inhibition of bacterial growth by cellular antibody. Cellular antibody was found to inhibit bacterial growth on agar plate or in normal monocytes by the aid of complement and lysozyme. Antibody of immune monocytes was transferred to live organisms of a virulent strain 116-54 as already described.[17] The sensitized bacteria thus obtained were washed twice with veronal buffer, and the suspension of sensitized bacteria, lysozyme and complement were mixed and incubated at 37°C. After incubation for 18 hr, the inhibition of colony formation on agar plate was scored.

RESULTS

Transfer of cellular immunity to nonimmune monocytes through the transfer agent. When recipient mice were injected intraperitoneally with transfer agent (TA), the peritoneal monocytes from such mice acquired cellular immunity and inhibited intracellular multiplication of bacteria, resisting cell degeneration caused by phagocytosis of a virulent culture of *S. enteritidis*. The results are shown in Figs. 1 and 2.

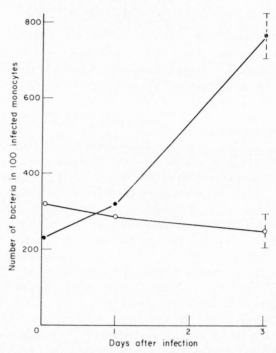

FIG. 1. Transfer of cellular immunity by TA and inhibition of intracellular multiplication of *S. enteritidis* 116–54. Symbols: ○, monocytes of mice which were treated with RNA extracted with phenol from the immunized monocytes; ●, monocytes of mice which were treated with RNA extracted with phenol from normal mouse monocytes.

In the earlier paper it was demonstrated that the cellular antibody was detectable in the peritoneal monocytes, spleen and lymphnodes of immunized mice. In this investigation, the relation between the formation of cellular antibody and the acquisition of cell resistance following treatment with TA was examined. As shown in Table 1, the cellular antibody was detectable in the monocytes of mice which had been treated with the RNA extracted with phenol from immune monocytes. As controls, the monocytes of mice treated with the phenol extracted RNA from normal monocytes were found to be negative in antibody formation.

Inhibition of bacterial growth by the cellular antibody produced in the monocytes of mice treated with TA. Antibody of immune monocytes, obtained from the peritoneal cavity of mice treated with TA, was transferred to live

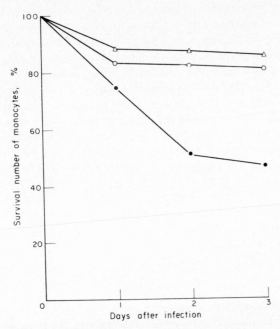

FIG. 2. Transfer of cellular immunity of TA and resistance against cell degeneration caused by infection. See the footnote of Fig. 1. △, normal monocytes without infection.

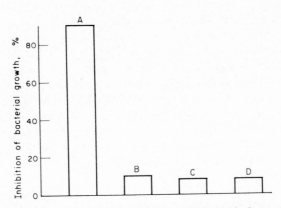

Fig. 3. Inhibition of bacterial growth by cellular antibody. A virulent strain 116-54 of *S. enteritidis* (1×10^8 organisms/ml) was used and treated with the peritoneal monocytes (1×10^8 cells/ml) of normal or immunized mice as described previously (17).The reaction mixture contained Ag (bacteria) treated with monocytes (or serum), complement, and lysozyme. A, treated with the monocytes of mice immunized with live vaccine of *S. enteritidis*; B, treated with the monocytes of mice immunized with heat-killed vaccine; C, treated with normal monocytes; D, treated with the serum of mice immunized with live vaccine.

organisms of a virulent strain 116-54, as specified under Methods, by mixed organisms of monocytes with a virulent culture. As shown in Fig. 3, the colony formation of bacteria was reduced to 2.0 to 1.1 % of the control. The *in vitro* treatment of bacteria with TA did not show such a decrease in colony formation even with the aid of complement and lysozyme.

In vitro transfer of cellular immunity to nonimmune monocytes through TA. The RNA extracts were prepared from the peritoneal monocytes derived from the immunized mice. When the nonimmune peritoneal monocytes were treated *in vitro* with TA, they inhibited intracellular multiplication of bacteria and resisted cell degeneration caused by phagocytosis of a virulent culture, resulting from the formation of cellular antibody. The representative results are shown in Fig. 4.

Pretreatment of TA with ribonuclease caused the loss of ability by TA to induce cellular immunity and resulted in the inhibition of antibody response. The transfer property of cellular immunity by TA was still retained when treated with deoxyribonuclease.

DISCUSSION

The cellular immunity in monocytes which was found in the infection of experimental animals with salmonellae was reported in infections of guinea pigs with *M. tuberculosis*,[20-23] of sheep with *Listeria monocytogenes*,[24,25] and of rabbit and guinea pig with *P. tularensis*.[26] Our results in the infection of mice with *S. enteritidis* have been confirmed recently by another laboratory.[27] It was concluded that the host resistance in immunity by live organisms is primarily dependent on the acquired immunity of monocytes which have enhanced intracellular destructive capacities. It was also demonstrated from this laboratory that monocytes, obtained from the peritoneal cavity of mice immunized with a live vaccine of *S. enteritidis*, contained specific antibody which reacted with live organisms even after absorption with formalinized antigens of the same organism. This antibody inhibited the growth of *S. enteritidis* 116-54 with the aid of complement and lysozyme, either on the nutrient agar plate or in the monocytes of normal mice.[17]

As for the cellular antibody in monocytes, there are two possibilities for its formation: (1) formation by plasma cells and adsorption onto monocytes, and (2) formation by monocytes.

An important role of plasma cells in antibody synthesis was demonstrated by many research workers[28,29] and has been firmly established by immunofluorescent studies[30] and by the single cell culture technique.[31,32]

It has been demonstrated from this laboratory that cellular immunity in monocytes is transferable to nonimmune monocytes. *In vivo* or *in vitro* treatment of monocytes with TA initiates the formation of cellular antibody in monocytes, and they acquire cellular immunity and inhibit intracellular multiplication of bacteria. The cellular antibody present in the immune monocytes

TABLE 1. CELLULAR ANTIBODY IN PERITONEAL MONOCYTES OF IMMUNIZED MICE

Monocytes obtained from mice	Absorption with formalinized (116-54)	Antigen for hemagglutination	Number of monocytes per reaction mixture ($\times 10^7$)							
			8	4	2	1	0.5	0.25	0.125	None
Immunized with live vaccine	before absorption	formalinized (116-54)	++	++	+	+	+	–	–	–
Immunized with formalinized vaccine			++	+	+	–	–	–	–	–
Immunized with live vaccine	after absorption	live (116-54)	++	+	+	+	–	–	–	–
Immunized with formalinized vaccine			–	–	–	–	–	–	–	–

was not found in the serum of mice during the whole life of mice after the last injection of live vaccine.

From these results it was strongly suggested that the cellular antibody present in the monocytes of immunized mice was produced by the monocytes themselves and not antibody transferred to the monocytes from immune sera.

Consequently, the present author proposes that the monocytes constitute an additional cell line to plasma cellular and lymphoid cell lines responsible for antibody formation.[33,34]

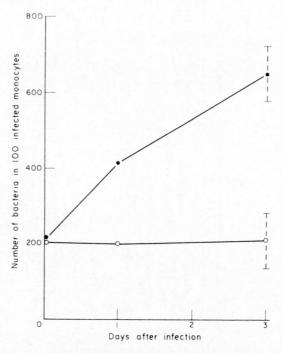

Fig. 4. *In vitro* transfer of cellular immunity by treatment with TA. The monocytes were obtained from the peritoneal cavity and suspended in culture medium as described previously.[17] For the transfer experiment with phenol extracted RNA, a number of cells equivalent to 10 per 1 recipient monocyte were incubated together. Symbols: ○, RNA extracted with phenol from immune monocytes; ●, RNA extracted with phenol from normal monocytes.

The biological activity of ribonucleic acid in immunity was reported from this laboratory[11–14] and by Fishman.[35] And the biological activity of RNA and ribosome in antibody formation has been studied by many research workers.[36–40] One thought the functional RNA in antibody formation and the other findings suggested the existence of RNA antigen complex which was extracted from organs of animals stimulated for antibody formation. Up to

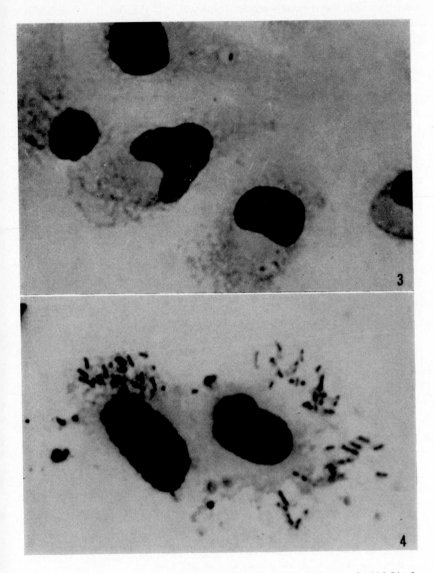

PLATE 3. Inhibition of intracellular multiplication of a virulent strain 116-54 of *S. enteritidis* by the monocyte of mouse immunized with live vaccine. Three days after infection.

PLATE 4. Destruction of monocyte of mouse immunized with killed vaccine by the intracellular multiplication of a virulent strain 116-54 of *S. enteritidis*. Three days after infection.

now it has not been clearly understood whether the functional RNA is a carrier of information for antibody formation or plays a role for antigenic information as a complex of RNA antigen or fragment thereof. Similarly, it has not been decided whether the transfer agent reported from this laboratory contains immunogenic substances or not.

REFERENCES

1. T. OKONOGI, M. FUKAI, S. MITSUHASHI, M. NAGAI and M. KAWAKAMI, *Japan. J. Exptl. Med.*, **29**, 71, 1959.
2. D. R. E. MACLEOD, *J. Hyg.*, **52**, 9, 1954.
3. W. W. C. TOPLEY, *Lancet*, **1**, 1337, 1929.
4. D. USHIBA, K. SAITO, T. AKIYAMA, M. NAKANO, T. SUGIYAMA and S. SHIRONO, *Japan. J. Microbiol.*, **3**, 231, 1959.
5. S. MITSUHASHI, M. KAWAKAMI, Y. YAMAGUCHI and M. NAGAI, *Japan. J. Exptl. Med.*, **28**, 249, 1958.
6. S. MITSUHASHI, In *Mechanisms of In ection* and *Immunity of Cytopathogenic Bacteria*. Ed. by K. Ando *et al.* Yamamoto Press, Tokyo, 1959.
7. S. MITSUHASHI, H. HASHIMOTO and M. KAWAKAMI, *Japan. J. Exptl. Med.*, **30**, 376, 1959.
8. S. MITSUHASHI, I. SATO and T. TANAKA, *J. Bacteriol.*, **81**, 863, 1961.
9. I. SATO, T. TANAKA, K. SAITO and S. MITSUHASHI, *J. Bacteriol.*, **83**, 1306, 1962.
10. K. SAITO, T. AKIYAMA, M. NAKANO and D. USHIBA, *Japan. J. Microbiol.*, **4**, 395, 1960.
11. S. MITSUHASHI, K. SAITO, I. SATO and T. TANAKA, *Proc. 16th Meeting of Japan Bacteriol. Ass.*, 1961.
12. S. MITSUHASHI and K. SAITO, *J. Bacteriol.*, **84**, 592, 1962.
13. K. SAITO and S. MITSUHASHI, *J. Bacteriol.*, **90**, 629, 1965.
14. I. SATO and S. MITSUHASHI, *J. Bacteriol.*, **90**, 1194, 1965.
15. S. MITSUHASHI, Reports from the Ministry of Education, Japan. UDC 616. 15-097, p. 352, 1963.
16. S. MITSUHASHI, M. KAWAKAMI and S. KURASHIGE, *Proc. Japan Acad.*, **41**, 635, 1965.
17. S. KURASHIGE, N. OSAWA, M. KAWAKAMI and S. MITSUHASHI, *J. Bacteriol.*, accepted for publication, 1967.
18. R. A. NELSON and H. C. WOODWORTH, Conference on complement. Walter Reed Institute of Research, Washington, D.C.
19. K. NISHIOKA, *J. Immunol.*, **90**, 86, 1963.
20. M. B. LURIE, *J. Exptl. Med.*, **75**, 247, 1942.
21. J. FONG, P. SCHNEIDER and S. S. ELBERG, *J. Exptl. Med.*, **105**, 25, 1957.
22. E. SUTER, *Amer. Rev. Resp. Dis.*, **83**, 535, 1961.
23. I. MILLMAN, *Amer. Rev. Resp. Dis.*, **85**, 30, 1962.
24. G. B. MACKANESS, *J. Exptl. Med.*, **116**, 381, 1962.
25. A. N. NJOKU-OBI and J. W. OSEBOLD, *J. Immunol.*, **89**, 187, 1962.
26. B. D. THORPE and S. MARCUS, *J. Immunol.*, **92**, 657, 1964.
27. F. M. COLLINS and M. MILNE, *J. Bacteriol.*, **92**, 549, 1966.
28. S. AMANO, M. HIRATA and Z. FUJII, *Medical Rev.*, **1**, 63, 1945.
29. A. FAGRAEUS, *J. Immunol.*, **58**, 1, 1948.
30. E. H. LEDVC, A. H. COONS and J. M. CONNOLLY, *J. Expt. Med.*, **102**, 73, 1955.
31. G. J. V. NOSSAL and J. LEDERBERG, *Nature*, **181**, 1419, 1958.
32. G. J. V. NOSSAL, A. SZENBERG, G. L. ADA and C. M. AUSTIN, *J. Exptl. Med.*, **119**, 485, 1964.
33. S. MITSUHASHI, *Proc. Japan Acad.*, **42**, 184, 1966.
34. S. MITSUHASHI, *Proc. Japan Acad.*, **42**, 280, 1966.

35. M. FISHMAN, *J. Exptl. Med.*, **114**, 837, 1961.
36. J. FONG, D. CHIN and S. S. ELBERG, *J. Exptl. Med.*, **118**, 371, 1963.
37. H. FRIEDMAN, *Science*, **146**, 934, 1964.
38. B. A. ASKONAS and J. M. RHODS, *Nature*, **205**, 470, 1965.
39. E. P. COHN, R. W. NEWCOMB and L. K. CROSBY, *J. Immunol.*, **95**, 583, 1965.
40. V. LAZDA and J. L. STARR, *J. Immunol.*, **95**, 254, 1965.

DISCUSSION

S. SENO: I agree with you on the point that macrophages have intracellular antibodies, though it does not neccessarily mean the antibody synthesis takes place in macrophages. According to my experience the ascites macrophages from a sensitized animal are swollen momentarily and degenerate when they come in contact with the antigen but those from non-sensitized animal do not. Now I have two questions, concerning the transference of RNase from macrophage to lymphocyte. The first question is, how is the RNA released from macrophages in a native state: by break-down of the cell membrane or is there any special pathway to be considered? The second question is what is the process of intake of native RNA by lymphocytes, when it is released in the native state from the macrophage. If the RNA is taken by phagocytosis, then RNA may be decomposed by active RNase in the phagocytic vesicle. There is a report suggesting the fusion of macrophage cytoplasm with that of the lymphocyte. But I have never encountered such a cytoplasmic fusion between lymphocyte and macrophage on tissue sections.

A. MITSUHASHI: (1) I cannot give you the physiology for the release of RNA. The RNA is obtained in the culture fluid of immune monocytes. This suggests that the RNA is excreted physiologically into the surrounding medium *in vitro* or *in vivo*. There is another possibility that the RNA is released from the monocytes when they are destroyed.

Many granulomas are seen in the liver, spleen and bone marrow of mice immunized with an attenuated strain of *S. enteritidis*. When these mice are challenged with a virulent strain for superimmunization, rapid disappearance of granulomas from such organs takes place, probably by the immunological response of monocytes, causing a rapid increase of antilethal resistance in mice.

(2) Practically in viral infection or fertilization, the DNA or RNA which is introduced into the cells, is not inactivated by the nucleases of the cells. I cannot give you the process by which the immune RNA is not inactivated by nucleases, but takes an immunological role in such cells.

As you mentioned, the fusion of cytoplasm may be one of the processes by which the immune RNA is transferred from the immune to the non-immune cells.

HISTOCHEMICAL
AND IMMUNOCHEMICAL STUDIES
ON SOME NUCLEASES

Shigeru Morikawa and Yoshihiro Hamashima

Department of Pathology, Faculty of Medicine, Kyoto University, Japan

It is well known that the nucleases, the depolymerizing enzymes of nucleic acids, are widespread in nature, being found in many animals, plants and microbiological forms. As for ribonuclease (RNase), it is revealed that there are at least two different groups of the enzymes, one of which has an optimal alkaline pH, named alkaline ribonuclease, and another has an acid one, named acid ribonuclease. Some differences besides optimal pH have been revealed in the enzymic properties and in the intracellular distributions or the tissue distributions. The physiological role of the intracellular RNases, though, has not been investigated sufficiently. Roth[1] and Utsunomiya et al.[2] have suggested that RNases may play a role in the synthesis of protein especially concerned with ribosomes. De Lamirande et al.[3] observed that a possible relation may exist between the ratio of alkaline and acid RNase activity of a tissue and its rate of cellular division. On the other hand, Herriott et al.[4] Alexander et al.,[5] and Connolly et al.[6] assumed that the ribonuclease activities of peripheral blood or leucocytes might protect a host from infection by pathogenic RNA virus. However, it may be said that the role of intracellular RNases and the relation between alkaline RNase and acid RNase are almost entirely unknown.

In the present paper, it is intended to clarify the relation between alkaline RNase and acid RNase in physiologic states. The distribution of these enzymes in some organs has been examined at the cellular level by the fluorescent antibody technique. Also, the distribution of cellular RNA has been investigated histologically by acridine orange staining.

MATERIALS AND METHODS

Preparation of antigens. Crystalline alkaline RNase and crystalline DNase of bovine pancreas (Sigma Chemical Co., U.S.A.) were used as the antigens. The greatest difficulty encountered in this investigation was in the preparation of the pure antigen of acid RNase and in the obtaining the specific antibody against acid RNase. Then, the isolation and purification of acid RNase was performed from the bovine spleen by a slight modification of

191

the method of Maver et al.[7] at our laboratory. Finally, spleen nuclease preparation was chromatographed on DEAE-cellulose (0.9 mEq/g. Brown Co., U.S.A.). In the procedure shown in Fig. 1, elution was accomplished by means of a convex exponential gradient, obtained with a mixer of the constant volume type. A gradient of sodium phosphate from 0.005 M (pH 8.0) to 0.15 M (pH 5.4) was begun at II, illustrated in Fig. 1. Most of the acid RNase activity was found in fraction D. The fraction of peak D was rechromatographed on a DEAE-cellulose column under the same conditions as described above. The identical fraction of fraction D on rechromatograph was used as the antigen of the acid RNase. RNases were assayed by a modification of the method of Kunitz.[8]

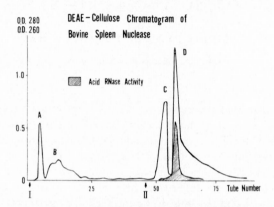

FIG. 1. Spleen nuclease preparation (400 mg in 4 ml) chromatographed on DEAE-cellulose column (1.1 × 52 cm). Initial buffer: 0.005 M sodium phosphate, pH 8.0. A convex gradient to 0.15 M sodium phosphate, pH 5.4, was begun at II. A constant volume (120 ml) mixing chamber was used. Flow rate: 10 ml per hour, with effluent collected in 5-ml fractions.

Preparation of antisera and conjugates. Anti-alkaline RNase antisera and anti-acid RNase antisera were obtained from rabbits and anti-DNase antisera from guinea pigs. Globulin fractions of the antisera were prepared by precipitating them three times with one-third saturation of ammonium sulfate. The anti-alkaline RNase globulin was conjugated with fluorescein isothiocyanate (Baltimore Biological Laboratory Inc., U.S.A.). The anti-acid RNase globulin was conjugated with tetramethyl rhodamine isothiocyanate (B.B.L.). The conjugates were purified by fractionation on DEAE-cellulose column (0.9 mEq/g. Brown, U.S.A.) as described by McDevitt et al.[9]

Assay for specificity of antisera. Specificity of antibody was examined by the double gel diffusion method of Ouchterlony and immunoelectrophoretic assay. A modification of the method of Hirschfeld[10] was used for immunoelectrophoresis.

Fractionation of the acid ribonuclease preparation by agar gel electrophoresis. Agar plates used were prepared in pH 8.6 veronal lactate buffer. The acid RNase preparation was applied into a slit (8 × 0.5 mm) which was cut out at the center of the plate. Electrophoresis was run at 8 mA per plate and 120 V for 30, 45 or 60 min. Then the agar plate was immediately cut out across the electrophoretic direction in small rectangular pieces, each 5 mm in width. Each of the agar blocks was immersed in 1.5 ml of 0.1 M acetate buffer pH 5.6 in a small test tube, and was placed in the cold (4 °C), with intermittent shaking, for 5 to 7 days. Each 1 ml of the extract was assayed for acid RNase activity.

Preparation of paraffin and frozen sections. The paraffin or frozen sections were obtained from the fresh organs of bovine pancreas, liver, spleen and thymus. For paraffin sections the following fixatives were used: (1) 95% cold ethanol, (2) 95% cold ethanol containing

5% glacial acetic acid, (3) 95% cold ethanol containing 1% glacial acetic acid, and (4) 10% formalin in phosphate buffered saline. For frozen sections: (1) 95% ethanol, (2) 95% ethanol containing 5% glacial acetic acid, (3) 95% ethanol containing 1% formalin, (4) acetone, and (5) 10% formalin in phosphate buffered saline.

Staining method for fluorescein-labeled conjugates. Following washing with phosphate buffered saline, sections were stained with fluorescein-labeled conjugate in a moist chamber at room temperature for 16 hr, or at 37°C for 2 hr. As for anti-acid RNase conjugate, the conjugate was absorbed with both normal bovine serum and fraction C of the chromatograph of the nuclease preparation (Fig. 1) by incubation at 37°C for 2 hr before use.

Acridine orange stain for RNA. Paraffin sections were used. The staining procedure was according to the method of Armstrong.[11]

RESULTS

Immunodiffusion studies. Anti-alkaline RNase rabbit antisera showed a single precipitin arc at the anodic site in the agar plate, against the crystalline alkaline RNase of the bovine pancreas which was migrated electrophoretically. In the immunoelectrophoretic pattern (Fig. 2a) of anti-acid RNase rabbit antisera, seven arcs were observed against either the nuclease preparation of the bovine spleen, or fraction D (Fig. 1). Two components of these arcs, which were observed near the well at the anodic site, were identified with the components of fraction C (Fig. 1) that showed no acid RNase activity. Three arcs appeared at the cathodic side of the agar plate were revealed due to the bovine serum proteins belonging to α or β globulins immunoelectrophoretically. So, after absorption of the anti-acid RNase antiserum with both fraction C and normal bovine serum by incubation at 37°C for 2 hr, the absorbed antiserum made only two arcs against fraction D in the immunoelectrophoretic pattern. Anti-acid RNase fluorescein-labeled conjugate also showed the same pattern as the absorbed antiserum did.

It was found, as shown in Fig. 2b, that both of the components of the two arcs observed after absorption of the antisera had acid RNase activities, from the results of the electrophoretic fractionation of the acid RNase preparation on agar gel, and the results of the immunoelectrophoresis which were performed under the same conditions simultaneously. From this fact it may be concluded that there are at least two acid RNases in the bovine spleen and that they are independent antigenically of one another. It may be confirmed, also, that the anti-acid RNase conjugate will react with two acid RNases of bovine only, after the absorption with both fraction C and bovine serum.

Cross-reactivity. The cross-reactivity between alkaline RNase and acid RNase could not be observed in the agar plate by double diffusion as illustrated in Fig. 3. Anti-alkaline RNase antiserum formed a single precipitin band between the crystalline alkaline RNase of the bovine pancreas and formed no band between the acid RNase preparation, or the crystalline DNase of the bovine pancreas. Anti-acid RNase antiserum absorbed formed two bands between the acid RNase preparation, corresponding to the immunoelectrophoretic pattern. No band was made between the other antigens examined. Anti-DNase antisera did not form any precipitin band between the RNases.

HISTOLOGICAL STUDIES

Pancreas. For fluorescent staining of alkaline RNases in the bovine pancreas, all the fixatives including ethanol, acetone or formalin described above were available in both frozen and paraffin sections. However, the distribution of pancreatic alkaline RNase showed some variation in the cytoplasm of acinar cells depending upon the staining procedure. In paraffin sections

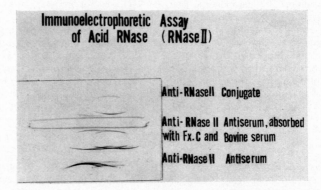

FIG. 2a. The upper and middle wells were filled with fraction D. The lower well was filled with the spleen nuclease preparation. After electrophoretic procedure, the antisera and conjugate illustrated above were filled into the trenches.

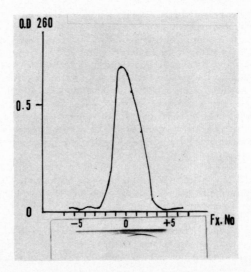

FIG. 2b. Acid RNase activity of fraction D of the spleen nuclease preparation, fractionated by agar gel electrophoresis. The peak of acid RNase activity did not form a symmetrical curve. The immunoelectrophoretic pattern showed below was procedured under the same conditions, simultaneously.

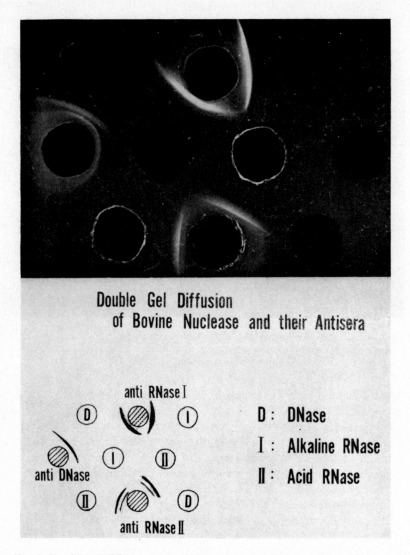

FIG. 3. Double gel diffusion study of the nucleases and their antisera. No cross-reactivity was revealed among them. The arc around the well filled with anti-DNase antisera may be due to the existence of denatured protein.

fixed with various fixatives or in frozen sections fixed with 95% ethanol or 10% buffered formalin, the zymogen granules at the apices of acinar cells showed specific fluorescence intensely (Fig. 4). On the other hand, the specific

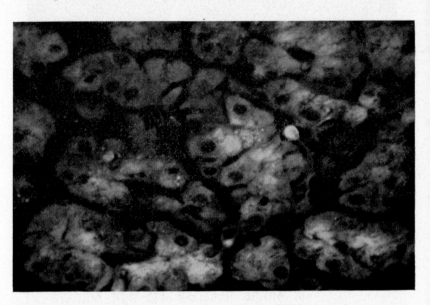

FIG. 4. This picture illustrates the location of alkaline RNase in bovine pancreas. Bright fluorescein deposits are demonstrated in the cytoplasm of the acinar cells. Paraffin section fixed with 95% ethanol.

fluorescence was observed as fine granules at the base of acinar cells and faint fluorescence was observed at the apex in frozen sections fixed with acetone. The nuclei of the acinar cells were not fluorescent in any sections. The cells in the islet of Langerhans and the epithelium of the ducts of the pancreas showed no fluorescence in any sections. A majority of acinar cells were observed containing alkaline RNase in zymogen granules, in accordance with the findings in paraffin sections.

The specific fluorescence of acid RNase was observed in the cytoplasm of several cells in the islet as fine granules and was observed diffusely in the cytoplasm of some mononuclear cells or leucocytes scattered in the stromal tissue of the pancreas. Occasionally, the nuclei of the cells in the islet showed the fluorescence of acid RNase slightly, but the nuclei of the acinar cells did not show the fluorescence entirely. The bright orange-red fluorescence of RNA by acridine orange stain was distributed mostly at the base of the acinar cells. Only slight fluorescence of RNA was observed at the apex of the acinar cells and the cytoplasm of the islet cells.

Liver. The staining procedure other than the frozen section fixed with acetone failed to show any fluorescence of alkaline RNase in hepatic cells.

In frozen sections fixed with acetone, slight specific fluorescence of alkaline RNase was observed in the perinuclear area of cytoplasm and in the nuclear membrane of hepatic cells. Sometimes, the nuclei also showed the fluorescence in frozen sections. The hepatic cells were not stained with anti-acid RNase conjugate excepting the nuclei, which were stained faintly sometimes in frozen sections fixed with acetone or 10% buffered formalin. The leucocytes in the sinusoid showed the specific bright fluorescence of acid RNase

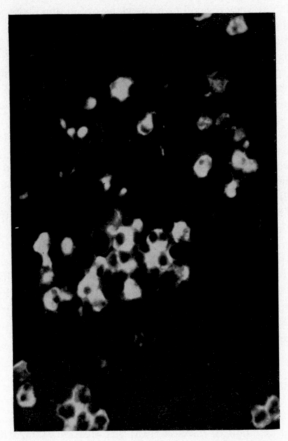

FIG. 5. The cells which show brightly fluorescent cytoplasm contain acid RNase. Bovine spleen; frozen section fixed with 10% buffered formalin.

in their cytoplasm. Only faint fluorescence was observed in Kupffer's cells in some sections. A moderate degree of fluorescence of RNA was observed as coarse granules throughout the cytoplasm of the hepatic cells.

Spleen. Acetone and 10% buffered formalin gave good results as fixatives for RNases in spleen sections. Some of the large mononuclear cells and leucocytes in the red pulp of the spleen were stained in their cytoplasms with anti-alkaline RNase conjugate. Specific fluorescence of acid RNase was also observed more intensely in the cytoplasm of large mononuclear cells, plasma cells, binucleated round cells and leucocytes which were scattered or in groups in the red pulp (Fig. 5). The fluorescence of alkaline RNase was observed rather diffusely in the cytoplasm of round mononuclear cells. On the other hand, the fluorescence of leucocytes was observed as small granules. The nuclei of the some cells in the red pulp were stained with anti-acid RNase conjugate sometimes, but no nucleus was stained with anti-alkaline RNase conjugate in the spleen.

The cytoplasm of the cells in the lymph follicles showed faint fluorescence with the exception of the reactive germinal centers, in which phagocytic cells and large mononuclear cells are scattered. In such germinal centers, some of the mononuclear cells showed the specific fluorescence of the RNases. Cells containing alkaline RNase were less numerous than those containing acid RNase in the bovine spleen. As a matter of course, all of the cells observed in a section were not fluorescent, but only a small portion of them showed specific fluorescence.

The cells containing RNA were observed mainly in the red pulp, and some of them corresponded with the cells containing RNases. But leucocytes showed no fluorescence for RNA by acridine orange stain.

Thymus. The same kinds of the cells as in the spleen were fluorescent in the medulla of the thymus by the anti-alkaline RNase conjugate or by the anti-acid RNase conjugate. The lymphoid cells in the cortex showed no fluorescence by any staining procedure. The nuclei of the thymic cells were not stained with either of the RNase conjugates. The number of cells containing acid RNase was higher than that of cells containing alkaline RNase in the bovine thymus. The fluorescence of RNA was observed in the cells scattered in the medulla of the thymus and its distribution was rather similar to that of the RNases.

DISCUSSION

Thus far, much information has accumulated regarding the fluorescent antibody technique including the investigations for enzymes.[12-14] It seems that the fluorescent antibody technique has an advantage over the other histochemical methods, in the case of the investigation of the distribution of the different enzymes that show same enzymatic activities, such as RNases, DNases, etc. As for RNase, it has been recognized that there are at least two different enzymes, namely the alkaline RNase and the acid RNase. The immunological cross-reactivity between the alkaline and the acid RNase is not revealed by the immunodiffusion studies, in spite of possessing the same depolymerizing activity to RNA. The crystalline alkaline RNase of the bovine pancreas is reported to be subfractionated into 3 or 4 fractions by ion-exchange cellulose column chromatography[15] or immunoelectrophoretic assay.[16] However, the anti-alkaline RNase antisera makes only one precipitin band against the enzyme by both of immunodiffusion studies in our experiment. The bovine acid RNase preparation was subfractionated or divided into two enzymes, each of which has acid RNase activity and has different antigenicity to one another, as illustrated in Figs. 2a, b. This result may correspond to the result of chromatography on the ion-exchange cellulose column of the bovine spleen preparation, which shows the two peaks of acid RNase activity reported by Maver et al.[7] The conjugates used in this investigation will show separately the location of one of the alkaline RNases and the location of one or two of the acid RNases.

Our observations of the localization of alkaline RNase in the pancreas are similar to those of Marshall[12] and Daoust,[17] who reported on the localization of RNases and did not mention anything about the acid RNase which we found in the islet of Langerhans and in the interstitial tissue, in this report. The variation of the distribution of alkaline RNase depending upon the staining procedure is seen to be an important problem. The localization of alkaline RNase in the base of acinar cells revealed in frozen sections fixed with acetone may correspond to the report of de Lamirande et al.[3] in which the RNases were found present in the mitochondrial fraction. Only the frozen sections fixed with acetone revealed the localization of alkaline RNase in hepatic cells. De Duve et al.,[18] Reid et al.,[19] and de Lamirande et al.[20] reported the presence of acid RNase activity in lysosomal fraction and in soluble fraction of the rat liver preparation. Though, in the present work, we only found the location of acid RNase in the infiltrating leucocytes intensely and sometimes in the nuclei slightly. The findings of Daoust[17] by the substrate film method, concerning the location of RNase in the spleen was confirmed by the fluorescent antibody technique in this investigation. Also, the localization of acid RNase in leucocytes, reported by Atwal et al.[21] was confirmed. The alkaline RNase, however, was revealed in leucocytes, too. The cellular localization of RNA was not always in accord with those of the RNases.

SUMMARY

1. The specific antibodies against the bovine alkaline and acid RNases were obtained from albino rabbits.
2. Immunological cross-reactivities among the bovine alkaline and acid RNases and the bovine DNase were examined by the immunodiffusion studies. And no cross-reactivity was proved.
3. The distribution of the alkaline and acid RNase in the pancreas, liver, spleen and thymus was investigated by the fluorescent antibody technique. The alkaline RNase was found mostly in the acinar cells of the pancreas, in the hepatic cells and in some of leucocytes and mononuclear cells in the spleen and the thymus. The acid RNase was contained mainly in the mononuclear cells, plasma cells, and leucocytes in

the spleen and thymus. Sometimes, the nuclei show the specific fluorescence of the RNase.

The distribution of cellular RNA was not always in accord with that of the RNases.

REFERENCES

1. J.S.ROTH, *J. Biophys. Biochem. Cytol.*, **7**, 443, 1960; **8**, 665, 1960.
2. T.UTSUNOMIYA and J.S.ROTH, *J. Cell Biol.*, **29**, 395, 1966.
3. G. DE LAMIRANDE and C.ALLARD, *Ann. N.Y. Acad. Sci.*, **81**, 570, 1959.
4. R.M.HERRIOTT, J.H.CONNOLLY and S.GUPTA, *Nature (London)*, **177**, 702, 1956.
5. H.E.ALEXANDER, G.KOCH, I.M.MOUNTAIN and O. VAN DAMME, *J. Exp. Med.*, **108**, 493, 1958.
6. J.H.CONNOLLY, R.M.HERRIOTT and S.GUPTA, *Brit. J. Exp. Path.*, **43**, 402, 1962.
7. M.E.MAVER, E.A.PETERSON, H.A.SOBER and A.E.GRECO, *Ann. N. Y. Acad. Sci.*, **81**, 599, 1959.
8. M.KUNITZ, *J. Biol. Chem.*, **194**, 563, 1946.
9. H.O.McDEVITT, J.H.PETERS, L.W.POLLARD, J.G.HARTER and A.H.COONS, *J. Immunol.*, **90**, 634, 1963.
10. J.HIRSCHFELD, *Sci. Tools*, **7**, 18, 1960.
11. J.A.ARMSTRONG, *Exp. Cell Res.*, **11**, 640, 1956.
12. J.M.MARSHALL, JR., *Exp. Cell Res.*, **6**, 240, 1954.
13. H.D.MOON and B.C.McIVUR, *J. Immunol.*, **85**, 78, 1960.
14. K.YASUDA and A.H.COONS, *J. Histochem. Cytochem.*, **14**, 303, 1966.
15. C.H.W.HIRS, S.MOORE and W.H.STEIN, *J. Biol. Chem.*, **200**, 493, 1953.
16. B.G.CARTER, B.CINADER and C.A.ROSS, *Ann. N. Y. Acad. Sci.*, **94**, 1004, 1961.
17. R.DAOUST, *Internat. Review Cytol.*, **18**, 191, 1965.
18. C. DE DUVE, B.C.PRESSMAN, R.GIANETTO, R.WATTIAUX and F.APPELMANS, *Biochem. J.*, **60**, 604, 1955.
19. E.REID and J.T.NODES, *Ann. N.Y. Acad. Sci.*, **81**, 618, 1959.
20. G. DE LAMIRANDE, C.ALLARD, H.C. DACOSTA and A.CANTERO, *Science*, **119**, 351, 1954.
21. O.S.ATWAL, J.B.ENRIGHT and F.L.FRYE, *Proc. Soc. Exp. Biol. Med.*, **115**, 744, 1964.

DISCUSSION

R.KINOSITA: Do you think that the alkaline and acid RNases referred to, would show the depolymerizing activity within the normal cells in which you have identified their presence?

Y.HAMASHIMA: We have no idea how to identify enzymic activity specifically in the cells of the section.

In our *in vitro* examination relative to the relationship between the enzymic activity and its antigenicity, it was found that the enzyme incubated with the antiserum shows a lesser activity than the enzyme incubated with normal serum.

S.SENO: Looking at your picture, I was impressed that some RNase detected by your method will be active, which seems to be contained in pino- or phago-cytotic vesicles, and that some others appearing diffuse in cytoplasms may be inactive in the presence of inhibitor. From the biological standpoint these two RNases, active and inactive ones, must be observed separately. And can you tell me whether or not there is any method of observing them separately.

Y. HAMASHIMA: No, I have no idea. I agree with your important note to make sure that there are two different types of active and inactive RNases in the cells, and also to check, morphologically, the presence of an inhibitor against RNase activities in the tissues or in cells.

H. TERAYAMA: You mentioned in your first slide that there is no acid RNase in the liver. But acid RNase is actually present in the liver, though most of it is found in lysosomes.

I wonder where you got or how you got the information about the absence of acid RNase in the liver.

Y. HAMASHIMA: I would like to answer the comment that Dr. Terayama made.

Present information indicates that only a few leukocytes in sinusoids of the liver, not parenchymal cells, contain acid RNase.

RNase in lysosomes of the liver which you mentioned is probably in these leukocytic lysosomes in the sinusoids, because the total amount of RNase in the liver is quite small.

But we know that it is difficult to find the best fixative for hepatic cells by means of the immunofluorescent technique.

DIVISION–LIMITING MORPHOGENETIC PROCESSES IN *TETRAHYMENA*

Erik Zeuthen and Norman E. Williams

This paper deals with the significance for later cell division in the ciliate protozoan *Tetrahymena pyriformis* (Fig. 1), of the oral primordium (OP), a system of cytoplasmic fibers and centrioles* which, in reaction with the plasma membrane and underlying cisternae, develops into the complex mouth structure (oral apparatus = OA) of this cell.

The oral apparatus and its primordium are useful morphogenetic markers of the position of the cell in its division cycle, which is also a morphogenetic cycle.† The autoradiographic data reported in this paper will be interpreted together with other results previously obtained for *Tetrahymena* populations with induced division synchrony. The analysis contributes to the standing discussion of the mechanisms by which temperature shocks induce division synchrony in *Tetrahymena* populations. We have, where possible, wanted to stress similarities rather than dissimilarities between *Tetrahymena* and mitotic cells.

* Use of the term "centriole" here requires comment. A "centriole" is defined as a granule at the focus of the astral rays (Mazia, *The Cell*, Vol. III, p. 116, 1960). The term "kinetosome" is usually used with reference to the ciliate cortex, and a kinetosome is defined as a granule at the base of a cilium. The structural similarity between centrioles and kinetosomes is now generally acknowledged. The difficulty of naming relevant structures, which in ciliates have no known role in nuclear division and which are not engaged as basal bodies for cilia, is traditionally solved by use of the designation "non-ciliated kinetosomes". However, in a previous study of the ciliate oral primordium,[3] we have shown (1) that the granules of this ectoplasmic system are foci for the formation of fibrillar arrays, (2) that this formation is functionally linked with cell and nuclear division, and (3) that some of these granules never form cilia,[1] i.e. never specialize as true kinetosomes. For these reasons, and also because we wish to suggest that the molecular mechanics of this cortical system may be similar to mitotic mechanisms in higher cells, we will use the term "centrioles" for the granular elements of the oral primordium of *Tetrahymena*.

† *Tetrahymena* is a polycentriolar cell, and amongst higher cells, it is one of the fastest and most hardy growers. *Tetrahymena* can be cultured bacteria-free in crude and fully defined media. Cultures are not readily infected with bacteria, because contaminants are phagocytized. Mass cultures can be easily and automatically synchronized for cell division. The cells are large (50 μ) and mobile by ciliary movement. They divide by binary fission, for lack of better knowledge described as amitosis, and they come apart immediately, so that a suspension culture consists of single cells well suited for electronic cell counting.

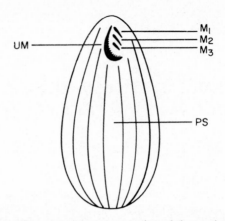

FIG. 1. *Tetrahymena*. Diagrammatic representation of the cortical morphology of *Tetrahymena* as revealed by silver nitrate preparations. The meridinal lines represent the somatic rows of cilia. The oral apparatus consists of three membranelles (M1, M2, M3) and an undulating membrane (UM). The oral primordium for the posterior division product appears in the area designated as PS (primordium site) (Williams and Zeuthen, 1966).

THE ORAL APPARATUS AND ITS PRIMORDIUM

The fine morphology of the OA has been examined in detail by Nilsson and Williams,[1] partly from sections of whole cells, partly from whole OAs isolated by lysis of cells in 1.5 M t-butanol (Williams and Zeuthen[2]). An isolated OA and its model is shown in Fig. 2. The OA consists essentially of three elements: pellicle, kinetosomes + centrioles, and fibrous material. The pellicle (plasma membrane with internally adhering flat cisternae, the pellicular sacs) is continuous over most of the structure, excepting the cilia and the food vacuoles, which are lined with plasma membrane. The kinetosomes of the mouth are arranged in triple rows (M1, M2, M3 in Fig. 2) which constitute the basal bodies for three triple rows of cilia in membranelles 1, 2 and 3, and in a single row which underlies the cilia in the undulating membrane (UM). At the base of the UM is a second (median) row of bodies, but these are centrioles in close contact with the plasma membrane. From this row and from membranelle 3 sprout thin fibrils which, as a 15 μ long "deep fiber" bundle (DF), reach the vicinity of the macronucleus. The OA is placed in the front of the cell (Fig. 1) and frontally to two blindly terminating kineties (ciliary rows). Before a newly divided *Tetrahymena* cell can again divide, a new mouth is formed. The OP normally first appears after the elapse of 40% of a generation's time. It always forms near the equator of the cell (at PS in Fig. 1) and in the open space between the two kineties which end behind the OA. It consists of many subpellicular centrioles arranged without order. Later steps in the morphogenesis of the new mouth are the arrangement of the subpellicular centrioles in rows; the establishment of direct con-

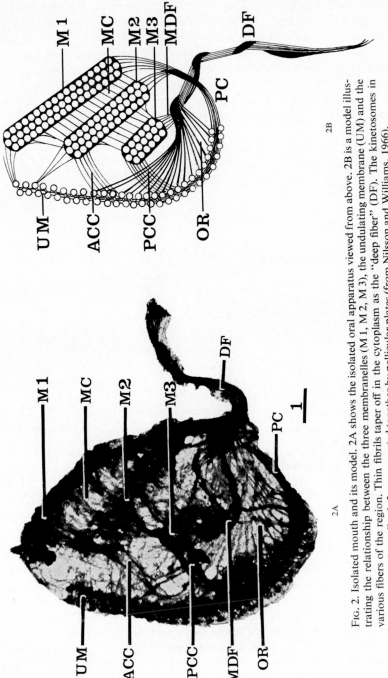

FIG. 2. Isolated mouth and its model. 2A shows the isolated oral apparatus viewed from above. 2B is a model illustrating the relationship between the three membranelles (M 1, M 2, M 3), the undulating membrane (UM) and the various fibers of the region. Thin fibrils taper off in the cytoplasm as the "deep fiber" (DF). The kinetosomes in membranelles 1–3 are cemented together by pellicular plates (from Nilsson and Williams, 1966).

tact, through openings in or between the pellicular sacs, between centrioles
and plasma membrane; the further differentiation of the centrioles into
ciliated kinetosomes; and the development of an extensive system of fibrils
between centrioles (or kinetosomes) and of fibers emanating from the mouth
and reaching other parts of the cell, illustrated, for example, by the above
mentioned deep fiber. Morphogenetically important materials precipitate
out at various places, cementing structures together. The fused pellicular
plates (Fig. 2) are cemented structures.

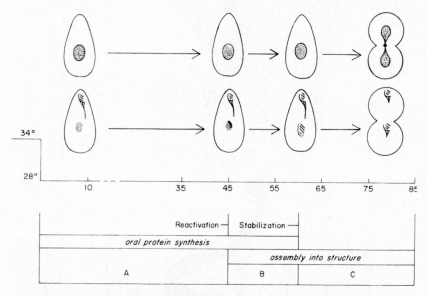

FIG. 3. Summary diagram of morphogenetic events in synchronously dividing
Tetrahymena. Development of the oral primordium as well as changes in the macronucleus and the old oral apparatus are indicated above the time axis. Time is in
minutes following the last heat shock of the synchronizing treatment. The critical
events, "reactivation" and "stabilization", divide the total time from the end of
treatment until division into three distinct periods (from Williams and Zeuthen,
1966).

The morphogenetic sequence in heat-synchronized cells is demonstrated
in Fig. 3. At the time EH (end of heat shocks) the primordium is present in
the synchronized cells, but its centrioles are placed without order and there
is no visible morphogenesis over the next 40 min. Then follows a period with
swift subpellicular ordering of the centrioles into rows. This phase is largely
over at EH + 60 min, but there is likely to be overlap with later morphogenetic events referred to above but not known in much detail.

Before order has yet appeared in the primordium, there are no filaments,
fibrils or fibers between the centrioles. When order is becoming established,
the centrioles can be seen to be connected by numerous filaments, 20 mμ

fibrils and thicker fibers. Thus, creation of order between centrioles is accompanied and (analogy: mitosis) probably conditioned by the appearance of fibers, some of which are of microtubular dimensions.[2]

ARE PROTEINS OF THE ORAL APPARATUS IN CONTINUOUS EXCHANGE WITH PROTEINS OUTSIDE THIS STRUCTURE?

The OA can be isolated from cells lysed with 1.5 M tertiary butanol of neutral pH and low ionic strength. When this is done with exponentially multiplying cells previously labelled with tritiated amino acids, the "hot" OAs can be radioautographed for labelled proteins (Williams and Zeuthen[3]). The radioautographic counts scored can only to a minor degree result from contamination with cytoplasmic proteins adsorbed onto the structure during the isolation: if "cold" mouthes isolated from one group of cells are swirled around in the lysate from hot cells, the cold mouthes label weakly (mean 4 counts/OA, Fig. 4) and this, we think, measures adsorption. Hot mouthes from the lysed cells are strongly labelled (mean 25–30 counts/OA, Fig. 4). The distribution indicated for the OAs isolated from hot cells is more or less typical of all experiments here to be described.

Cells were grown under conditions (heat shocks) which preserve the OAs present at the beginning of the incubation, but block the formation of new ones. The heat shocks induced conditions for a later synchronous division.[4]

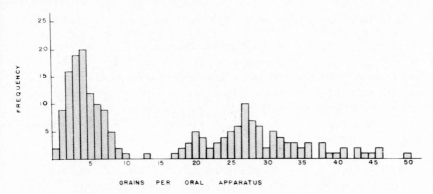

FIG. 4. *Right*: Autoradiographic counts per OA isolated from exponentially multiplying cells incubated with tritiated histidine (10 μC/ml) for 4 hr. *Left*: Counts for cold OA's exposed to the lysate from which the OA's to the right were isolated.

The cells were allowed to incorporate labelled amino acids for 10 min periods, initiated 10 min before the times shown by the points in Fig. 5. They were then lysed and autoradiographed for amino acid label in the isolated OPs (curve I) and OAs (curve II). The data are presented without subtraction of the background measured in Fig. 4. For the first 40 min (28°) after EH there is morphogenetic arrest of the OP (Fig. 3), but between 15 and 40 min there is incorporation at increasing rates (Fig. 5, curve I). The rates (counts/OP × 10 min) reach a peak value at 55–65 min, then drop. In general the incorporation in the old (and larger) oral apparatus (curve II) follows the same pattern as the curve for the primordium, but it neither rises so quickly before,

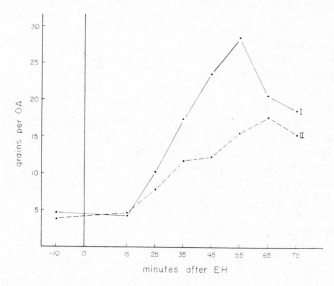

FIG. 5. Autoradiographic counts per OP (I) and OA (II). Results of pulse-labelling of the cells for 10-minutes periods with tritiated histidine (14 μC/ml). Label added 10 minutes before each point.

nor drops so much after the peak. Therefore, the two curves come close at the time when the primordium is maturing to a fully formed mouth, in all respects indistinguishable from the old oral apparatus.

It will be noticed from Fig. 5 that the resumption of primordial morphogenesis is not marked as a discontinuity on curve I, and the observations reported thus far are more indicative of a fast turnover of the proteins in both the old mouth and in the primordium than of growth of any of the structures.

In collaboration with Mr. Ole Michelsen we have asked the question whether the OA is a center of protein synthesis. The tentative answer is "No", because when butanol-isolated OAs were incubated with tritiated amino acids, an ATP-regenerating system, and with supernatant from cells lysed by freeze-thawing, the autoradiogram showed incorporation of amino acids into protein only over lumps of cytoplasm, not over fairly clean mouthes. Turnover as referred to above may therefore mean that proteins synthesized extraorally continue to move into the mouth structures, and again move out, either after some change that makes them unfit for combination in the structure or without such change.

In the next experiment the proteins of exponentially multiplying cells were labelled with tritiated histidine for 12 hours. The cells were then washed twice in unlabelled medium, and cold histidine was added in large excess (chase). The population was sampled over three generations, beginning at the time when log growth, stopped for some time by the washing, has been resumed. The results are plotted in Fig. 6, without correction for the expected unspecific contamination (cf. Fig. 4). Proper mathematical analysis has not yet been made, but we consider that the distribution curves for counts per OA are monomodal for all time points sampled. Therefore, the results may be interpreted to suggest that protein-label is chased from an old mouth, and that proteins synthesized generations ago show up in identical new structures.

The dilution of label on the growing population of OAs less than corresponds to the increase in number of OAs. The overall shape of the curve suggests, either ineffective chasing over the first generation's time after the chase, or the presence at the initiation of the chase—or at the somewhat later time of resumption of logarithmic multiplication—of extraoral hot

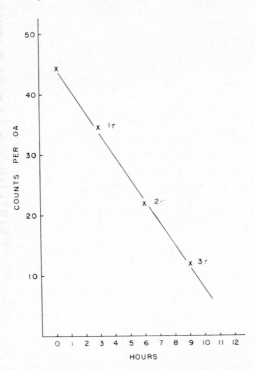

FIG. 6. A population was grown for 12 hr with triated histidine 10 μC/ml. Then the label was removed with 2 washes in a medium which was diluted 1:4 (OAs isolate better in this medium). The OAs were isolated immediately after log growth was resumed (0 hr) and then at one generation intervals for 3 generations.

proteins specified for incorporation into new mouthes. In the two latter cases an extraoral pool of oral precursor proteins, present when the chase begins, or when log-growth is resumed, would gradually move into structures as they form, and the radioactivity of the extraoral pool would be diluted with cold precursor proteins, formed later. The shape of the curves (Fig. 6) and the suggested monomodality of the curves for distribution of counts per OA even 3 generations after the chase to us supplies some support for the view that the oral structures are in, or approach, equilibrium with extraoral precursor proteins specified for the structures. It is implied that pool size could be considerable in exponentially multiplying cells. The next experiment supplies evidence that it is large in synchronized cells.

EXTRAORAL PROTEIN PRECURSORS FOR ORAL STRUCTURES

In the following experiments, tritiated amino acid was supplied to cells exposed to heat shocks in the program which synchronizes division. The growth medium was replaced with inorganic medium (no label) after heat shock 5. Chaser amino acid was added at the end of shock 7 (EH) to stop further synthesis of radioactive proteins. The mouthes were then sam-

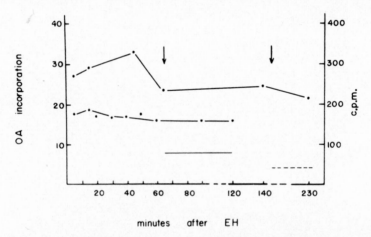

FIG. 7. Upper curve: Cells were synchronized with 7 heat shocks in nutrient medium (2% proteose-peptone, 0.1% liver extract) to which was added tritiated leucine (10 μC/ml), just prior to the first heat shock. The cells were transferred to a "cold" inorganic medium after shock 5, and chaser-leucine was added after shock 7 (1500 × the concentration of the label).The OA's were isolated at times of points. Arrows indicate times of maximum division, 1 and 2. Ordinate shows autoradiographic counts per OA. *Middle and lower curves* are based on experiments under identical conditions by Crockett *et al.*[5]. The label (C[14] phenylalanine) is retained in the cell protein (curve with points). If the OA's were in isotope equilibrium with the bulk of the cell proteins we would have expected the isolated OA's to have the counts shown by the levels or initial levels of the three curves shown. Our experimental (upper) curve is quite different and requires the interpretation given in text.

pled through two divisions. In the present case the fresh medium was inorganic, and cellular and organelle multiplication took place under starvation conditions. There were two reasons for the choice of starvation conditions in this experiment: (i) the chase was expected to be maximally effective because the amino acid pool is smaller in starving than in growing cells,[5] and (ii) data were already available[5] for rates of protein synthesis under the conditions of this experiment, which reproduced those under which these previous data were collected.

The results are shown in Fig. 7. The counts per old OA went up till EH + 45 min, and then down. At division 1 (arrow) the new and old mouthes are morphologically and radioautographically indistinguishable, and the same is true at division 2 (arrow) in which perhaps three out of four mouthes are new. Even with reasonable correction for background, we find that the dilution of the label on the mouth structures is far smaller than expected from the observed multiplication of these structures, and we therefore interpret our results to indicate that at EH proteins specified for fitting the mouth structures are present extraorally in large amounts.

DISCUSSION

When individual *Tetrahymena* cells are subjected to heat shocks of the type used in the synchronizing treatment, to cold shocks or to certain inhibitors of protein synthesis, division is generally delayed for a period of time in excess of the duration of the treatment. These delays have been called "setbacks". The division is only delayed, however, if the agent takes effect prior to a point late in the cell cycle, called the "physiological transition point", after which the cells are insensitive to these agents. Furthermore, the extent of the set-back is age dependent, i.e. the division of older cells (prior to the physiological transition point) is delayed more than that of younger cells. This phenomenon provides the basis for the induced division synchrony.

Cann[6] has developed a kinetic model for *Tetrahymena* synchronization

$$A \rightarrow B \rightarrow C$$
$$\downarrow$$
$$X$$

The model bears evolutionary resemblance to an earlier reaction scheme of Rasmussen and Zeuthen[7] from which it takes over the concept that preparation of cell division is by a shifting balance between synthesis (A → B) and removal of an intermediary product (B) in a side reaction (B → X). New is the concept that division is initiated only when the product (C) of two sequential reactions has accumulated to a critical concentration.

Cann's model will account for set-backs obtained with heat, cold, *p*-fluorophenylalanine and (in part) with mercaptoethanol if it is assumed that all agents slow reaction A → B, slightly accelerate reaction B → C and greatly stimulate or even initiate reaction B → X. The model accounts for the transition point as "that age of the cell beyond which it is not possible, by short time exposure to supraoptimum temperature, to interfere with the attainment of the critical concentration of C in the same time as required by cells progressing undisturbed toward division".

Studies of oral morphogenesis in *Tetrahymena* have shown that this developmental system has properties strikingly similar to those of the critical system postulated on the basis of the set-back data. This particular ciliate cell forms a second oral apparatus before division is initiated. Early development of this organelle is blocked by temperature shocks and certain metabolic inhibitors (Fig. 3, period A). After the system has been left undisturbed at 28° for some time the block is followed by morphogenetic reactivation, and for a limited time prior to the transition point (stabilization) the above treatments induce resorption of the reactivated oral primordium (Fig. 3, B). Finally, the time of the physiological transition is also a time of developmental stabilization of the primordium: shock treatments (as defined above) after the transition point will not interfere with the further development of the primordium into a new mouth (Fig. 3, C).

In a previous paper[2] we stressed that the oral primordia in cells examined immediately after the synchronizing treatment lack subpellicular fibers, and also that fibers make their first appearance at the time when oral morphogenesis is resumed ("reactivation", Fig. 3). One is therefore free to guess at which time macromolecules (fiber precursors) begin to aggregate into fibers. Clearly, this event occurs before, and perhaps well before, fibers can be demonstrated with the electron microscope. Because shock treatments cause set-backs early after the synchronizing treatment, long before the oral primordium is reactivated, and continue to have effect for some time after the reactivation, the simplest hypothesis we can make is that the synchronizing treatment blocks the polymerization of precursors into fibrous material (late step in sequence A → B in model), and that it also depolymerizes fibers, visible and invisible, already present (side-reaction B → X). At least for shock exposures to heat and to mercaptoethanol[8] it is easy to see that these two effects may go together. The two treatments will tend to prevent the formation of, or break, respectively cohesive and sulfur bonds. Thus, the subpellicular fiber system of the oral primordium may represent the primary site of action within the morphogenetic system for the treatments which simultaneously cause set-backs and delay or reverse morphogenesis. Returning to Cann's model we can now suggest that B represents, not an unknown macromolecule, but a known macromolecular aggregate of time-dependent complexity, and intermediate on the structural pathway to a stable mouth.

We propose (cf. diagram of Fig. 8) that the structurally complex oral primordium and apparatus contain a minimum of three components: whole centrioles and at least two kinds of proteins in the microtubules and in bundles of tubules and filaments. Morphogenesis of the oral primordium depends on a more or less precisely running aggregation of these elements into an initially very labile structure. Across the middle of Fig. 8 we have shown this pictorially (follow the heavy arrows). Disorder (1, left) gives place to order (2, middle), and order precedes a second kind of disorder (3, right).

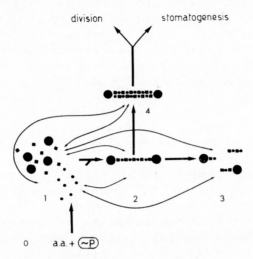

FIG. 8. Illustration of arguments presented in the text. The *heavy* arrows indicate the path of morphogenetic assembly of structures essential for division and for oral morphogenesis. Step (1 → 2) is limited by the synthesis of more of the reactants (1) or by one of the reactants (text). It may also depend on synthesis of catalytic factor(s), enzyme(s). The latter possibility would demand revision of the details in the lower left of the scheme, which revision is rather obvious, and is not shown. The *thin* arrows refer to the possibility (text) that proteins in structure are in continuous exchange with proteins outside structure.

This reflects the view that the assembly of macromolecules into a definable structure requires that the building blocks are picked in some order, and it suggests that before its stabilization (2 → 4) a labile structure like the oral primordium tends to fragment by breakage at random places, which process is greatly accelerated by e.g. heat shocks. When it breaks, the structure is likely first to liberate pieces (3, right) which cannot be directly reutilized, because they are differently specified than the original building blocks (possibility (i)). Liberation of such original blocks in quantity seems possible (ii), but to us less likely, simply because so many more bonds must be broken for their liberation. The decision between these two possibilities, (i) and (ii), is important, because it greatly influences our stand on the question about the nature and function of those proteins which must be synthesized *de novo*[7] after the heat shocks to permit new structures to be assembled from, or in part from, protein precursors present in the cell at the time EH. Is this fraction enzymic with catalytic function(s), or does it contain elements which must be incorporated into the growing structures?

Let us assume that the developing and mature structures have low affinity for some of their proteins which will therefore always be present in quantity outside the structure,—growing, mature or broken. In our experiments and analysis these proteins would show themselves as "extraoral precursors". Proteins with low affinity to the structures would not be revealed in our experiments and analysis, but they could perhaps to a higher degree be responsible for assembling the structures from their macromolecular constituents and for holding structures, once formed, together. It might be true for both types of proteins that synthesis after EH is necessary to shift the balance from macromolecular disorder (Fig. 8, left) towards structural order (Fig. 8, middle). The situation described would be a likely consequence if we accept possibility (i), according to which structures broken, e.g. with heat, fail to liberate elements which can be easily reutilized for making new structures.

The most extreme consequence of possibility (ii) would be that all building blocks held in structure would be liberated by heat shocks for immediate reutilization. This would not explain why the cell is short of essential proteins after a series of heat shocks, and we would have to postulate selective damage to catalytic factors (polymerizing enzymes?) or, already modifying the premises under which we now discuss, to fractions of proteins which have been part of structure. After EH there would then be need for the selective synthesis either of specific enzyme(s), or only of specific fraction(s) of the proteins which enter the structure. It will be clear that from the experiments reported we cannot make a final decision between these various possibilities.

If synthesis in synchronized cells of more protein triggers the polymerization of this with other extraoral proteins, or of extraoral proteins already available, into labile fibres (B in the model), then reaction A → B (model) can be resolved into two steps: macromolecular synthesis and aggregation (0 → 1 → 2 in Fig. 8). Step 0 → 1 is sensitive to inhibitors of protein synthesis (puromycin[9]) and of oxidation[10] and oxidative phosphorylation[11,12] amongst which we can count cold.[13,14] Step 1 → 2, polymerization, is physical; it involves the formation of loose bonds, and it requires correct conformational properties of its components. It is likely to be adversely affected if synthesis (0 → 1) goes on in the presence of parafluorophenylalanine, which is incorporated into the proteins of *Tetrahymena* about as efficiently as phenylalanine.[15]

Structural breakdown of the developing OP can be induced by contamination of one or more of the structural proteins or relevant enzymes with parafluorophenylalanine, or by change in the balance between polymerizing and depolymerizing reactions involved in the structural build-up. The essential observation is that a number of influences or agents cause total structural breakdown when applied before a stabilization point. This is no longer true after the stabilization point has been passed. It is around this time that the developing primordium begins to incorporate amino acid label at the slower rate characteristic of the mature mouth (cf. Fig. 5). If—as argued on p. 208—incorporation largely measures exchange of proteins, then morphogenetic stabilization correlates with increased stability of bonds in the macromolecular association which is the developing mouth. This suggests that the morphogenetic stabilization involves macromolecular interactions superimposed on those in the OP. The latter view has been foreshadowed in a discussion by Mazia and Zeuthen,[8] dealing with the significance of S-bonds in the structural build-up of then unknown structures essential for later division in *Tetrahymena*. Stabilization by extensive side-to-side fusion of microtubules (cf. diagram Fig. 8(4)) would satisfy Cann's model and harmonize with the known morphology of the oral primordium[2] and apparatus.[1] Behnke and Forer[16] have shown that in crane fly spermatid tails, free cytoplasmic microtubules are less stable to heat and cold than are the two central microtubules, and these again are less stable to heat than the peripheral double tubules in the 9 + 2 flagellar pattern of the developing tail. The stabilization of the oral apparatus is likely to result from even more complex interactions between fibrils than those seen in a flagellum.

In conclusion, we presently consider that heat shocks produce set-backs and induce division synchrony by (i) blocking or slowing aggregations of macromolecules into cellular structures (reaction A → B in Cann's model), and (ii) by dispersing structures which have not yet passed the point of morphogenetic stabilization (acceleration or opening of reaction B → X in model). The diagram (Fig. 8) has been drawn to suggest that OP-proteins are thereby lost to the cell because they are tied up in dispersed and useless fractions of oral structures. This creates shortage of those proteins which are present extraorally in limiting amounts. These proteins, or some of them, or specific enzymes involved in fitting them together, must be synthesized *de novo*, and they may represent the "division proteins" of Rasmussen and Zeuthen.[7,17] These views can be harmonized with those of Watanabe (this symposium).

Synchrony induction by heat shocks has been found to be more efficient in *Tetrahymena* than in any other cell studied. This may reflect that the cellular phase with labile structural build-up of the oral apparatus is of relatively much longer duration than the assembly phase of the mitotic apparatus in mitotic cells.

It is not difficult to accept the idea that the proteins of the subpellicular oral fibre system are critical for oral morphogenesis in *Tetrahymena*. It is not as easy to accept that this system is the key to understanding cell division in *Tetrahymena*. However, there are ways in which we might interpret the significance of oral fiber proteins for cell division. First, these proteins or their fibres may be required for division, because they control the formation and/or functioning of division mechanisms within the nucleus and fission line. Secondly, the oral fiber proteins may represent one of a class of proteins which are important for division and which have the same general properties. That is, there may be other "division proteins" which accumulate over the cell cycle, can be removed by temperature treatments and exposures to an amino acid analogue, and are no longer essential or subject to removal after the point of physiological transition or stabilization. Such proteins may be involved in nuclear division, cytokinesis or other aspects of cell division. With regard to the applicability of this idea to cells other than *Tetrahymena* it may be pointed out that present indications are that the elements of the *Tetrahymena* oral fiber system are microtubules.[10] They are therefore similar, at least in ultrastructure, to spindle fibers.

REFERENCES

1. J. R. NILSSON and N. E. WILLIAMS, *Compt. Rend. trav. Lab. Carlsberg*, **35**, 119, 1966.
2. N. E. WILLIAMS and E. ZEUTHEN, *Compt. Rend. trav. Lab. Carlsberg*, **35**, 101, 1966.
3. N. E. WILLIAMS and E. ZEUTHEN, Second International Conference on Protozoology, London, Excerpta Medica International Congress Series No. 91, 1965.
4. O. SCHERBAUM and E. ZEUTHEN, *Exp. Cell Res.*, **6**, 221, 1954.
5. R. L. CROCKETT, P. B. DUNHAM and L. RASMUSSEN, *Compt. Rend. trav. Lab. Carlsberg*, **34**, 451, 1965.
6. J. R. CANN, *Compt. Rend. trav. Lab. Carlsberg*, **33**, 431, 1963.
7. L. RASMUSSEN and E. ZEUTHEN, *Compt. Rend. trav. Lab. Carlsberg*, **32**, 333, 1962.
8. D. MAZIA and E. ZEUTHEN, *Compt. Rend. trav. Lab. Carlsberg*, **35**, 341, 1966.
9. G. G. HOLZ, L. RASMUSSEN and E. ZEUTHEN, *Compt. Rend. trav. Lab. Carlsberg*, **33**, 289, 1963.
10. L. RASMUSSEN, *Compt. Rend. trav. Lab. Carlsberg*, **33**, 53, 1963.
11. K. HAMBURGER and E. ZEUTHEN, *Exp. Cell Res.*, **13**, 443, 1957.
12. K. HAMBURGER, *Compt. Rend. trav. Lab. Carlsberg*, **32**, 359, 1962.
13. J. FRANKEL, *Compt. Rend. trav. Lab. Carlsberg*, **33**, 1, 1962.
14. E. ZEUTHEN, In *Synchrony in Cell Division and Growth*, p. 99. Ed. by E. Zeuthen, Interscience Publishers, New York–London–Sydney, 1964.
15. E. ZEUTHEN and L. RASMUSSEN, *J. Protozool.*, **13** (Suppl.), 29, 1966.
16. O. BEHNKE and A. FORER, *J. Cell Sci.*, **2**, 169, 1967.
17. E. ZEUTHEN, First IUB/IUBS International Symposium, Stockholm. In *Biological Structure and Function*, Vol. II. Ed. by T. W. Goodwin and O. Lindberg, Academic Press, London–New York, 1961.
18. D. R. PITELKA, *J. Protozool.*, **8**, 75, 1961.

DISCUSSION

Y. HOTTA: (1) Is your key-protein, namely OA protein, a single protein or a group of proteins?

(2) How clearly could you separate the new OA from the old?

(3) How do you think about the universality of OA protein? Is there any information on finding OA protein in the organisms which do not have OA?

E. ZEUTHEN: (1) We do not know, but I refer to Dr. Y. Watanabe's talk later in this symposium.

(2) Dr. Williams can distinguish between old and new oral apparatuses until 65 min after the termination of the last heat shock.

(3) The changes in SH-protein in water-soluble and 0.6 M KCl soluble proteins (Watanabe and Ikeda, 1965) are much the same as found in sea urchin eggs. I expect that proteins which constitute "mitotic" fibres, and cortical proteins located at the place of the furrow, are involved in *Tetrahymena* division as in normal mitotic division. Whether the two types are essentially similar or different I dare not suggest. Our analysis points to the former, presumably universal type of proteins as essential for division.

K. IZUTSU: Do you think that the fibrous structure seen between the kinetosomes is homologous to the spindle fibers in higher animals structurally and functionally?

E. ZEUTHEN: We need more good electron-microscopic pictures before we can stress structural homology with spindle fibers. The basis on which we suggest that our fibers serve to orient centrioles relative to each other is the same as in mitosis: connecting fibers appear at the time when the centrioles move.

Y. WATANABE: The findings of fiber formation in oral primordium after EH plus 45 min are very interesting. I would like to ask you regarding the solubility of the fiber in the oral apparatus. Did you try to solubilize the isolated oral apparatus by 8M urea or by some other agents capable of depolymerizing protein complex?

E. ZEUTHEN: We have not yet studied the solubility properties of the oral primordium.

N. KAMIYA: I am wondering if the proteins forming the fibrils are ATP-sensitive. How do the fibrils behave in a glycerol model on addition of ATP or Ca^{++}?

E. ZEUTHEN: Also this has not yet been done. I need point out that just to observe the microscopic detail of the isolated primordium is a task. However I do believe that possibly the isolated primordium could be further studied along the lines you have suggested.

NUCLEAR CONTROL
OF CELL SPECIALIZATION—
VIEWED FROM THE STUDIES
OF DIFFERENTIATION
OF MAMMALIAN ERYTHROID CELL*

SATIMARU SENO, MASANOBU MIYAHARA and KOZO UTSUMI

Department of Pathology and Institute for Cancer Research, Okayama University
Medical School

STUDIES of DNA, RNA and protein syntheses by using bacterial cells
have greatly contributed to the establishment of an up-to-date concept
of gene activation and cell specialization mechanisms.[1,2] Observations on
the cell specialization of higher organisms in culture as well as *in vivo* revealed a similar mechanism as that found in bacteria.[2,3,4] This is not surprising
when we consider that the genes of all kinds of organisms from viruses to
higher animals and plants are composed of the same four common bases.
Synthesis of the DNA-like RNA, or messenger RNA, is essential for cell
specialization, and this is the very key to open the door for cell differentiation with somatic protein synthesis, by the aid of ribosomes and transfer
RNA.

In the development of multicellular organisms, however, differing from
the case of bacteria, the cells divide and multiply themselves prior to their
specialization, without synthesis of somatic protein. This is also the case in
some regenerating tissues of adult organisms, e.g. red cell formation from
their erythroid precursors. The decaded cells are compensated through the
repeated cell divisions of the precursor cell followed by specialization.

As described by Holtzer, the cell division or DNA synthesis will be a
biological process antagonistic to cell specialization or somatic protein synthesis.[5] Generally, the undifferentiated cells divide actively and the specialized cells reduce or lose their mitotic activities. This means that the
somatic cells composing the multicellular organism have developed some
control mechanism between the gene duplication and its expression, by which
the fertilized egg develops into the multicellular adult organism and the latter
uniformly maintains its cellular components. The problem is whether the

* Supported by a grant from the Japanese Government, Ministry of Education.

long life m-RNA is synthesized in the early stage of differentiation and yet the somatic protein synthesis is suppressed[4] or the m-RNA is synthesized only in the later stage of specialization.[6]

Here, on the bases of the results obtained from the studies of erythroid cell differentiation in rabbits, we plan to demonstrate that the m-RNA for somatic protein is formed at the early stage of differentiation but the somatic protein synthesis is controlled by the nucleus.

It is believed that the differentiation of the stem cell to the youngest precursor of erythroblast, proerythroblast, is triggered by the erythropoietin, which is produced in the kidney.[7] There is much evidence, however, that the stem cell will receive the information for differentiation from the reticuloendothelial cell, a kind of macrophage found in the center of an erythroblastic islet in the bone marrow. Damage to the macrophage results in anemia of the normochrome type, in which the red cell number in the circulating blood decreases without any abnormalities in the cell specialization process and hemoglobin synthesis, e.g. aplastic anemia in man.[8] A similar kind of anemia can be induced experimentally in animals by intravenous injections of blocking agents for the reticuloendothelial system, like India ink,[9] polyvinylpyrrolidone[10] and colloidal silver.[11] It is uncertain whether the inducer is the erythropoietin itself or some other substance produced in the reticuloendothelial cell, stimulated by erythropoietin, but the observations suggest that if the differentiation of the stem cell is triggered once, the specialization of the proerythroblast to red cells proceeds automatically, if provided with the materials needed for the specialization.

Concerning the cell specialization process the proerythroblast matures to red cells through four cell divisions. At each cell division both the nuclear and cell volumes reduce themselves by about one half, progressively,[8,12] and the cells are denucleated after the terminal cell division, forming anucleate red cells.

According to the steps in grade of specialization and cell division, the specialization process may be divided into five stages: pro-, early basophilic, late basophilic, polychromatic and orthochromatic erythroblasts. Each stage of specialization can be distinguished in smeared cells by the nuclear and cell diameters. By Giemsa stain the somatic protein, hemoglobin, is stained red and appears first in the polychromatic stage. In normal hemopoiesis the denucleation occurs in the orthochromatic stage by the extrusion of the nucleus without any karyolytic process.[13]

As is well known, in acute anemia of man a number of large red cells may appear in the circulating blood, i.e. macrocytosis. Such a macrocytosis can be induced in the rabbit by several injections of phenylhydrazine-HCl, and it was reasonably supposed that the macrocytes would be formed by early denucleation,[14] as the cell volume of erythroblast reduces itself at each cell division and no cell division takes place after denucleation.[15] Recently, it has been established that the macrocytosis is actually formed by early denucleation skipping one or two terminal cell divisions. In anemic animals some big reticulocytes, which do not synthesize RNA[17,18] and do not divide further, proved to contain a quantity of RNA, the level of which is comparable to those of polychromatic and late basophilic erythroblasts.[16]

In the recovery stage of phenylhydrazine-induced anemia the mean corpuscular volume becomes as much as twice the normal, indicating that almost all the cells are denucleated at the polychromatic stage skipping one cell division. These macrocytes contained a large quantity of hemoglobin, nearly twice the normal (Fig. 1). This fact suggests that the m-RNAs required for the synthesis of the expected amount of hemoglobin had to be prepared by the stage of basophilic erythroblast, at the latest.

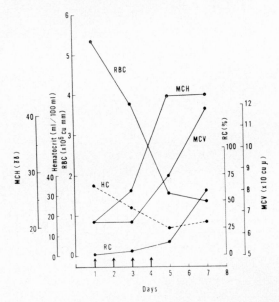

FIG. 1. Changes in red cell values in phenylhydrazine anemia. Note rapid increase in cell volume and in hemoglobin content per cell. RBC: red cell count. RC: reticulocyte count. MCH: hemoglobin content per cell. Arrows: phenylhydrazine-hydrochloride injections (0.5 cc of 2.5% solution per kg per day).

The RNA synthesis of the erythroblast observed by flash labeling with H^3-uridine *in vitro* and radioautography revealed that the RNA synthesis in the nucleus is most marked in younger precursors, proerythroblasts, probably including the stem cell, and decreased exponentially with the advance of cell specialization reaching the minimum level at poly- and orthochromatic stages (Fig. 2). This RNA synthesis was inhibited in the presence of actinomycin D, slightly at $10\,\gamma$/ml and completely at $50\,\gamma$/ml, and will be the synthesis of m-RNA. These findings are consistent with those of Swift and associates,[19,20] Torelli,[21,22] and Goldwasser and collaborators,[23] and with the findings of Sassa.[24] They found that m-RNA for hemoglobin is synthesized in the early stages of differentiation[19-23] and the enzymes for heme synthesis appear shortly after the stimulation of bone marrow with erythropoietin,[24] which is believed to induce the differentiation of the stem cell to proerythroblast.

The observation of heme absorption at 406 mµ on the smeared cells from the normal rabbit bone marrow revealed that heme appears as early as the

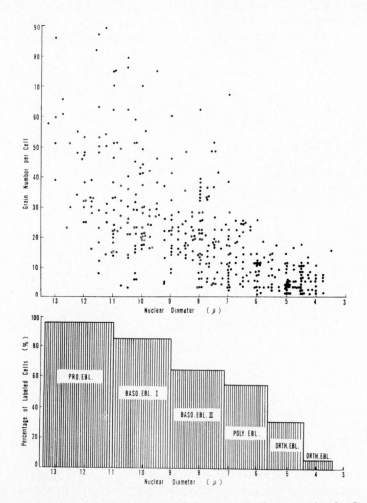

FIG. 2. RNA synthesis of the erythroblasts observed by radioautography after flash labeling with H³-uridine. Bone marrow cells were incubated with H³-uridine (20 µc/ml) for 1 hr at 37°C, smeared, treated with 0.1% DNase, covered by stripping film Kodak AR-10 and exposed for 9 days. Each filled circle shows the grain count found on one cell and the column shows the mean percentage of labeled cell number to the total cell number (labeling index) in each maturation stage.

basophilic stage. Heme content per cell estimated by microspectrophotometry at 406 mμ increased with the advance of cell specialization stages.[8,25]
However, only a slight increase was observed during the nucleated stages,
and a marked elevation of hemoglobin level was always observed after
denucleation.[25] Similar results were obtained on the bone marrow smears
of blood depleted animals, which included macrocytes denucleated at an early

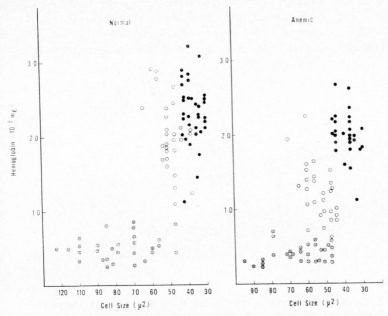

FIG. 3. Hemoglobin contents of erythroid cells from normal (RBC: 580×10^4/mm^3)
and blood depleted (RBC: 325×10^4/mm^3) rabbits, measured by microspectrophotometry. Filled circles: matured red cells, open circles: reticulocytes, crossed
circles: erythroblasts.

stage (Fig. 3). These data indicate clearly that heme or hemoglobin synthesis is somehow arrested in the nucleated stages and enhanced by denucleation, showing the nuclear control of heme synthesis in spite of the presence of
m-RNAs. This nuclear effect on the somatic protein synthesis suggests that
some substance to suppress the somatic protein synthesis is liberated from
the nucleus to the cytoplasm.

Immature reticulocytes shortly after denucleation are nothing but the
cytoplasm of erythroblasts enveloped by the cytoplasmic membrane. They
have mitochondria, ribosomes, an endoplasmic reticulum, lysosomes and the
Golgi apparatus.[8] Maturation means the loss of all these components and
can be measured by the decrease in the *substantia granulofilamentosa* or
RNA, whose degradation proceeds inversely proportionately to the synthesis of hemoglobin.[26]

The maturation test of reticulocytes *in vitro* in an oxygenated environment revealed that they mature to red cells in about 8 to 10 hr losing their *substantia granulofilamentosa*[27,28] or RNA[16] (Fig. 4a). The maturation time was nearly the same in reticulocytes from normal and anemic animals induced by phenylhydrazine, and for those denucleated in orthochromatic and polychromatic stages skipping cell division (Fig. 4b). Since the hemoglobin content of the red cell formed by early denucleation can be double of that denucleated at orthochromatic stage, early denucleation shortens the period required for the synthesis of the expected amount of hemoglobin per unit volume of cytoplasm, skipping the time expected for the next nucleated stage.

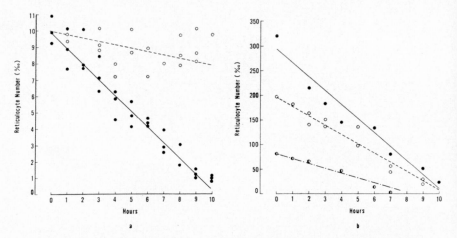

FIG. 4. Maturation of reticulocytes *in vitro*. (a) Hourly decrease in reticulocyte number in normal rabbit blood under deoxygenated (open circles) and oxygenated (solid circles) conditions. Observations on 3 animals in each and show the inhibited maturation by deoxygenation. (b) Hourly decrease in the number of reticulocytes from the rabbit in the recovery stage of phenylhydrazine anemia, 5 to 14 days after the phenylhydrazine injection (0.05 cc/kg), observation on 4 animals under oxygenated conditions. Note that the reticulocytes matured with almost the same time lapse as in (*a*).

The maturation test of reticulocytes in the presence of 2,4-dinitrophenol and antimycin A, and in an aerobic environment showed a marked retardation of maturation with the suppressed degradation of RNA (Fig. 5) and the *substantia granulofilamentosa*[29,30] (Fig. 4a). These data suggest that maturation of the reticulocyte is a process dependent largely on energy supplied by oxidative phosphorylation. The analysis of reticulocytes incubated for 3 hr at 37°C for the maturation test revealed that ATP level fell markedly in the presence of 2,4-dinitrophenol and antimycin A (Fig. 6.). This indicates the possibility that the mechanism of the nuclear control of the maturation of the cytoplasm or somatic protein synthesis with RNA degradation is closely related to the suppression of energy metabolism of the cytoplasm.

Therefore, it may be supposed that histone is responsible for the control of energy metabolism. The arginine-rich histone from calf thymus proved to be a strong uncoupler for oxidative phosphorylation, shown by observing the

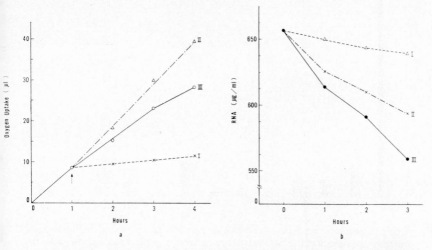

FIG. 5. Oxygen consumption (a) and RNA degradation (maturation) (b) of reti-culocytes *in vitro* at 37 °C in the presence of 2,4-dinitrophenol and antimycin A, the observations on circulating blood from the rabbit with phenylhydrazine anemia. (a) Oxygen consumption of reticulocytes. Reticulocytes: 50% of whole red blood cells. Suspending medium: 1/5 M tris-HCl buffer solution (pH 7.40) added with each volume of Gey's solution and normal rabbit plasma. 0.72 ml of packed cells was suspended in 1.08 ml of the medium. Arrow shows the time when the reagents were added. (b) RNA degradation in reticulocyte suspension. Reticulocytes: 60% of whole red blood cells. RNA was isolated by the slightly modified method of Schmidt and Thannhauser[29] and estimated by the orcinol reaction. I: Incubated with anti-mycin A (5 γ/ml). II: Incubated with 2,4-dinitrophenol (10^{-4} M). III: Control with physiologic saline solution. For the precise method refer to the paper of Miyahara *et al.*[29]

effect on rat liver mitochondria *in vitro*.[31,32] At a concentration of 10^{-3}% arginine-rich histone suppressed ATP formation of isolated mitochondria in the presence of ADP, with a marked respiratory release (Fig. 7). This char-acteristic of arginine-rich histone suggests the possibility that a small amount of histone is released from the nucleus to the cytoplasm, lowers the ATP level in the cytoplasm and suppresses the synthesis of somatic protein.

There is no evidence in any cell that the histone is released from the nucleus to the cytoplasm, but the observation of Prescott on *ameba proteus* suggests this possibility.[33] He observed the disappearance of the nuclear protein labeled with H^3-amino acids after cutting off the cytoplasm thirty times, once a day for 30 days, through which the cell division and DNA synthesis was completely inhibited.

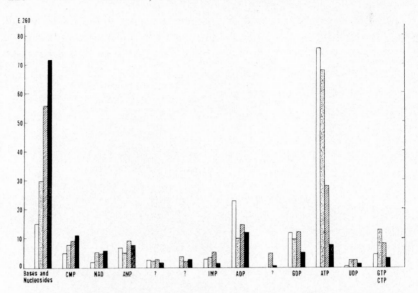

Fig. 6. Changes in amount of acid-soluble phosphorus compounds in reticulocyte suspension (reticulocytes: 60% of whole red blood cells). The columns show the amounts after 3-hr incubation with 2,4-dinitrophenol (10^{-4} M, hatched columns), antimycin A (5γ/ml, solid columns) and with physiologic saline solution (dotted columns). Open columns show the values at the initiation of incubation.

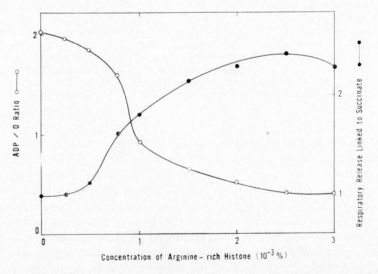

FIG. 7. Effect of arginine-rich histone on the oxidative phosphorylation (ATP formation by adding ADP) and the respiration of rat liver mitochondria in the media containing succinate. Each point was given from the curves drawn by oxymeter by adding ADP. For the precise method refer to the paper of Yamamoto et al.[31,32]

SUMMARY

As has been pointed out in the beginning of this paper, cells composing the multicellular organisms divide and multiply themselves before cell specialization in the regeneration process as well as in the embryonic development. This may mean that m-RNA is synthesized only at later stage of cell differentiation. Observations on the erythroid cell differentiation and specialization process, however, indicate that m-RNA synthesis for somatic protein occurs in the early stage of differentiation but somatic protein synthesis is largely suppressed till the end stage of specialization. Studies on the cell specialization process before and after denucleation disclose that the nucleus is responsible for the control of cell specialization. And the character of arginine-rich histone as a strong uncoupler for the oxidative phosphorylation suggests the possibility that the histone is the very controller for somatic protein synthesis which is largely dependent on the energy liberated by oxidative phosphorylation.

REFERENCES

1. F. JACOB and J. MONOD, *J. Mol. Biol.*, **3**, 318, 1961.
2. J. D. WATSON, *Molecular Biology of the Gene*, W. A. Benjamin, Inc., 1965.
3. W. BEERMAN and U. CLEVER, Chromosome puffs. In *Scientific American*, **120**, No. 4, p. 50, 1964.
4. J. D. EBERT, *Interacting Systems in Development*, Holt, Rinehart and Winston, Inc., 1965.
5. H. HOLTZER, In *General Physiology of Cell Specialization*, p. 80. Ed. by D. Mazia and A. Tyler, McGraw Hill, New York, 1963.
6. K. SCHERRER, L. MARCAUD, F. ZAJDCL, I. M. LONDON and F. GROS, *Proc. Nat. Acad. Sci.*, **56**, 1571, 1966.
7. J.-P. NAETS and A. HEUSE, In *Erythropoiesis*, p. 98. Ed. by L. O. Jacobson and M. Doyle, Grune and Stratton, New York, 1962.
8. S. SENO, *Acta Pathol. Jap.*, **7** (suppl.), 741, 1957.
9. K. MATSUOKA, *Acta Med. Okayama*, **19**, 161, 1965.
10. T. TOYAMA, *Acta Med. Okayama*, **19**, 307, 1965.
11. E. KOMIYA, S. NAKAMURA and T. MURAKAMI, *Kumamoto Igakkaizasshi*, **5**, 1, 1929.
12. S. SENO, *Acta Haemat. Jap.*, **27**, 718, 1964.
13. M. AWAI, S. OKADA, J. TAKEBAYASHI and S. SENO, *Acta Haemat.*, **39**, 193, 1968.
14. F. STOLMAN, JR., D. HOWARD and A. BELAND, *Proc. Soc. Exptl. Biol. Med.*, **113**, 986, 1963.
15. L. M. LOWENSTEIN, *Exp. Cell Res.*, **17**, 336, 1959.
16. S. SENO, M. MIYAHARA, O. OCHI, H. ASAKURA, K. MATSUOKA and T. TOYAMA, *Blood*, **24**, 582, 1964.
17. S. SENO, M. MIYAHARA, O. OCHI, K. MATSUOKA, T. TOYAMA and T. SHIBATA, *Acta Med. Okayama*, **17**, 253, 1963.
18. P. PINHEIRO, C. P. LEBLOND and B. DROZ, *Exp. Cell Res.*, **31**, 517, 1963.
19. J. A. GRASSO, J. W. WOODARD and H. SWIFT, *Proc. Nat. Acad. Sci.*, **50**, 134, 1963.
20. J. A. GRASSO and J. W. WOODARD, *J. Cell Biol.*, **31**, 279, 1966.
21. U. TORELLI, G. GROSSI, T. ARTUSI, G. EMILIA, I. R. ATTIYA and C. MAURI, *Acta Haemat.*, **32**, 271, 1964.
22. U. TORELLI, D. QUAGLINO and C. MAURI, *Acta Haemat.*, **35**, 129, 1966.
23. S. B. KRANTZ and E. GOLDWASSER, *Biochim. Biophys. Acta*, **103**, 325, 1965.

24. S.SASSA, *Acta Haemat. Jap.*, **29**, 45, 1966 (in Japanese).
25. S.SENO, T.ODA and K.UTSUMI, *Symp. Cell. Chem.*, **8**, 223, 1958.
26. H.G.SCHWEIGER, E.SCHWEIGER and I.VOLLERTSEN, *Biochim. Biophys. Acta*, **76**, 482, 1963.
27. S.SENO,K. KAWAI, S.KANDA and K.NISHIKAWA, *Mie Med. J.*, **4** (suppl.), 1, 9, 1953.
28. S.SENO,K. KAWAI, S.KANDA and K.NISHIKAWA, *Mie Med. J.*, **4** (suppl.), 1, 19, 1953.
29. M.MIYAHARA, S.SENO, K.HAYASHI and O.OCHI, *Biochim. Biophys. Acta*, **95**, 598, 1965.
30. S.SENO, M.MIYAHARA, M.AWAI, M.INOUE, K.UTSUMI, G.YAMAMOTO and T.-H.MURA-KAMI, *Symp. Cell. Chem.*, **15**, 207, 1965.
31. G. YAMAMOTO, K.UTSUMI, T.H.MURAKAMI, M.MIYAHARA and S.SENO, *Symp. Cell Chem.*, **15**, 217, 1965.
32. K.UTSUMI, and G.YAMAMOTO, *Biochem. Biophys. Acta*, **100**, 606, 1965.
33. D.M.PRESCOTT, In *Cell Growth and Cell Division*, p. 111. Ed. R.J.C.HARRIS, Acad. Press, New York, 1963.

DISCUSSION

H.BUSCH: (1) An inhibitory effect has been found of DNA-like RNA present in chromatin on uptake of amino acids into protein in the Nirenberg system. This effect may be related to the regulatory role of the nucleus in blocking cell protein synthesis.

(2) The 2a histone fraction, an AL or arginine–lysine histone, has been found by Dr. Schwartz in our department, to block mitochondrial oxidation. Its relation to the suggestion of Prof. Seno is that histones may regulate oxidation, these effects may be mediated by very small amounts of proteins.

S.SENO: I did not realize that nuclear DNA-like RNA has an inhibitory effect on protein synthesis, but it seems to be probable that this RNA is also correlated to the control of hemoglobin synthesis which is enhanced after denucleation in rabbit erythroblasts.

Concerning the AL histone, I have not observed the inhibitory effect on mitochondrial oxidative phosphorylation, but ly sine-rich histone was much less in the effect.

H.TERAYAMA: I wonder if there are mitochondria in the cells related to erythropoiesis. If there are, I would like to know how they behave during the course of erythropoietic differentiation, or if not, do you believe that nuclei are a site responsible for the oxidative phosphorylation?

S.SENO: Erythroblasts and reticulocytes have actual mitochondria, though their number of mitochondria seems to be reduced successively with the advance of cell specialization. Energy will mainly be used for protein synthesis. Actually, the incorporation of H^3-leucine into protein was very high at the proerythroblastic stage and decreased with the advance of cell differentiation. But it does not necessarily mean hemoglobin synthesis. Hemoglobin synthesis ensues at rather later stages of differentiation. Besides this, I would like to say that the degradation of RNA in reticulocyte is an energy-requiring process and the energy supplied by mitochondria will also be used for the activation of RNase.

III. CELL MULTIPLICATION
AND DIFFERENTIATION

THE REGULATION
OF DNA REPLICATION IN CHROMOSOMES
OF HIGHER CELLS*

J. Herbert Taylor

Institute of Molecular Biophysics and Department of Biological Sciences,
Florida State University, Tallahassee, Florida, U.S.A.

THE first evidence that the sequence of replication of DNA in a chromosome is highly regulated was obtained in autoradiographic studies of the root cells of *Crepis* (Taylor, 1958) and somatic cells of the Chinese hamster (Taylor, 1960). The latter illustrates the point best. Established cell lines from male and female embryos were grown *in vitro* where H^3-thymidine could be readily supplied to the cells for a short time and then removed and replaced by a medium containing unlabeled thymidine. Samples of the unsynchronized cell population were fixed at hourly intervals thereafter and the incorporation of H^3-thymidine into various chromosomes was examined in autoradiographs. It was clear that all parts of most chromosomes were not being labeled by such a pulse exposure to the isotope.

The most striking variations in labeling of large blocks of DNA occurred in the sex chromosomes, but was by no means restricted to these. For example, in male cells, the first labeled chromosomes to reach metaphase (those labeled at the end of *S*, DNA synthetic phase), invariably had more tritium in the X- and Y-chromosomes than in the autosomes. On the other hand, the last labeled cells to reach metaphase (those labeled in early *S*) had little if any label in the Y-chromosome or in the short arm of the X. In female cells labeled early in *S*, one X was almost wholly unlabeled while the other had tritium only in the short arm. If labeling occurred late in *S* in female cells, the whole of one X and the long arm of the other contained tritium. To summarize, in cells derived from a male the Y-chromosome and the long arm of the X were replicated in the last half of *S*, while the short arm of the X-chromosome was replicated in the first half of *S*. In cells derived from a female the whole of one X and the long arm of the other was re-

* This work was supported in part by a contract between the Division of Biology and Medicine, U.S. Atomic Energy Commission and the Florida State University. The assistance of Philip Miner with the gradient centrifugation studies is gratefully acknowledged.

A travel grant from the National Science Foundation is gratefully acknowledged.

231

plicated in the last half of S, while only the short arm of one X was replicated in the first half of S. Since then, this out of phase replication of the X- and Y-chromosomes as well as sectors of autosomes has been established in many different species of both animals and plants.

That the pattern is consistent from cell generation to cell generation for a given type of cell was indicated by autoradiography, but was perhaps more convincingly demonstrated for all of the DNA by Mueller and Kajiwara (1966). They labeled a partially synchronized culture of HeLa cells in early S and noted that the same DNA molecules labeled in early S were the first to be labeled in a subsequent S phase. Evidence is accumulating that this late replicating DNA is relatively inactive in genetic transcription (Klevecz and Hsu, 1964) and in the expression of genes located in these regions (Davidson, Nitowsky and Childs, 1963). However, a particular sector of a chromosome may change its replication pattern and presumably its rate of transcription during cellular differentiation. For example, the portions of the sex chromosomes, which replicate late in somatic cells of the Chinese hamster, were found to replicate early in spermatogonial cells where these regions might be reasonably expected to be active (Utakoji and Hsu, 1965).

The basis for this correlation between late replication and genetic activity is not yet understood, but the widespread occurrence of the phenomenon and its changes during differentiation suggest that a major role in the regulation of gene action is involved. The late replicating regions of chromosomes usually show a tendency to remain condersed during interphase and frequently exhibit a differential condensation during prophase compared to the early replicating regions. The correlation between condensation patterns and apparent absence of genes in some chromosomes or chromosomal regions led to the coining of the term "heterochromatin" for this material. However, the term now has limited usefulness since it has been employed to describe so many differential behaviors of chromosomes in forms where the function or lack of function could not be investigated. In any case, the state of activity of these and other regions of chromosomes is likely to be a matter of variation among differentiated tissues. If the material did not function at some stage in the life cycle it would probably be eliminated by the pressure of natural selection.

Before we can make much progress in understanding the significance of these variations in replication and condensation, much more must be learned about the control of DNA replication at the molecular level and the actual mechanisms involved in replication *in vivo*. Studies of isolated enzymes have provided a number of possibilities, but none of the systems appear to be complete enough or refined enough to give us a clear picture of the replication and integration of DNA into a chromosome. On the basis of autoradiographic evidence, we have for a long time assumed that chromosomes consist of units of DNA which can be replicated independently. These hypothetical units have recently been referred to as replicons (Taylor, 1963; 1964). Only recently have we obtained evidence concerning their size in higher cells and suggestions concerning their structure. The remainder of the paper will be devoted to these questions. What is the size of the replicons and what features mark the beginning and possibly the end of a replicon. These considerations do not require that we know the arrangement of replicons within a chromosome, but much of the present evidence suggests that they are tandemly linked, perhaps with no non-nucleotide spacers as we had originally supposed. A recent model (Fig. 1) based on the idea that a whole chromosome arm may be one continuous piece of DNA has been presented (Taylor, 1966; 1967).

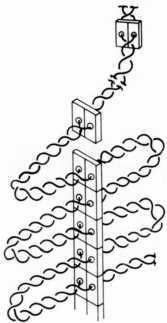

FIG. 1. A diagram of a segment of the chromosome model used as a basis for discussion. The model consists of a single DNA duplex with pairs of replication guides attached at intervals which separate the replicons. The replication guides could either be permanently united end to end to form a column or temporarily coupled. However, it should be noted that the hypothetical replication guides must have a polarity if they are not permanently united. In order to maintain semi-conservative segregation of strands, all of the replication guides attached to a single strand would have to reunite after separation in the same column or at least segregate together (after Taylor, 1966).

One way to find the maximum size of a replicon is to measure the rate of growth of new DNA during replication. With this rate, the length of DNA in a chromosome and the total time required for replication, the size of a replicon can be given an upper limit. To measure the rate of growth, the

density label BUdR (bromouracil deoxyriboside) was used for Chinese
hamster cells grown in culture. By supplying the nucleoside at a concentration of 10^{-5} M in the medium along with FUdR (fluorouracil deoxyriboside) at a concentration of $2x\ 10^{-6}$ M, the cells can be induced to almost
completely substitute BU for thymine in the newly formed strands of DNA.

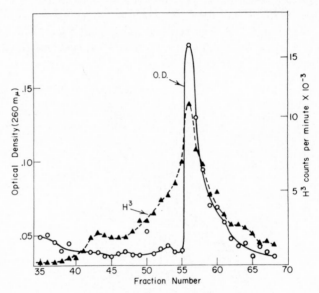

FIG. 2. Sedimentation profile of DNA in a CsCl gradient. DNA from Chinese hamster cells grown for 10 min in a medium with BUdR-H^3 and isolated by a procedure
that maintains a high molecular weight.

The FUdR is a potent inhibitor of thymidylate synthetase and since in these
cells the pool of thymidylate is small, a quick depletion is possible. The
average density of the unsubstituted DNA in CsCl is about 1.70 and the
hybrid molecules with one strand fully substituted with BU have a density
of about 1.75. When cells were pretreated for 10 min with FUdR to deplete
the pool of thymidylate and then changed to a medium with BUdR-H^3
for 5 or 10 min, the rate of growth was revealed. Figure 2 shows a profile of
DNA centrifuged in a solution of CsCl with an average density of about
1.70 for 48 hr at 37,500 rpm in a S3w9 Spinco rotor. The gradient solution
was dripped out in the usual way and the optical density and radioactivity of
each 3 drop fraction determined. The DNA was isolated by a technique
which yields very large pieces, probably with an average molecular weight
of more than 50×10^6. It may be seen that no fully substituted particles were
formed. However, when the DNA was broken by shearing (shaking over
chloroform: isoanyl alcohol and passage through Bio-Gel P-60), it can be
seen (Fig. 3) that most of the BU labeled DNA was isolated by breakage

from the unsubstituted DNA and now banded at a density typical of the BU-thymine hybrid. However, when DNA from cells grown for 5 min in BUdR-H^3 was treated the same way, little if any fully substituted DNA was obtained (Fig. 4). Therefore, pieces large enough to form pure hybrid particles have grown in 10 min, but not in 5 min.

If we assume that the breakage of DNA by shear is random, the size of the BU substituted stretches of DNA can be estimated from the size of the pieces (molecules) in the CsCl solution and the ratio of the BU in fully substituted pieces to that in the partially substituted pieces. Very briefly, the reasoning is as follows.

If the stretches of BU-DNA in the chromosome are a very small proportion of the total and do not exceed in length that of the broken pieces in solution, the probability of getting pure BU hybrid is nearly zero. Some will, of course, be possible if the smallest pieces which can be banded are smaller than the stretches of BU, but for our purposes this fraction can be ignored.

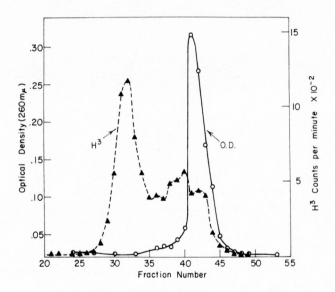

FIG. 3. The profile of another sample of the DNA shown in Fig. 2 sheared to a lower molecular weight.

If the stretches of BU-DNA are just two times the length of the pieces broken out by shear, one-half of the BU will appear in particles of pure hybrid and one-half will appear in impure particles (short stretches of BU hybrid attached to unsubstituted DNA). The time at which this 1:1 ratio is attained is somewhere between 5 and 10 min, probably 6–8 min. The particle weight of the pieces produced by shear has not yet been measured, but we know from the work of others that a maximum size is probably 4–6 × 10^6. If we use the larger number and estimate that pieces two times the molecular weight of the particles in the gradient are synthesized in 6 min, the rate of growth is one micron per minute (one micron of DNA has a molecular weight of approximately 2×10^6 and the pieces which are two times the size of those in the gradient, are estimated to have a molecular weight of 12×10^6). If we use the lower number for the molecular weight (4×10^6) and 8 min as the time required to synthesize lengths two times that size, the rate would be one half micron per minute.

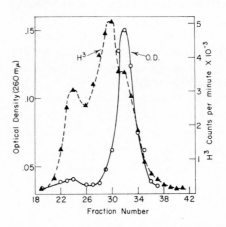

FIG. 4. A profile of DNA from cells grown for 5 min in a medium with BUdR-H^3. DNA sheared to a molecular weight similar to that shown in Fig. 3.

The time of replication for the long arm of the X-chromosome of the Chinese hamster can be estimated from autoradiographic studies to be about $3\frac{1}{2}$ hr (Taylor, 1960). The length of this arm is given by Hsu and Zenzes (1964). It is 3 microns of a total chromosome length of 124 microns (diploid complement). One long arm would be 2.43% of the total DNA which has been estimated to be 7×10^{-12} g (Huberman and Riggs, 1966). This would be equivalent to 51,000 microns of DNA in the long arm of the X. If each replicon grows for the total time required for replication of this chromosome arm (200 min), the average replicon will have a length of 200 microns (mol. weight *ca.* 400×10^6) and there would be about 255 of these in the long arm of the X (51,000 ÷ 200). The estimate of 255 is of course a minimum figure for the number of replicons since each replicon probably does not replicate for the full $3\frac{1}{2}$ hr. Likewise, the size of 400×10^6 daltons per replicon is a maximum size. Other data not presented here lead me to predict that the size is much smaller, possibly no more than $12-24 \times 10^6$ daltons, but the details cannot yet be given. This rate of growth of DNA in the hamster cells measured by the method described above compares favorably with that estimated by autoradiography for the DNA of HeLa cells by Cairns (1966).

The nature of the initiation points can be surmised from recent data obtained on substitution of BU into DNA of *Vicia* root cells (Haut and Taylor, 1967; Taylor, 1966). Details cannot be given, but a brief description of the data and its implication will be mentioned.

When *Vicia* roots are grown in solutions containing an inhibitor for thymidylate synthetase (aminopterin at 10^{-6} M) and BUdR (10^{-5} M) a semi-conservative replication pattern can be demonstrated. However, an unusual feature was also observed. In addition to the fully substituted hybrid DNA obtained after a few hours growth in BUdR-C^{14}, two slightly-heavy labeled peaks appeared (Fig. 5). Evidence from subsequent studies with hamster cells indicate that these two bands of partially BU-substituted material are produced by replicons interrupted in their growth by a deficiency of thymidylate. Apparently the exogenous supply of bromodeoxyuridylate is insufficient to replace the thymidylate in *Vicia* cells. The bimodality in banding of this partially substituted DNA can be explained by assuming a bias in the distribution of the BU in the two new molecules during the

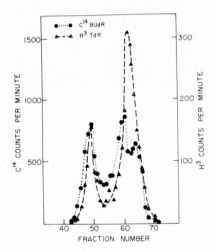

FIG. 5. A profile of DNA isolated from roots of *Vicia faba* which had been growing continuously in BUdR-C^{14} for several hours.

replication of the first part of the replicon. It largely disappears when the pieces of DNA become fully substituted. The bimodality can also be demonstrated by growing roots in solutions of BUdR-C^{14} without an inhibitor of thymidylate. In this case there is no fully substituted DNA, but the BU labeled DNA bands in two peaks similar to the incompletely substituted DNA shown in Fig. 5 (see Fig. 4, Taylor, 1966). The interpretation is that the initiation region of replicons is composed of DNA rich in A-T pairs, in which most of the A is in one strand and the T in the complementary one. In addition, the BU is more readily utilized in the initiation region than in the remainder of the replicon. If this were not so, the limited substitution in the absence of a thymidylate inhibitor should not produce an easily detectable bias. The suggestion might be that a different enzyme produces the polymer at the initiation region (primer) than along the remainder of the replicon (Taylor, 1966). The size of the region would need to be at least 100 nucleotide pairs in length to produce the observed differences in density. Consistent with this interpretation is the fact that the relative amount of label in the lighter of the two bands is always less than in the heavier.

If some initiation regions were rich in G-C pairs the labeling experiments above would not necessarily reveal them. A mechanism for controlling the time of replication could involve several devices which cannot yet be specified. Two polymerases active at different times in S, could recognize replicons with variations in the initiator regions. Alternatively primer DNA's necessary to initiate synthesis at complementary sites (small molecules of poly dA, poly dT, poly dC, poly dG, or any other special sequence) could be produced either early or late in the S but not in the other part of the DNA synthetic phase (Taylor, 1966). At present there is no way to choose between these and other possible mechanisms for the control of the replication or the function of the various regions of DNA.

REFERENCES

CAIRNS, J. (1966) *J. Mol. Biol.*, **15**, 372.
DAVIDSON, R.G., NITOWSKY, H.M. and CHILDS, B. (1963) *Proc. Nat. Acad. Sci. (Wash.)*, **50**, 481.
HAUT, W.F. and TAYLOR, J.H. *J. Mol. Biol.*, 26, 389.
HSU, T.C. and ZENZES, M.T. (1964) *J. Nat. Cancer Inst.*, **32**, 857.
HUBERMAN, J.A. and RIGGS, A.D. (1966) *Proc. Nat. Acad. Sci. (Wash.)*, **55**, 599.
KLEVECZ, R.R. and HSU, T.C. (1964) *Proc. Nat. Acad. Sci.*, **52**, 811.
MUELLER, G.C. and KAJIWARA, K. (1966) *Biochem. Biophys. Acta*, **114**, 108.
TAYLOR, J.H. (1958) *Exp. Cell Res.*, **15**, 350.
TAYLOR, J.H. (1960) *J. Biophys. Biochem. Cytol.*, **7**, 455.
TAYLOR, J.H. (1963) *J. Cell. Comp. Physiol.*, Suppl. 1, **62**, 73,
TAYLOR, J.H. (1964) In *Symp. of the Society for Cell Biology*, Vol. 3. Ed. by R.J.C.Harris, Academic Press, New York.
TAYLOR, J.H. (1966) In *3.Wissenschaftliche Konferenz der Gesellschaft Deutscher Naturforscher and Ärzte*, Semmering bei Wien, 1965. Ed. by P.Sitte, Springer-Verlag, Heidelberg.
TAYLOR, J.H. (1967) In *Molecular Genetics*, Part II. Ed. by J.H.Taylor, Academic Press, New York.
UTAKOJI, T. and HSU, T.C. (1965) *Cytogenetics*, **4**, 295.

DISCUSSION

Y. TAKAGI: Your experiments, together with many experiments performed in various laboratories, indicate that the replication of chromosomes proceeds in one direction from its initiating point along the strands of the DNA molecule. I would like to ask you whether you have heard any report indicating that there are two types of DNA polymerase, that is, commencing DNA replication from 3'- and 5'-terminals of the DNA molecule.

J. H. TAYLOR: No, I do not believe anyone has isolated a DNA polymerase which adds to the 5'-terminus of DNA chains. Perhaps, if the replicons are short enough, the polymerase may work only in one direction, leaving one strand single for a short time.

S. SPIEGELMAN: I should like to detail some implications of your interesting findings and the model you derive from them. They predict that the DNA "duplicore" (the enzyme(s) concerned with making DNA duplicates) has the ability to recognize and prefers certain sequences, like the RNA replicases we have been studying. In contrast, a DNA polymerase whose major function is to repair damage by filling in gaps, will not possess this recognition mechanism. If it did damage in a certain region of the DNA, this would not be reparable, or would be so with difficulty. This line of reasoning predicts that the "duplicase" activity would be sensitive to fragmentation of the DNA to levels such that many pieces do not have "initiating sequences". On the other hand, the "repair" polymerase might be quite insensitive to such modification of the template. This could provide a method for distinguishing between "duplicase" and the "repair" enzyme.

H. BUSCH: Does all chromosomal synthesis begin at one end and if so, are there necessarily multiple initiating sequences required as well as multiple sites of enzymatic action?

J. H. TAYLOR: Synthesis of DNA in chromosomes of all higher cells appears to begin at many sites at nearly the same time. These sites appear to be characteristic for a particular chromosome, but are not often at the ends. The first instance observed of variations in synthesis along a chromosome happened to be in root cells of *Crepis*, where the late replicating DNA is localized near the centromeres. The cells labeled in late *S* sometimes showed a gradient of decreasing labeling from the end to the centromere. This was evidently due to the peculiar distribution of late replicating DNA in these particular chromosomes.

The only chromosomes which appear to replicate from one end to the terminus are found in bacteria, *E. coli* and *B. subtilis*.

THE MECHANISM OF ANTIMITOTIC
ACTIVITY OF DAUNOMYCIN

A. DI MARCO and R. SILVESTRINI

Istituto Nazionale per lo Studio e la Cura dei Tumori, Milano,
Gruppo di Ricerche Oncologiche del C.N.R.

IN STUDIES on DNA duplication and replication of the whole chromosomal structure some useful information may be offered by substances able to interfere with different stages of these processes. Our experiences concern one product of the *Str. peucetius* metabolism named by us Daunomycin (Da)[1] (Fig. 1). In the formation of complexes of this substance with DNA, shown by different physico-chemical methods, an intercalation of the planar anthracene-like chromophore between DNA adjacent base pairs[2-3] could play a role.

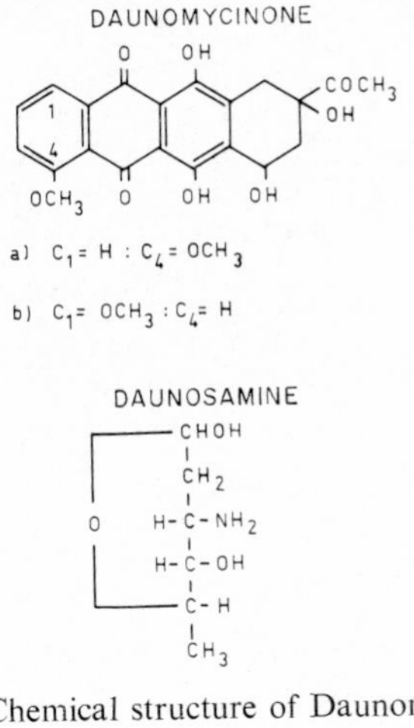

FIG. 1. Chemical structure of Daunomycin

However, the amino group present in the sugar residue should be considered responsible for a large increase in the bond's strength, as evidenced by the viscosity changes brought about in DNA solution by Da and not by its N-acetyl derivative.[4] A shift in the fusion temperature of DNA caused by Da indicates the formation of extra bonds between the two DNA strands, and a consequent increase in the probability of rebuilding of the original

241

native structure at low temperature. Denatured DNA and monofilamentous Ec9 phage of DNA do not show such behaviour.[5]

The inhibitory effect of Da on DNA-dependent RNA synthesis and on DNA replication could be considered as mere consequences of this binding. The interference exerted on the incorporation of labelled precursors into RNA should be related to the inhibition exerted by Daunomycin on the activity of DNA-dependent RNA polymerase.[2,6] As observed for actinomycin, the degree of enzyme inhibition increases with the antibiotic concentration and can be comparatively antagonized by DNA. The effect on DNA synthesis was observed:

(1) *in vitro* with DNA-dependent DNA polymerase;[6] here, a strong dependence on DNA concentration in the medium is observed, and
(2) on the incorporation of nucleic acid precursors into DNA by HeLa cells cultures[7] (Fig. 2).

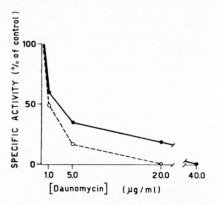

FIG. 2. Effect of increasing concentrations of Daunomycin on the incorporation of d-^{14}C adenine into DNA (o– – –o) and RNA (•——•) of HeLa cells.

A marked interference of Da with ^{3}H thymidine incorporation into DNA of *in vitro* HeLa cells was also observed by autoradiography.[8] Trying to understand the molecular mechanism of Daunomycin interference with DNA replication, two different possibilities should be considered:

(1) the binding of the antibiotic to DNA causes a steric hindrance to the formation of the hypothetical DNA–DNA polymerase complex, and
(2) taking for granted that strand separation is a necessary requirement to the function of the catalytic system,[9] the observed tendency of Daunomycin to tie together the two strands of DNA molecule should inhibit the very beginning of the polymerization reaction.

The latter possibility is supported by the considerable reduction in the ability to impair the DNA synthesis by Da-N-acetyl derivative, which has a reduced capacity to bind to DNA with no effect on DNA solution viscosity.

The experimental evidence of an inhibitory effect of Daunomycin on the replica of double-stranded DNA phage, but not of the monofilamentous Ec9 phage of DNA,[5] is also suggestive of the same mechanism. How far is the interference of Da with the mitotic activity of the cell dependent on the observed inhibition of DNA synthesis?

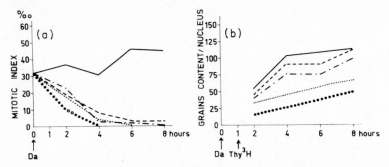

FIG. 3. Mitotic index (a) and ^{3}H thymidine incorporation (b) in HeLa cells, after treatment with D aunomycin. Control ———; 0·1 μg/ml – – – –; 0·2 μg/ml – · – · –; 0·5 μg/ml ······; 1 μg/ml ••••••.

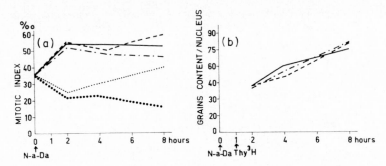

FIG. 4. Mitotic index (a) and ^{3}H thymidine incorporation (b) in HeLa cells, after treatment with N-acetyl derivative of Daunomycin. Control ———; 5 μg/ml – – – –; 10 μg/ml – · – · –; 20 μg/ml ······; 40 μg/ml ••••••.

In this connection, it must be firstly recalled that direct phase contrast observations showed that Da can stop the mitotic activity of the single cells when added up to 20 min prior to the prophase stage.[10] Furthermore, on HeLa cells a remarkable inhibition of M.I. can also take place with slightly active Da doses on ^{3}H thymidine incorporation into DNA (Fig. 3).

The same is true for the N-acetyl derivative of Da (Fig. 4); in this case the reduction of mitotic activity divorced by any effect on DNA synthesis leads to cells with increased content of radioactive DNA (Figs. 5, 6). As the ability of this compound to bind to DNA is considerably reduced, the possibility of a true statmokinetic effect of Da, unrelated to the binding to DNA, should be considered.

Trying to look into Daunomycin interference with different phases of the regenerative cycle, we synchronized slide cultures of rat fibroblasts following the procedure indicated by Xeros, Bootsma and Firket[11-13] (see Methods).

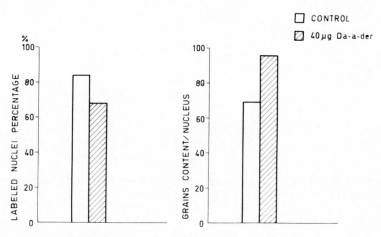

FIG. 5. Thymidine incorporation into control cultures and cultures treated for 24 hours with 40 μg/ml of Da-N-acetyl derivative.

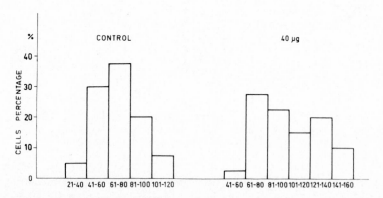

FIG. 6. Frequency of cells showing different grain contents in control cultures and in cultures treated for 24 hours with 40 μg/ml of Da-N-acetyl derivative.

This is based on the findings of the inhibitory effect of these high concentrations of thymidine on DNA synthesis.[14-17]

After removal of excess thymidine the cells proceed with DNA synthesis from the point in the S period at which they had been halted.[18]

In agreement with this statement, the treatment in the experiment reported, (Fig. 7) covers a wide portion of the synthetic phase but owing to the tenacious binding of Da to DNA the antibiotic is probably present in the cells from the moment of addition onwards. The figure shows that the treatment of the cultures with a concentration of 0.2 µg/ml Da produces a marked drop in number of cells which enter mitosis. That the impairment in mitotic ability of these treated cells could be a consequence of the interference with the

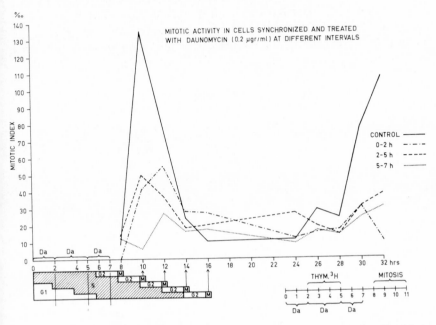

FIG. 7. Action of Daunomycin on all division in synchronized cultures of rat fibroblasts.

DNA's replication is shown by the reduction in the proportion of labelled interphase cells in the group which was treated and labelled at the same time. Mitotic anomalies and chromosomal aberrations (such as three-group metaphase, laggard chromosomes, chromosome breaks, chromatid breaks with ring formation, true anaphasic bridges) are frequent in cultures taken at different intervals from the treatment. These anomalies are also present in cells taken at the second mitotic wave, and are probably an expression of permanent damage undergone by cells which escaped the mitotic block.

How far could the inhibition of RNA synthesis be related to Da antimitotic activity? As protein synthesis is ultimately dependent on messenger and ribosomal RNA, it is evident that an inhibition of RNA synthesis would in the long run impair the reproductive capacity of the cell. But the sense of the posed question is whether this antibiotic does inhibit the synthesis of RNA molecules more specifically related to DNA duplication and more broadly to cell reproduction.

As such, one could consider messenger RNA responsible for the initiation of the synthesis of the enzymes required for DNA duplication (thymidine kinase, DNA polymerase, etc.) or for the synthesis and assembly of spindle proteins.

To analyze the possible relationship between the RNA synthesized in the different times of the cycle, and the subsequent DNA synthesis and cell division, we used mice fibroblast slide cultures, synchronized with the indicated procedure, but treated with Da at different intervals from the end of the mitotic wave. The partial loss of synchrony generally observed in the second mitotic division was thought to have only limited relevance in a study of the

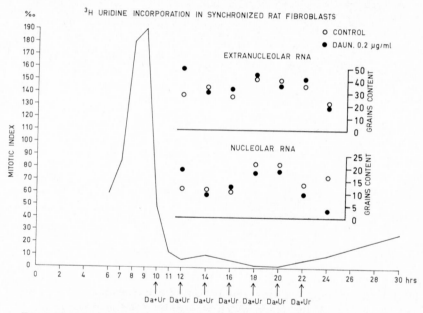

Fig. 8. Nuclear and extranucleolar incorporation of ^{3}H uridine by synchronized cultures of rat fibroblasts. The cells were treated with Daunomycin over 2 hr intervals, and ^{3}H uridine incorporation was evaluated over the same intervals.

events which precede the beginning of DNA synthesis. The grain counts of nucleolar and extranucleolar RNA per cell, after a 2 hr incubation with ^{3}H uridine shows the presence of a definite peak of both values but in particular of nucleolar label from the 6th to the 12th hour after the end of the mitotic wave (Fig. 8).

Thereafter, in coincidence with the beginning of the S phase, the RNA nuclear activity drops to the values characteristic for the beginning of the

G_1 phase. Taylor[19] and Robbins[20] failed to observe a periodical RNA synthesis in synchronized cultures.

It must, however, be stressed that both these authors measured the rate of synthesis on samples of cell population taken from the culture. Another point which deserves consideration is the utilization, in our experience, of diploid cells at the first *in vitro* transplant. A concentration of Da which exerts no influence on the RNA nuclear synthesis during the middle of the G_1 phase has, however, two noticeable and opposite effects at the beginning and at the end of the G_1. In fact, a stimulation of uridine incorporation is observed in the treatment with Da from the 9th to the 11th hour.

A similar effect was repeatedly observed in asynchronous cultures with low concentrations of this drug, as well as with actinomycin D and mitomycin, but it could not be obviously related to the events of the mitotic cycle.

In the present case, a relationship could be looked for between the binding of the drug to some regulator genes and the de-repression of the transcription process.

In this experiment it should be noted that a light reduction in nucleolar activity has no effect on the subsequent DNA synthesis.

In the next experiment Da concentration was therefore increased to 0.5 µg/ml (as shown in Fig. 9). The cultures were treated with Da during 2 hr periods from the 9th to the 21st hour, after the removal of excess thymidine, in the presence of ^3H uridine. A duplicate set of cultures, treated

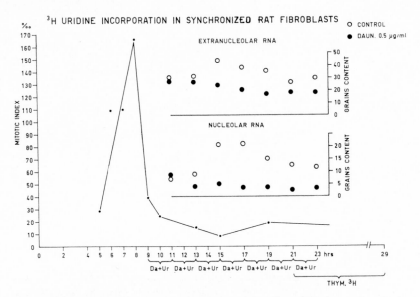

FIG. 9. Nucleolar and extranucleolar incorporation of ^3H uridine by synchronized cultures of rat fibroblasts. The cells were treated with Daunomycin over 2 hr intervals, and ^3H uridine incorporation was evaluated over the same intervals.

with Da during the indicated intervals, were exposed to ^{3}H thymidine from the 21st to the 29th hour.

A nearly complete suppression of the nucleolar incorporation is shown, and a considerable but not so deep reduction of the extranucleolar one. These results are in agreement with previous observations[21] showing a different degree of susceptibility to Da of uridine incorporation into RNA observed in nucleolar and extranucleolar areas, and with the analogous previous results obtained by Perry[22] working with actinomycin.

Further experiments carried out in our laboratories on HeLa cells demonstrated that Da exerts a considerable inhibition on newly synthesized RNA (extracted by the phenol-sodium-dodecyl-sulphate method according to Hiatt[23]) which sediments out in a sucrose gradient in the 28 S and 16 S zones.

On the contrary, an RNA aliquot not extracted by the indicated procedure appeared less sensitive to the action of Da and actinomycin.

A reduction of thymidine labelling appears in cells treated from the 13th hour onwards and becomes stronger in next intervals until the 19th hour (Fig. 10). The deepest inhibition of both RNA and DNA synthesis is achieved with treatments effected in the central part of the G_1 period (the

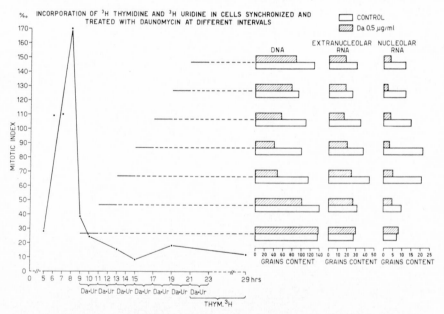

FIG. 10. Action of Daunomycin on the incorporation of ^{3}H uridine and ^{3}H thymidine by synchronized cultures of rat fibroblasts. After the first mitotic wave, cells were treated with Daunomycin over 2 hr intervals. The ^{3}H uridine incorporation was measured over the same intervals. ^{3}H thymidine was added at the 21st hour with respect to the time of removing excess thymidine.

one that shows the extra synthesis of RNA). Could this indicate an essential role of the RNA which is synthesized at this time for the DNA replica?

As the Da–DNA bond is not readily reversible, could the observed hot points of sensitivity mean that the transcription of the "replicon" follows an ordered sequence in time?

A definitive answer to these questions cannot be obtained by this kind of experience only, but it will serve to stress the importance of a deeper insight into the biochemical events which occur in the G_1 stage.

METHODS

Explants from the legs of 24–48 hr old rats were put on a coverslip covered with a very thin layer of chicken plasma.

Cultures were grown in Porter flasks in a medium consisting of:

40% Hanks' balanced salt solution,
40% placental cord serum, and
20% chick embryo extract.

After 30 hours' growth the cultures were incubated for 24 hr in the presence of 2 mM thymidine, then for 16 hr in normal medium and finally 24 hr in the presence of 2 mM thymidine.

Starting from this time, samples of cultures were drawn for mitotic activity determination. The cultures were treated with Daunomycin at different doses according to the scheme described and incubated respectively in the presence of ^{3}H thymidine (s.a. 160 mc/mM), 2 μc/ml, and ^{3}H uridine (s.a. 1.2 C/mM), 5 μc/ml.

The cultures were fixed in Bouin's fluid, washed in TCA 5% at 4°C and then in thymidine and uridine carrier. The coverslips were stuck on histologic slides by peligom and submitted to the ARGraphic technique of stripping film. After an exposure time of 3 days the autoradiographes were developed in D19b Kodak for 5 min at 18°C, fixed in F5 Kodak for 10 min at 15°C, and then washed in tap water and distilled water.

After drying, the cultures were stained in Mayer hematoxylin at 4°C.

The determination of labelled precursor incorporation was evaluated by counting the number of reduced silver grains per nucleus, after ^{3}H thymidine incubation, and per nucleolus and extranucleolar area after ^{3}H uridine incubation.

Each value reported in the figures was obtained from the observation of four cultures and represents the medium value from 80–100 cells.

ACKNOWLEDGEMENT

The author thanks Mrs. R. T. Dasdia for her technical assistance.

REFERENCES

1. A. GREIN, C. SPALLA, A. DI MARCO and G. CANEVAZZI, *Giorn. Microbiol.*, **11**, 109, 1963.
2. D. WARD, E. REICH and I. H. GOLDBERG, *Science*, **149**, 1259, 1965.
3. W. FULLER, Personal communication, 1965.
4. E. CALENDI, A. DI MARCO, M. REGGIANI, B. SCARPINATO and L. VALENTINI, *Biochem. Biophys. Acta*, **103**, 25, 1965.
5. E. CALENDI, R. DETTORI and M. G. NERI, *Giorn. Microbiol.*, in press.
6. G. HARTMANN, H. GOLLER, K. KOSCHEL and W. KERSTEN, *Biochem.*, **341**, 1126, 1964.
7. A. RUSCONI and E. CALENDI, *Biochim. Biophys. Acta*, **119**, 413, 1966.
8. R. SILVESTRINI, A. DI MARCO, S. DI MARCO and T. DASDIA, *Tumori*, **49**, 399, 1963.
9. F. I. BOLLUM, In *Progress in Nucleic Acids Research*, Academic Press, 1963.
10. A. DI MARCO, M. SOLDATI, A. FIORETTI and T. DASDIA, *Cancer Chemioth. Rep.*, **38**, 39, 1964.
11. N. XEROS, *Nature*, **194**, 682, 1962.
12. D. BOOTSMA, L. BUDKS and O. VAS, *Exp. Cell Res.*, **33**, 301, 1964.
13. H. FIRKET, *Compt. Rend. Soc. Biol.*, **158**, 1408, 1964.
14. N. R. MORRIS, P. REICHARD and G. A. FISHER, *Biochim. Biophys. Acta*, **68**, 93, 1963.
15. R. B. PAINTER and R. E. RASMUSSEN, *Nature*, **201**, 409, 1964.
16. T. T. PUCK, *Cell. Biol.*, **19**, 57, 1963.
17. A. DI MARCO, P. BARBIERI, L. FUOCO and A. RUSCONI, *Acta U.I.C.C.*, XVI, 1960.
18. T. T. PUCK, P. SANDERS and O. F. PETERSEN, *Biophys. J.*, **4**, 441, 1964.
19. E. W. TAYLOR, *Exptl. Cell Res.*, **40**, 316, 1965.
20. E. ROBBINS and M. SCHARFF, *Cell Synchr.*, 353, 1964.
21. A. DI MARCO, R. SILVESTRINI, T. DASDIA and S. DI MARCO, *Riv. Ital. Istoch.*, **11**, 211, 1965.
22. R. R. PERRY, *Exptl. Cell Res.*, **29**, 400, 1963.
23. H. H. HIATT, *J. Mol. Biol.*, **5**, 217, 1963.

DISCUSSION

F. H. KASTEN: (1) I found this presentation an extremely interesting one and I wish to compliment Dr. Di Marco on this study. In our laboratory, Mr. Strasser and I have been using the excess thymidine method for studying nucleoprotein synthetic patterns during the synchronized cell cycles of human CMP tumor cells in culture, and have reported some of our results elsewhere. One thing we saw with the aid of time-lapse photography was that in the presence of excess thymidine, in addition to an increase in cellular and nuclear sizes, abnormalities are induced in some cells. These include boiling at the cell surface, unusual vibratory movements, and the presence of some rounded cells which appear to be blocked in mitosis. The fact that such anomalies are induced by thymidine suggests that we need to view the synchronized cycle which follows with caution and be on the alert for abnormal morphologic and biochemical patterns. In our hands, double-thymidine blockage for 24 hr periods leads to a few mitotically blocked cells in a rounded state at the beginning of the S phase but no other lesions including mitotic abnormalities at the mitotic burst which occurs 10 hr later. Can you tell us whether there is any evidence of cellular abnormalities in your material, especially since I think you used 4-day treatments with thymidine?

(2) It appears that your drugs were added in the period between 10 and 21hr after the initiation of the synchrony. Is this the G_1 period for these cells?

(3) My last question is a general one and addressed to any of the biochemists who are accustomed to isolating nucleoli and performing nucleic acid analyses on this material. I think it is very important to re-emphasize the cytologic difficulties inherent here. The nucleolus has no membrane enveloping it, so that any nucleic acid constituents which are not attached to granular or filamentous structures of the nucleolus must fill the interstitial regions or vacuoles, probably in a gel-like form. In view of this situation, I wonder about the reliability and proper interpretation possible using the mass isolation procedures. This point, although of general importance, is raised with particular reference to the question of the origin of 18 S RNA.

A. Di Marco: (1) In the cultures treated by thymidine, to get the synchronization, there is an increase in the number of mitotic anomalies. In parallel experiments these rise to 5% of all mitoses in comparison to 2.5% in control, non-synchronized cultures.

(2) The drug treatment was performed at the moment when there was observed (in control slides) a drop in the mitotic figures; in the experiment reported in Fig. 9 the antibiotic was added from the 9th to 11th hour onwards for 2 hr periods until the 21st to 23rd hour. Following the evaluation in unsynchronized cultures of rat fibroblasts the G_1 phase lasts 10–12 hrs. As the overall regeneration time in our system is unchanged (22–24 hr), it could be speculated that the treatment was performed during G_1 phase; however it should be remarked that we have not checked the actual absence of DNA synthesis in this period.

M. Izawa: (3) Evidence from rapid labelling and specificity of molecular species and/or nucleotide compositions of synthesized RNA in nucleolus studies on mass isolated nucleoli strongly suggests that it is a site of independent RNA synthesis in the nucleus. The RNA found in isolated mass nucleoli is, therefore, not a simple contaminant derived from other parts of the nucleus. This means that the mass isolation procedure is reliable for biochemical study of the RNA synthesis in the nucleoli. However, we cannot exclude completely a possibility that some RNA components might be lost from nucleoli during the isolation. Concerning this point we have to look carefully for whether or not nuclear 18 S RNA is localized really in.

ACTION OF MITOMYCIN C ON CULTURED SALAMANDER CELLS–A PHASE-CONTRAST AND CYTOCHEMICAL STUDY*

FREDERICK H. KASTEN and CREED WOOD†

Pasadena Foundation for Medical Research, Pasadena, California; Department of Anatomy, University of Southern California School of Medicine, Los Angeles, California; Department of Pathology, Loma Linda University, School of Medicine, Loma Linda, California

THE selective effects of mitomycin C (MC) on deoxyribonucleic acid (DNA) has been demonstrated in bacterial systems (Shiba et al., 1958; Kersten and Rauen, 1961; Reich, 1961; Sekiguchi and Takagi, 1960). The antibiotic is reported not to affect the synthesis of RNA or protein in microorganisms (Reich and Tatum, 1960; Sekiguchi and Takagi, 1960; Reich and Franklin, 1961). In cultured mammalian cells and ascites tumor cells, MC inhibits DNA synthesis and exerts little or no action on protein and RNA synthesis (Kuroda and Furuyoma, 1963). At the morphologic level in mammalian cells, MC has been shown to inhibit mitosis, induce pycnosis in interphase cells, and produce cytoplasmic vacuolization, nucleolar lesions, giant cells, and chromosomal aberrations (Kuroda and Furuyoma, 1963; Reich et al., 1961; Kimura, 1963; Kobayashi, 1960; Lapis and Bernard, 1965; Nowell, 1964; Cohen and Shaw, 1964; Oledzska–Slotwinska et al., 1964). A concise and recent summary of some biochemical and biological effects of MC is available (Szybalski and Iyer, 1964). On the basis of experimental data, a number of modes of action of MC have been proposed to account for its biological activity. In one model, MC is said to act by activating DNase (Nakata et al., 1961; Kersten, 1962) to cause depolymerization of DNA (Shiba et al., 1959; Reich et al., 1961; Kersten and Rauen, 1961). There is evidence for stimulation of DNA polymerase activity in mammalian cells

* This investigation was supported in part by U.S.P.H.S. Research Grants CA 07991 from the National Cancer Institute and NB 03113 from the National Institute of Neurological Diseases and Blindness, an Institutional Grant to the University of Southern California from the American Cancer Society, and a grant from the Greenville Foundation. The assistance of C. George Lefeber in preparing the plates is gratefully acknowledged.

† Supported by an N.I.H. Program for Physical Biology, Grant No. 2T1GM945 to Pomona College. Present address of C.W.: Yale University School of Medicine, New Haven, Connecticut.

(Magee and Miller, 1962) but not in bacteria (Nakata *et al.*, 1961; Sekiguchi and Takagi, 1960). MC has also been shown to increase activities of acid hydrolases in ascites tumor cells (Niitani *et al.*, 1964), which implicates lysosomes in the direct or indirect scheme of action of MC. On the basis of *in vitro* studies of interaction between MC and DNA, it is suggested that DNA synthesis is blocked by demonstrable *in vitro* crosslinking of complementary DNA strands (Szybalski and Iyer, 1964). This is supported by the work of Matsumoto and Lark (1964).

In view of diverse responses elicited by MC in mammalian cells, the primary lethal mode of action of this antibiotic is obscure. To aid in evaluating this problem, a combined cytologic and cytochemical investigation was carried out on a previously unstudied system; namely, cultured amphibian cells. The large size *in vitro* of Salamander epithelial lung cells and their nuclei and chromosomes provides ideal material for following the sequential action of MC by phase-contrast microscopy, autoradiography, and cytochemical staining of fixed cells. Results will be presented which demonstrate an early inhibition of DNA synthesis. This is followed by a loss of DNA in discrete nuclear loci, and then the solubilization of histone protein and reaggregation into multiple granules at these loci. Direct effects of MC on mitotic chromosomes are also shown.

MATERIALS AND METHODS

Cultures and Treatment

Small pieces of lung tissue from a Western salamander, *Taricha granulosa* supplied from Oregon by Dr. James Kezer, were explanted under cellophane in the Rose chamber and grown at room temperature in Eagle's MEM medium supplemented with 10% calf serum. Under these conditions, sheets of large, ciliated epithelial cells emigrate from the explant in 5–7 days and display mitotic activity during the following two weeks. Cultures could be maintained during this entire period at a constant pH without a change of medium. All experimental treatments and observations were done on such healthy cells. Additional cytologic observations have been reported (Seto and Pomerat, 1965; Kasten, 1965a).

Phase-contrast, Time-lapse Cinematography

Routine still phase-contrast microscopy was used to follow the evolution of lesions induced in living cells by MC. Other cultures were followed by phase optics using 16 mm time-lapse filming at a taking rate of 1–2 frames per minute. Such experiments were filmed for 2–3 days after the addition of MC. A description of the multiple cine units employed in this laboratory and their modes of operation have been described in detail (Lefeber, 1963).

Autoradiography

Some cultures were treated with H^3-thymidine at 1 μc/ml for 24 hr and then treated with MC at 5γ/ml for 4, 18, and 24 hr. In other experiments, cultures were exposed first to MC at 5γ/ml for 5, 18, and 24 hours. At the end of the required treatment time with H^3-thymidine and MC, cultures were washed in Gey's balanced salt solution (BSS) to remove serum protein prior to fixation in Carnoy's acetic–alcohol (1:3). Other details of specimen preparation and film handling and processing are given elsewhere (Kasten *et al.*, 1965). In these experiments, preparations were left under film for 9 days.

Cytochemical Staining and Cytophotometry

Routine staining was done using Jacobson's multiple staining method. Nucleic acids were stained with gallocyanin chromalum (Einarson, 1951), azure B bromide at pH 4.0 (Flax and Himes, 1952), methyl green pyronin and acridine orange or AO (Mayor, 1961). DNA was stained selectively by the Feulgen reaction (Feulgen and Rossenbeck, 1924) using the de Tomasi recipe for Schiff's reagent (de Tomasi, 1936) and the fluorescent-Feulgen reaction using the auramine O-SO$_2$ reagent (Kasten, 1959; Kasten, 1960). Proteins were detected with mercuric bromphenol blue (Mazia *et al.*, 1953).Naphthol yellow S (Deitch, 1955) was employed to stain proteins containing the dibasic amino acids, lysine, arginine, and histidine. Other selective protein stains included alkaline Biebrich scarlet for basic proteins (Spicer, 1962), alkaline fast green for histones (Alfert and Geschwind, 1953) and ammoniacal silver for histones (Black and Ansley, 1964). Cytophotometry of DNA was carried out on Feulgen-stained nuclei as previously described (Kasten, 1965b).

RESULTS

By means of time-lapse filming, it was possible to show that MC can act directly on dividing chromosomes. In one sequence, two treated cells in prophase were each observed to progress to metaphase. After a prolonged metaphase, both cells regressed to interphase in tetraploid states. Direct effects of MC on metaphase chromosomes were documented in another cell. In this case, the culture had been exposed to MC at 1 γ/ml for 4 hr by the time the dividing cell was in metaphase (Fig. 1). Except for one chromosome which was outside the metaphase plate, the chromosomes themselves appeared normal. Four hours later, the division had reverted to a prophasic condition and the individual chromosomes had been disrupted into a series of dense, beadlike granules (Fig. 2). Eventually, this cell developed a nuclear membrane and attained an interphase condition. Further studies are needed to decide whether other stages of mitosis are sensitive to the drug.

By means of thymidine autoradiography, it was possible to label newly synthesized DNA in interphase nuclei (Fig. 3). Pretreatment of such cells with MC at 5 γ/ml for as little as 5 hr completely blocked DNA synthesis. On the other hand, cells were labeled normally when preincubated with labeled thymidine and then treated with the drug, up to 24 hr. New mitoses were not observed after 24 hr exposure to MC; this inhibition was probably due to the blockage of DNA synthesis.

Probably the most interesting effect of MC is seen in living interphase nuclei by phase-contrast optics. Untreated nuclei vary greatly in size and each usually contains two prominent nucleoli, various dense heterochromatic regions or a clear nucleoplasm, and some irregular granules (Fig. 4). Following addition of MC, there is no marked visible change for about 5 hr (5 γ/ml). At this time, some areas of the nucleoplasm become less dense and form multiple phase-light lacunae, each about 1 μ in diameter. At 8–9 hr, minute, dense beads begin to appear within the lacunae. These become more distinct with the passage of time and enlarge into prominent granules (Fig. 5). While the nucleolus may become slightly reduced in size, it remains prominent. After careful study by direct microscopic viewing and by time-lapse filming, it is clear that the newly-formed beads and granules do not originate from the nucleolus but appear *de novo* at multiple sites in the nucleoplasm.

PLATE I

FIG. 1. Metaphase chromosomes following treatment with MC at 1γ/ml for 4 hr. Except for one chromosome which has been lost from the spindle, there are no evident chromosomal abnormalities. Phase-contrast.

FIG. 2. Same field as in Fig. 1 after 8 hr treatment with MC. Note beading of chromosomes and reversion to prophasic condition. Typically, such metaphase damaged cells fail to complete mitosis and revert to interphase in tetraploid condition. Phase-contrast.

FIG. 3. Autoradiograph displays 5 labeled nuclei. Cells were exposed to H^3-thymidine for 24 hr and then treated with MC at 5 γ/ml for 24 hr. While MC blocks DNA synthesis, the prelabeled nuclei retain much of their DNA as shown here. Hematoxylin stain.

FIG. 4. Typical interphase nucleus of cultured Salamander epithelial lung cell. Note two prominent nucleoli (nu) and several smaller regions of a dense heterochromatic nature. Phase-contrast.

FIG. 5. Nucleus following treatment with MC at 5 γ/ml for 22 hr. Abundant numbers of newly formed dense granules are scattered throughout the nucleus. Arrow directs attention to a heterochromatic region which persists. Phase-contrast.

FIG. 6. At higher magnification, a nucleus is shown which was exposed to MC for 48 hr and then placed in normal medium for 24 hr. An assortment of dense, rounded granules of various sizes is shown, each of which arises *de novo* from the nucleoplasm. A prominent, unaltered nucleolus (nu) persists. Phase-contrast.

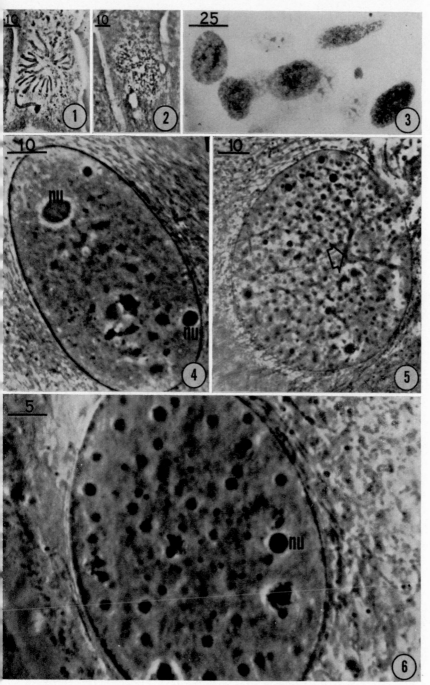

PLATE I

NAM

Also, they are apparently not derived from heterochromatic regions, which appear normal during early stages of granule formation (Fig. 5). In unusually favorable film sequences, the nucleoplasm is seen to lose its characteristic density and appears almost transparent. In this state, the granules seem to be "floating" in the fluid nucleoplasm and frequently move from side to side, as if in response to internal fluid movements. By 48 hr, all sizes of granules are observed, ranging from 0.4 to 3 μ in diameter, and number as many as 50 within a single nucleus (Fig. 6). The dense nucleolus persists during the entire evolution of the granules. At lower concentrations of MC (1 γ/ml), the same series of events takes place, with the aberrant granules first being seen at about 30 hr. Throughout the exposure to MC, mitochondria appear normal and there are no evident cytoplasmic disturbances. Cilia continue to beat at the surface of the cells. After 3–4 days in MC, cells begin to degenerate. Cells and nuclei become compressed and thick, preventing optimal observation by phase optics. Most of the cells die and form aggregates of debris.

The chemical nature of the nuclear lesions was evaluated by cytochemical staining. The characteristic nuclear distribution of DNA in untreated cells is especially well shown with the fluorescent-Feulgen stain (Fig. 7). Here we see the prominent heterochromatic regions and a single nucleolus with a rim of nucleolar-associated DNA. After 48 hr in MC when multiple granules are known to be present, such granules fail to stain for DNA. Also, the DNA normally present in these regions is absent, leading to the appearance of hollow spaces (Fig. 8). In another nucleus of the same figure,

---

PLATE II

FIG. 7. The normal intranuclear DNA pattern is shown in this untreated cell. DNA is present in the form of a dense mat of filaments, two especially dense regions of heterochromatin (h), and a ring of nucleolar-associated material around the nucleolus (nu). Fluorescent-Feulgen stain.

FIG. 8. After 48 hr treatment with MC, DNA is lost from specific foci, leaving multiple spaces (sp). These are prominently displayed in the upper nucleus. The lower nucleus also reveals a few DNA-deficient foci (arrows). Granules are present within the spaces but lack DNA and fail to be seen in this preparation. The persistence of heterochromatin (h) is suggestive that this is more resistent to MC-degradation than euchromatin. Fluorescent-Feulgen stain.

FIG. 9. One of the earliest responses to MC in cultured Salamander cells is a loss of nuclear protein at discrete foci, as illustrated here (arrows). Treated with MC at 5 γ/ml for 4 hr. Mercuric bromphenol blue stain.

FIG. 10. After 48 hr exposure to the drug, there is protein reaggregation and deposition in the previously depleted foci. Mercuric bromphenol blue stain.

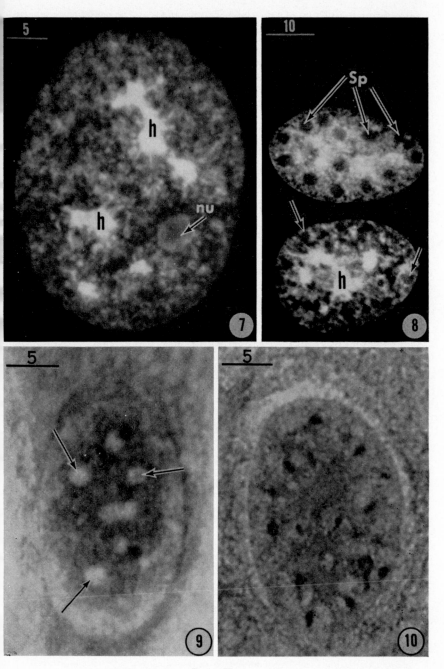

PLATE II

heterochromatic blocks of DNA appear unchanged in the presence of several abnormal nuclear spaces. The apparent loss of DNA in the regions occupied by granules was verified by Feulgen cytophotometry. After 24 hr treatment with MC at 1 γ/ml, there was an average loss of 35% DNA per nucleus in the population. The granules likewise failed to stain for DNA with the routine Feulgen stain, azure B bromide, methyl green, and gallocyanin chromalum. AO did not produce the characteristic yellow–green fluorescence of DNA in the granules. The absence of RNA in the granules was also verified with the appropriate stains. When MC-treated cells (4 hr at 5γ/ml) were stained for protein with mercuric bromphenol blue, there was a negative reaction (Fig. 9); this yielded a picture similar to that seen after staining for DNA—a group of unstained spaces within the nucleus. Later, after granules were present, these stained intensely with the bromphenol blue stain (Fig. 10). More specific protein stains revealed that the granules contained proteins rich in dibasic amino acids (naphthol yellows S) and were of a highly basic nature. Positive staining for histones were obtained with ammoniacal silver (Fig. 11), alkaline fast green (Fig. 12) and alkaline Biebrich scarlet (Fig. 13).

Finally, two other features of the action of MC on amphibian cells should be mentioned. One is the fact that the drug induces giant cell formation in a small percentage of cells, probably as a consequence of failure of some cells to complete mitosis and their subsequent reversal to interphase. Second, the drug acts directly on interphase nuclei to produce budding and loss of DNA in the cytoplasm. An example of this effect is shown in Fig. 14. The field of

---

PLATE III

FIG. 11. A fine granular distribution of histones is shown in this nucleus of an MC-treated cell (5 γ/ml) after 48 hr exposure. Arrows direct attention to dense concentrations of histones within 3 different nucleoli. The positioning of the nuclear border and identification of nucleoli are made possible by a composition phase-contrast photomicrograph of the same cell. Ammoniacal silver stain for histones.

FIG. 12. The abnormal deposits of protein previously seen in DNA-deficient foci (Figs. 8, 10) are shown in this nucleus to consist of histones. MC treatment at 5 γ/ml for 48 hr. Alkaline fast green stain.

FIG. 13. With another specific histone stain, the same abnormal pattern of histone deposition is shown—as was illustrated in Fig. 12. Alkaline Biebrich scarlet stain.

FIG. 14. After long exposure to MC (1 γ/ml for 72 hr) nuclear budding and fragmentation occurs. Such extranuclear DNA fragments, which fluoresce yellow–green, are seen in the upper cell. Large numbers of normal nucleated erythrocytes are seen, as well as a few larger epithelial cells. Acridine orange stain, fluorescence photomicrograph.

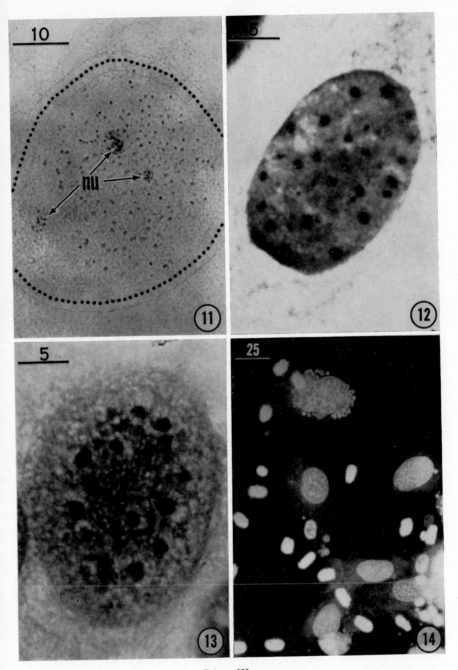

PLATE III

MC-treated epithelial cells and nucleated erythrocytes was fluorochromed with AO. The nucleus in the upper cell of this figure has been disrupted by the drug and displays budding and extranuclear DNA fragments.

DISCUSSION

Our results demonstrate that MC induces a variety of biochemical and cytologic lesions in cultured amphibian cells, some of which have been reported in mammalian cells and other lesions which have not been previously reported. In the order of their appearance, these lesions include the following:

1. Blockage of DNA synthesis.
2. Loss of approximately 35% DNA at multiple foci in each nucleus.
3. Direct effects on metaphase chromosomes to produce beading or reversal to interphase.
4. Inhibition of mitosis.
5. Intranuclear deposition of phase-dense histone granules in foci previously depleted of DNA.
6. Giant cell formation.
7. Nuclear budding and extrusion of DNA.

In the brief time remaining, it is not possible to discuss all the ramifications of these results. For this reason, I will direct most of my remarks to the unusual and new findings, namely, histone reaggregation resulting from MC treatment (Nos. 2 and 5 listed above).

The relatively rapid loss of DNA in interphase nuclei is undoubtedly due to loss of preexisting DNA rather than to inhibition of DNA synthesis. The reason for this conclusion is that only a fraction of the cell population is synthesizing DNA at a given time according to thymidine autoradiography, yet practically all interphase cells respond to MC treatment by loss of DNA from small 1–3 μ spaces scattered largely around the margin of each nucleus. This DNA fraction is apparently not derived from heterochromatin, judging from the persistence of heterochromatic blocks seen in living and stained cells. Since the heterochromatic blocks are considered to be genetically inactive DNA regions, it appears that such regions are more resistant to degradation by mitomycin C than genetically active DNA. The molecular basis for this difference with MC is not known although it is likely that the degradation itself occurs by depolymerization (Shiba et al., 1959; Kersten, 1962). In a discussion of the reasons underlying the selective fractionation of nucleohistones with salt solutions, Chargaff (1955) comes to the conclusion that nucleic acid chains relatively rich in guanylic and cytidylic acids are detached from histones more easily than those in which adenylic and thymidylic acids predominate. Is it possible that our results and those summarized by Chargaff have a common basis relating to inherent physicochemical differences between euchromatin and heterochromatin? In our experiments, the simultaneous appearance in each nucleus of multiple spaces lacking DNA indicates that the extracted DNA fragments are removed from many chromosome sites.

With the loss of DNA, the previously associated histone becomes temporarily solubilized, as judged by its failure to stain in the regions devoid of DNA. This histone fraction is not lost from the nucleus but reaggregates into large, individual visible beads which fill each space. These are visible in living cells and their evolution is seen to proceed from a barely visible size by light microscopy to large granules. While nucleoli persist in these altered nuclei, it is evident from the time-lapse studies and from staining results that the granules are not derived from nucleoli but arise *de novo* from molecular aggregations of free histones. While there has been no previous report that we are aware of what relates MC to nucleohistone degradation and subsequent histone aggregation, it is reasonable to speculate that the failure to observe this response in mammalian cells is due to the fact that the latter contain 30 times less DNA (or nucleohistone) per nucleus than Salamander cells (Vendrely, 1955), and the histone aggregations would not be easily visible by light microscopy. In this connection, there is a report by Lapis and Bernhard (1965) which deals with MC effects on the nucleolar fine structure of KB cells. These authors observed the presence of small, intense bodies or satellites in the nucleoplasm which are detectable by light microscopy. While pepsin did not digest these granules, no further cytochemical stains at the light microscope level were employed. Further studies are needed to evaluate this problem.

Our other findings on blockage of DNA synthesis, inhibition of mitosis, direct effects on metaphase chromosomes, giant cell formation, and nuclear budding with extrusion of DNA confirm previous reports based on MC treatment of mammalian cells. The fact that MC induces reversion of some metaphase Salamander cells to interphase suggests that this is one way in which proliferating malignant cells in the tumor-bearing patient temporarily escape damage from the drug. The reverted tetraploid interphase cells, in a few cases where they were followed by time-lapse filming, were observed to eventually develop typical nuclear lesions with multiple, phase-dense histone beads. Our observations and interpretations agree with those of Nowell (1964) and Cohen and Shaw (1965), who conclude that MC causes chromosome aberrations at a specific stage in the cell cycle. The participation of histones in protecting DNA from MC-induced damage is also suggested. It is difficult for us to interpret our findings on the basis of a single major mode of action of MC at the molecular level, and conclude that MC prevents DNA synthesis by crosslinking complementary chains (Szybalski and Iyer, 1964; Matsumoto and Lark, 1964). Normal nucleohistone changes during the normal cell cycle probably account for the variety of primary and secondary responses observed following drug treatment.

REFERENCES

1. M. ALFERT and I. GESCHWIND, *Proc. Nat. Acad. Sci.*, **39**, 991, 1953.
2. M. M. BLACK and H. R. ANSLEY, *Science*, **143**, 693, 1964.
3. E. CHARGAFF, In *The Nucleic Acids*, vol. 1, pp. 307–371. Ed. by E. Chargaff and J. N. Davidson, Academic Press, New York, 1955.
4. M. M. COHEN and M. W. SHAW, *J. Cell Biol.*, **23**, 386, 1964.
5. M. M. COHEN and M. W. SHAW, In *In Vitro*, vol. 1, pp. 50–66. Ed. by C. J. Dawe, Waverly Press, Baltimore, 1965.
6. A. D. DEITCH, *Lab. Invest.*, **4**, 324, 1955.
7. J. A. DE TOMASI, *Stain Techn.*, **11**, 137, 1936.
8. L. EINARSON, *Acta Path. Scand.*, **28**, 82, 1951.
9. R. FEULGEN and H. ROSSENBECK, H.-S. *Zts. f. Physiol. Chem.*, **135**, 203, 1924.
10. M. FLAX and M. H. HIMES, *Physiol. Zool.*, **25**, 297, 1952.
11. F. H. KASTEN, *Histochemie*, **1**, 466, 1959.
12. F. H. KASTEN, In *International Review of Cytology*, vol. **10**, pp. 1–100. Ed. by G. H. Bourne and J. F. Danielle, Academic Press, New York, 1960.
13. F. H. KASTEN, *Amer. Zool.*, **5**, 648, 1965 a.
14. F. H. KASTEN, *Stain Techn.*, **40**, 127, 1965 b.
15. F. H. KASTEN, F. F. STRASSER and M. TURNER, *Nature*, **207**, 161, 1965.
16. J. KAWAMATA, *Nature*, **183**, 1056, 1959.
17. H. KERSTEN, *Biochim. Biophys. Acta*, **55**, 558, 1962.
18. H. KERSTEN and H. M. RAUEN, *Nature*, **190**, 1195, 1961.
19. Y. KIMURA, *Gann*, **54**, 163, 1963.
20. J. KOBAYASHI, *Cytologia*, **25**, 280, 1960.
21. Y. KURODA and J. FURUYOMA, *Cancer Res.*, **23**, 682, 1963.
22. K. LAPIS and W. BERNHARD, *Cancer Res.*, **25**, 628, 1965.
23. C. G. LEFEBER, In *Cinemicrography in Cell Biology*, pp. 3–26. Ed. by G. G. Rose, Academic Press, New York, 1963.
24. W. E. MAGEE and O. V. MILLER, *Biochim. Biophys. Acta*, **55**, 818, 1962.
25. I. MATSUMOTO, and K. G. LARK, *Exp. Cell Res.*, **23**, 192, 1964.
26. H. D. MAYOR, *Tex. Rep. Biol. Med.*, **19**, 106, 1961.
27. D. MAZIA, P. A. BREWER and M. ALFERT, *Biol. Bull.*, **104**, 57, 1953.
28. Y. NAKATA, K. NAKATA and Y. SAKOMOTO, *Biochem. Biophys. Res. Comm.*, **6**, 339, 1961.
29. H. NIITANI, A. SUZUKI, M. SHIMOYAMA and K. KIMURA, *Gann*, **55**, 447, 1964.
30. P. C. NOWELL, *Exptl. Cell Res.*, **33**, 445, 1964.
31. H. OLEDZSKA–SLOTWINSKA, R. BASSLEER and S. CHEVREMONT–COMHAIRE, *Chemotherapia*, **8**, 210, 1964.
32. E. REICH, *Biochim. Biophys. Acta*, **53**, 132, 1961.
33. E. REICH and R. M. FRANKLIN, *Proc. Nat. Acad. Sci.*, **47**, 1212, 1961.
34. E. REICH, R. M. FRANKLIN, A. J. SHATKIN, and E. L. TATUM, *Fed. Proc.*, **20**, 154, 1961.
35. E. REICH and E. L. TATUM, *Biochim. Biophys. Acta*, **45**, 608, 1960.
36. M. SEKIGUCHI and Y. TAKAGI, *Biochim. Biophys. Acta*, **41**, 434, 1960.
37. T. SETO and C. M. POMERAT, *Copeia*, No. 4, 415, 1965.
38. S. SHIBA, A. TERAWAKI, T. TAGUCHI and J. KAWAMATA, *Biken's J.*, **1**, 179, 1958.
39. S. SHIBA, A. TERAWAKI, T. TAGUCHI and J. KAWAMATA, *Nature*, **183**, 1056, 1959.
40. S. S. SPICER, *J. Histochem. Cytochem.*, **10**, 691, 1962.
41. W. SZYBALSKI and V. N. IYER, *Fed. Proc.*, **23**, 946, 1964.
42. R. VENDRELY, In *The Nucleic Acids*, vol. 2, pp. 155–180. Ed. by E. Chargaff and J. N. Davidson, Academic Press, New York, 1955.

DISCUSSION

H. BUSCH: Smetana and I published a method for staining acidic nucleo-proteins in Cancer Research this February (1966). I wonder whether there may be an increase in acidic proteins in these complexes, i.e. is there a histone-acidic protein complex?

F. H. KASTEN: This is a good point to check.

J. H. TAYLOR: Do the dense bodies stain with fast green at high pH without removal of DNA?

F. H. KASTEN: The standard alkaline fast green technique for histones re-quires the prior removal of DNA in order to achieve histone staining. How-ever, it is not essential to remove DNA in order to demonstrate histones in the nuclear granules. They stain as well for histones with the alkaline Biebrich scarlet technique and the ammoniacal silver procedure, neither of which involves removal of DNA.

H. ISHIZAKI: Does the total amount of histone differ from the normal cells?

F. H. KASTEN: This is a good question. We have not made any quantita-tive histone measurements as yet and it is certainly important to know the DNA/histone ratios in individual nuclei of these treated cells.

Y. HOTTA: Two questions concerning the temporal arrest at metaphase: (1) What percentage of cells could be arrested in metaphase? and how long do they survive?

(2) In our laboratory, we found the sticky chromosome in anaphase and/or telophase. Does your metaphase arrested cell never go into anaphase?

F. H. KASTEN: (1) I have no data on this point. Cells which reverted to interphase appear morphologically viable but were not followed long enough to see if they could divide again.

(2) The several examples we followed by time-lapse photography were all in metaphase or late prophase at the time of treatment. They never proceeded past metaphase and all eventually regressed to prophase and interphase to give apparently viable polyploid cells. Our sample was small, however, so that it is possible that some arrested metaphases may go ahead and cleave.

P. FITZGERALD: Dr. Taylor wanted to know if you tried the Alfert stain without prior hydrolysis of DNA?

F. H. KASTEN: No.

ADDENDUM

Since this paper was presented in October, 1966, the author, in collaboration with Jeannette Huber, has extended these studies. Briefly, we found upon further examination that nucleoli of MC-treated amphibian cells become progressively smaller but do not contribute structurally to emerging histone granules. The granules develop from a fine chromosomal network. Following DNA inhibition and granular histone aggregation, RNA synthesis is reduced, especially in nucleoli, but protein synthesis continues un-abated. These results will be published in more detail in the near future.

EXPRESSION OF SOME ANTIGENIC CHARACTERISTICS IN EMBRYONIC KIDNEY CELLS CULTURED *IN VITRO**

TOKINDO S.OKADA

Laboratory of Development Biology, Zoological Institute, Faculty of Science,
University of Kyoto, Kyoto, Japan

CHANGES in differentiative traits of the cells cultured in monolayer have been a subject of many studies since an introduction of the technique of cell (or tissue) culture in cell biology.[1] A loss of several tissue-specific characteristics of tissue cells takes place in most cases with these studies. The earlier conclusion which was exclusively based upon morphological criteria has now been extended to the studies utilizing chemical, enzymological or immunological techniques for identifying the specificities. An apparent loss, however, is caused sometimes by the selective proliferation of non-specific cell types in culture conditions. In such a case as kidney which is naturally a composite of the several different types of tissue cells, it is particularly difficult to determine whether the loss may be due to cellular changes occurring within progenies of the cells originally belonging to the same tissue type or merely to a selective process. Weiler,[2] and Fogel and Sachs[3] adopted the former alternative using some antigenic characteristics for marking the tissue-specificity of hamster kidneys. In the present communication, the results of the same subject using chicken embryonic kidneys will be reviewed. In these studies, which have been carried out in the author's laboratory for the last few years, efforts have been made to correlate the changes in antigenic components with the loss of histoformative capacity of the cells. For details of the techniques, refer to the author's earlier publications.[4−7]

ANTISERA, DIAGNOSTIC REAGENTS FOR IDENTIFYING THE SPECIFICITIES

Cytoplasmic particulate materials obtained from adult chicken kidneys by ultracentrifugation at "microsomal" level are washed with EDTA and treated with DOC. A yellowish pellicle (K45) can be precipitated from the

* This investigation was supported in part by a PHS research grant RG-9469 from the Division of General Medical Sciences, U.S. Public Health Service.

DOC-soluble fraction by salting-out with ammonium sulfate of between 25 and 45% saturation (Fig. 1). This pellicle, when injected into rabbits after dialysis against water, elicits a highly kidney-specific antiserum.[8] After

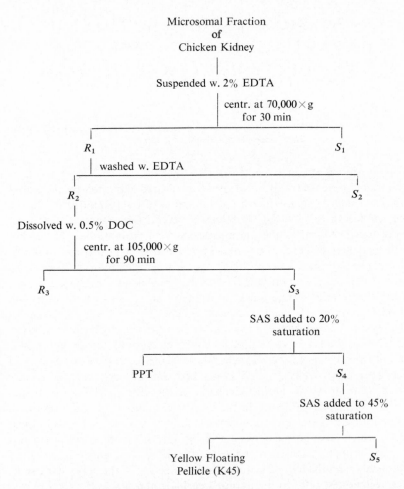

Microsomal Fraction
of
Chicken Kidney

Suspended w. 2% EDTA

centr. at 70,000×g
for 30 min

R_1 S_1

washed w. EDTA

R_2 S_2

Dissolved w. 0.5% DOC

centr. at 105,000×g
for 90 min

R_3 S_3

SAS added to 20%
saturation

PPT S_4

SAS added to 45%
saturation

Yellow Floating S_5
Pellicle (K45)

FIG. 1. Flow-sheet of isolation of "kidney-specific" antigen, K45, from the microsomal fraction. EDTA: sodium salt of diethylene aminotetraacetic acid, DOC: sodium deoxycholate, SAS: saturated solution of ammonium sulfate.

proper absorption with some heterologous antigens this antiserum (anti-K45) can be used as a diagnostic reagent for detecting "kidney-specificity", particularly by means of fluorescent antibody staining.[9]

Embryonic chicken kidney cells cultured in monolayer (MCK) for 6–7 days were injected into rabbits. An antiserum obtained was conjugated with fluorescent dye and absorbed with freshly dissociated kidney cells (FDK) until the conjugate did not stain FDK positive any longer (anti-MCK). This was used to discover whether a "new" specificity would appear in MCK.

K45 ANTIGENS IN MONOLAYER CULTURES OF KIDNEY CELLS (MCK)

In both embryonic and adult kidneys anti-K45 conjugated with fluorescein-isothiocyanate (anti-K45-FITC) stained specifically only the inner borders of cells of the proximal convoluted tubules (Fig. 2a). In FDK of the 14-day embryonic metanephros, only an apical cap of the cell fluorescence appeared

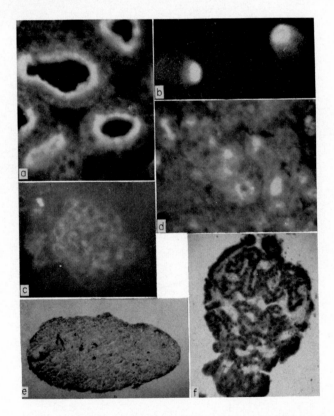

FIG. 2. "Kidney-specific" antigens in the different culture conditions. a: histological section of the embryonic metanephros, b: freshly dissociated cells, c: monolayer cultured cells, d: histological section of the reaggregate of b, e: histological section of the reaggregate after 4-days' organ culture, f: histological section of the reaggregate of the monolayer cultured cells. a, b, c, d: fluorescent photos stained with anti-K45-FITC. e, f: ordinary histological staining. b, d from Okada;[6] c from Okada.[7]

(Fig. 2b). In spreading MCK the florescence occurs diffusively in the cyto-plasm, especially in perinuclear areas (Fig. 2c). The living FDK adsorbs anti-K45-FITC in the apical cup if it can be treated with the latter for 30 min, without killing the cells. Such a surface staining does not occur in MCK.

Changes of K45-content in MCK were determined in the following manner. The results obtained by the quantitative precipitation test indicate a sharp decrease of K45 per unit amount of proteins with the culture day (Fig. 3a). To ascertain how much of this decrease is due to "selection",

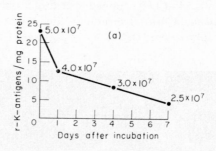

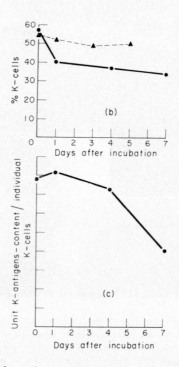

Fig. 3. Loss of K45 from the monolayer cultured kidney cells. From Okada.[7] The numbers in Fig. 3a represent the cell number necessary for preparing 1 ml of the DOC-extracts with ca. 1.0 mg protein each. K-antigens: K45 in the text.

a correction was made based upon the data obtained by the fluorescent staining method. Fortunately, anti-K45-FITC can stain, though weakly, MCK at 7 days, when the experiments were usually terminated. Changes in the percentages of the positive cells in MCK at the different stages of culture are given in Fig. 3b. Thus, an average content of K45 per individual positive cell can be calculated from the data given in both Figs. 3a and b, as is illustrated in Fig. 3c. An actual decrease in the antigens of the cells derived from the original K45-containing cells can be accepted from this result.

AN INCREASE OF "CULTURE" ANTIGENS (C-ANTIGENS)

Anti-MCK-FITC absorbed does not stain FDK, but does stain MCK. In order to see which fraction of the population of MCK contains this type of antigen to combine with the absorbed anti-MCK (C-antigens), 5-day MCK was divided into two aliquots, the one for staining with anti-CK, the other for anti-K45. The results indicate that at least 20% of MCK should contain both types of antigens simultaneously. Thus, an increase of C concomitant to a decrease of K45 in monolayer culture is not solely due to a selective change of the different cell types, but a regulative process in a single cell or in the cells derived from the same tissue types must be involved in this phenomenon.

TABLE 1. SPECIFICITIES OF THE 5-DAY CULTURED KIDNEY CELLS REVEALED
BY ANTIBODY STAINING METHOD

Experiment	Anti-K45-FITC		Anti-CK (absorbed with 0-day cells)	
	+	−	+	−
I	269	270	148	69
II	150	128	193	61
III	92	107	160	70
Total	511	505	501	200
	(50.1%)	(49.9%)	(72%)	(28%)

From Okada.[7]

There is no reason to assume C antigens as "culture-specific" and as actual "new" ones to appear only in MCK. C antigens are simply a mixture of antigens which are indetectable in the original tissue cells in a given technique. As to the chemical nature of these C antigens, the data suggest that Forssman type antigen is involved are now in our hands.[10] These were obtained from tests of the cytotoxicity of the antisera. Several kinds of rabbit antisera which were respectively prepared against different cellular fractions separated from the homogenate of adult chicken kidneys or of embryonic kidneys, were each added to the culture medium. They are not all strongly cytotoxic or not at all so within the range of the concentrations tested in the presence or absence of complement. On the other hand, anti-MCK was strongly cytotoxic even in the absence of complement. The cytotoxity of this antiserum was not "organ-specific"; lung cells of the embryonic chickens are equally sensitive. At the same time the antiserum has a hemagglutinin activity of very high titre for sheep red cells (SRB). An exhaustive absorption of anti-MCK with SRB completely removed its cytotoxicity. Thus, it may be surmised that such an antigen, perhaps Forssman antigen, that is common to SRB and elicits a cytotoxic antibody, may be increased in MCK. The statement confirms an earlier conclusion by Fogel and Sachs obtained with hamster kidney cultures.[3] FDK also should contain this type of antigen, because it is sensitive to the cytotoxic effect of anti-MCK.

CHANGES IN HISTOFORMATIVE CAPACITY

If FDK, instead of inoculating for monolayer-spreading culture, is allowed to reaggregate, reconstitution of the original architecture of kidney takes place in organ culture (Fig. 2e). The process was observed by means of the fluorescent antibody technique using anti-K45-FITC. The technique permitts us to follow the movement and displacement of the particular cell type (cells containing K45: K-cells) derived from the proximal convolution tubules. Sorting-out of these K-cells from other kidney cells (dark cells) starts after 24 hr of culturing, and small vesicles consisting of only the fluorescent cells were observed (Fig. 2d). In these vesicles the fluorescent cap was always directed toward the lumen.

MCK, 4 days old, in which K45 can no longer be detected on the cell surface but is diffusively distributed over the cytoplasm, is allowed to reaggreate. Sorting-out of the fluorescent cells from the dark cells does not occur and an amorphous mass without any distinct histological architecture was obtained after organ culture on top of the nutrient agar (Fig. 2f). From these results, we speculate that a localization of K45 antigen in the apical cytoplasm (including apical surface) of the K-cell may be a key principle for recognizing its counterpart.

SEPARATION OF SEMI-PURE K-CELL POPULATION
FROM THE WHOLE KIDNEY CELLS

For further study of the cellular changes occurring in MCK, it is a necessary prerequisite to start cell culture with a more homogeneous population consisting mostly of K-cells. This was carried out by means of differential trypsinization of the kidney fragments. The procedure is given in Fig. 4. The first dissociate (S_1) consists of wholly dark cells, when examined by the immunofluorescence method using anti-K45-FITC. Small cell clusters from about 10–20 cells are obtained in the most trypsin-resistant fraction R4. More than 90% of these clusters consist entirely of fluorescent cells.

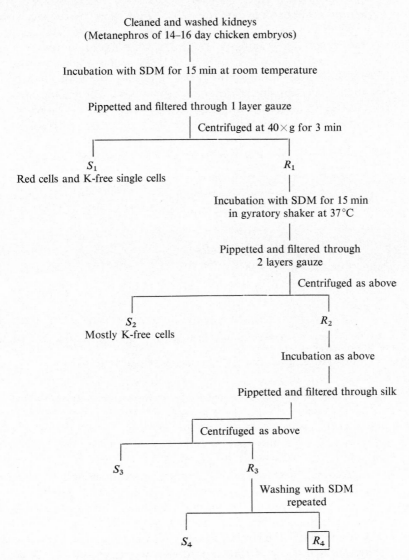

SDM: 0.15% trypsin and 20 mg collagenase in 100 ml CMF with 3 ml fowl serum.

FIG. 4. Flow-sheet of separation of tubular cell populations from freshly dissociated embryonic kidney cells.

EFFECT OF ACTINOMYCIN ON CELLULAR CHANGES
IN CULTURED KIDNEY CELLS

The results summarized on pages 271–2 show that the cellular changes occurring in the monolayer cultured cells involve, besides the decrease of K45, an increase in C-antigens. There are some indications which suggest that an increase in Forssman antigens in MCK is due to an active synthesis of these molecules.[11] If a synthesis of C antigens is stimulated in MCK, an inhibition of the synthetic process occurring in these cells might cause a prolonged preservation of the original specificity expressed by the presence of K45. To examine this possibility the effect of actinomycin (ACM) was studied.

TABLE 2. EFFECT OF ACTINOMYCIN ON THE LOSS OF K45-ANTIGEN

Days after culturing†	3-day	6-day	9-day		
			No treatment	ACM-treatment‡ at 3-day	ACM-treatment‡ at 6-day
No. of positive colonies { Exp. 1	46/50	8/12	Mostly negative	8/9	9/15
Exp. 2	31/32	8/15		8/12	8/16
% of positive colonies	92.4	43.2	nearly 0	76.8	57.6

† Culture medium: Ham's F10 supplemented with 5% calf serum and Coon's L-fraction of embryo extract.
‡ 1.5 µg ACM per plate (with 3 ml medium) for 30 min.

R4-fraction which was obtained from 14–15-day chicken embryonic kidneys and mostly consisting of the clusters of K45-containing cells, was used for this particular study. Approximately 1000 clusters of R4 were plated on each dish with 6 cm diameter. After 3 days, when each cluster was transformed into a colony of about 20–30 spreading epithelial cells, actinomycin-S was added at a concentration of 1.5 µg per dish containing 3 ml of culture medium. Treatment lasted for 30 min at 37°C and the cultures were transferred into fresh medium after thorough washing. Some pilot experiments showed that such a treatment inhibits an incorporation of H³-uridine into RNA to about one fourth the level of the control, when revealed with autoradiography. The specificity of thus treated cells was examined in a 5-day culture by means of the fluorescent staining with anti-K45-FITC. The results are summarized in Table 1. The 3-day colonies, which were exposed to ACM-effect, were mostly K45-containing ones (the positive colonies), and each colony consisted of the positive cells. In the control 6-day cultures, colonies consisting of thin and well-spreading cells, were all negative. On the other hand, many of the ACM-treated colonies contained positive cells, particularly in their central parts. No sign of deterioration was witnessed in these treated cells.

SUMMARY AND CONCLUSION

Changes in the differentiative traits in the cells spreading in monolayers were followed in the primary culture of the embryonic chicken kidneys. The "kidney-specific" antigens (K45) separated from the deoxycholate-soluble fraction of kidney microsomes were adopted for markers. When the cultured cells start to spread, K45 distributes over the cytoplasm and disappears from the cell surface, while in the intact tissue cells, the antigen is localized along the inner border of the cells of the proximal tubules including their surfaces. Antigen content per cells becomes gradually decreased thereafter.

Concomitant with these changes, another unidentified group of antigens (C) was increased during the first several week of culture. The results from the cytotic effect of the antiserum against the spreading cells suggest that some of these antigens are non-specific (in both tissue and species characters) and may probably be Forssman-type antigens.

These antigenic changes are paralleled with the loss of the histoformative capacity of these cells. The cells in which an oriented distribution of K45 in their original apical parts were lost, did not reconstitute a tubular tissue organization any more.

As revealed by immunofluorescence observations K45 is present only in the members of the proximal tubules. The method for separating these "positive" tubular elements from other negative cell members of kidneys was described. Some preliminary experiments indicate that the treatment with actinomycin was effective in preserving K45 in the monolayer culture of these tubular elements.

REFERENCES

1. E.H.DAVIDSON, *Adv. in Genetics*, vol. **12**, p. 144. Ed. by E.W.Caspari and J.M.Thoday, Academic Press, 1964.
2. E.WEILER, *Exptl. Cell Res.*, Suppl. **7**, 244, 1959.
3. M.FOGEL and L.SACHS, *Exptl. Cell Res.*, **34**, 448, 1964.
4. T.S.OKADA, and A.G.SATO, *Exptl. Cell Res.*, **31**, 251, 1963.
5. T.S.OKADA, *J. Embryol. Exp. Morph.*, **13**, 285, 1965.
6. T.S.OKADA, *J. Embryol. Exp. Morph.*, **13**, 299, 1965.
7. T.S.OKADA, *Exptl. Cell Res.*, **39**, 591, 1965.
8. T.S.OKADA, and A.G.SATO, *Nature*, **197**, 1216, 1963.
9. T.S.OKADA, *Exptl. Cell Res.*, **33**, 584, 1964.
10. T.MATSUZAWA, *Ann. Meeting Jap. Zool. Soc.*, 1965 (oral communication).
11. L.SACHS, In *New Perspectives in Biology*. Ed. by M.Sela, Elsevier, Amsterdam, 1964.

DIFFERENTIATION
OF THE CELL-FREE MATERIALS
WITH AGGREGATE-FORMING ACTIVITY
FROM EMBRYONIC CHICK LIVER CELLS*

YUKIAKI KURODA

National Institute of Genetics, Misima, Japan

UNDER suitable conditions trypsin-treated dissociated embryonic cells tend to adhere to each other giving rise to a histoformative structure. A rotation-mediated culture permitted analysis of the process of aggregation under controllable, replicable, and potentially quantitative conditions.[1]

In embryonic chick retina cells a progressive age-dependent decline in the aggregation index of the populations was found.[2,3,4] An age-dependent decline in cohesiveness of dissociated cells was also observed in embryonic chick liver cells[5,6,7] and lung cells.[8] This decrease in cohesive competence of cells from older embryos may reflect differentiation-accompanying changes in cellular biosynthetic patterns.

This paper deals with age-dependent changes in aggregability of dissociated liver cells, with the characteristics of the aggregate-forming activity in cell-free preparations obtained from liver cell cultures from younger and older embryos and with an attempt to induce the aggregation of liver cells from the older embryo by the cell-free preparation from the younger embryo.

MATERIALS AND METHODS

Methods for obtaining viable cell suspensions of embryonic chick liver and for obtaining aggregation of dissociated cells in rotation culture have been previously described.[1] Livers from 7-, 8-, 10-, 14-, 18-, and 20-day chick embryos were dissociated by treatment with 1% trypsin solution in CMF (calcium-, and magnesium-free Tyrode's solution) for 15 min at 37°C, and by flushing 10–15 times through the tip of a fine pipette. The culture medium consisted of Eagle's basal medium supplemented with 1 mM L-glutamine, 10% bovine serum, 2% chick embryo extract, and penicillin-strep-

* This investigation was supported by research grant RF-00058 from the National Institute of Health, United States Public Health Service.

277

tomycin at concentrations of 50 units ard 50 μg per ml medium, respectively. After filtration through a Swinney filter with a triple thickness of lens papers, cell suspensions each containing about 3×10^6 cells in 3 ml standard culture medium were rotated in 25 ml Erlenmeyer flasks on a gyratory shaker at 70 rpm at 38 °C for 24 hr.

Cell-free supernatants were prepared as follows. Dissociated liver cell suspensions from 7-day and 18-day chick embryos were shaken in a reciprocating shaker at 100 rpm for 2 hr at 38°C. These cell suspensions were centrifuged at 3000 rpm for 10 min and the supernatants filtered through a Swinney filter with a millipore filter to remove cells and cellular debris. Filtrates were tested for their effect on cohesion of freshly dissociated liver cells in a gyratory shaker at 70 rpm at 28°C for 24 hr.

RESULTS

Age-dependent Aggregation of Liver Cells

Dissociated liver cells obtained from a 7-day embryo formed one or two large spherical aggregates with smooth surfaces after 24 hr of cultivation in a gyrating flask. Liver cells dissociated from progressively older embryos

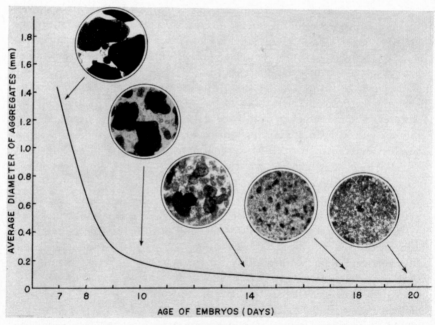

FIG. 1. Relationship between age of embryos and the average diameter of the cell aggregates. Twenty-four-hour cultures in standard medium rotated at 70 rpm at 38 °C. Explanations in text.

gradually lost their aggregability and formed smaller and numerous aggregates under the same culture conditions. Dissociated liver cells from 20-day embryos lost their aggregability almost completely and a large number of cells remained free as single cells after 24 hr of rotation culture. The relationship between age of embryos and average diameter of the cell aggregates is shown in Fig. 1.

On microscopic examination of stained sections aggregates obtained from 7-day embryonic liver cells were seen to consist predominantly of hepatic parenchyma; a sinusoid was formed with some cells arranged around it (Fig. 2a). Higher magnification showed that there is a lumen surrounded by several parenchymal cells in the aggregate (Fig. 2b). In aggregates formed from a 10-day embryonic liver cells parenchymal and sinusoid structures were still observed, although the mutual cohesion of each cell appeared to be looser than aggregates of 7-day embryonic liver cells (Fig. 2c). The reconstruction of the parenchymal structure was observed in aggregates obtained from 14-day embryonic liver cells. Histological observation of 24-hr rotation culture of 18-day embryonic liver cells showed that numerous minute cell clusters consisted of a few cells while many cells remained free as single cells (Fig. 2d).

To determine whether the aggregation-promoting factor acts on liver cells from younger embryos or the aggregation-inhibiting factor acts on liver cells from older embryos, mixed suspensions of liver cells from younger and older embryos were rotated. Aggregates obtained from mixed suspensions of 7-day and 18-day embryonic liver cells showed to have an average diameter intermediate between aggregates obtained from liver cells of each age alone of the embryos. Histological examination on relatively large aggregates obtained from such mixed suspension showed that the aggregates consisted predominantly of presumed 7-day embryonic liver cells and a few scattered 18-day embryonic liver cells (Fig. 2e). Difference in the size of the cells and the staining feature of cytoplasm were observed between both cells from 7-day and 18-day embryos.

These facts indicate that the aggregation-promoting factor, even if it is present in the liver cells from 7-day embryo, may not act on those from 18-day embryo, and that the aggregation-inhibiting factor, even if it is present in the liver cells from 18-day embryo, may not act on those from 7-day embryo.

Cell-free Supernatant with Aggregation-promoting Activity

At 28°C freshly dissociated liver cells of a 7-day embryo showed a reduced aggregation pattern after rotation culture under otherwise the same conditions (Fig. 2f). When the cells were rotated at 38°C for 2 hr, transferred to 28°C, and rotated for 22 hr, they formed relatively large aggregates. Cell-free supernatant was prepared from 2-hr reciprocation cultures at 38°C of 7-day embryonic liver cells and tested for its activity on aggregation of freshly dissociated liver cells from 7-day embryo at 28°C. The cell-free supernatant thus obtained showed to have an activity of considerably enhancing aggregation pattern (Fig. 2g).

Histologically the aggregates consisted of hepatic parenchyma (Fig. 2h) which appeared to be a comparable structure to that seen in aggregates formed at 38°C without the supernatant. The supernatant prepared from 7-day embryonic liver cells had an aggregation-promoting activity at 28°C

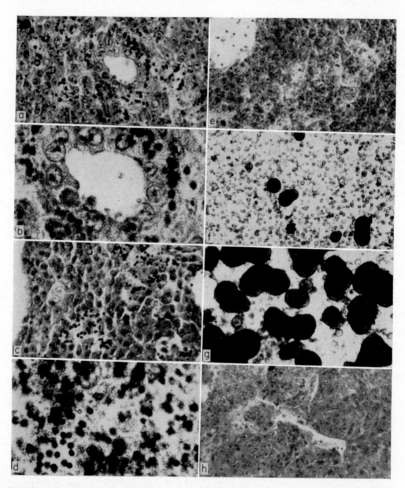

FIG. 2. a. Section of an aggregate of 7-day embryonic liver cells. ×300. b. Higher magnification of the section shown in Fig. 2a. ×750. c. Section of an aggregate of 10-day embryonic liver cells. ×300. d. Section of cell clusters of 18-day embryonic liver cells. ×750. e. Section of a co-aggregate formed from a mixed suspension of liver cells of 7-day and 18-day embryos. ×300. f. Aggregates of 7-day embryonic liver cells rotated at 28 °C for 24 hours. ×30. g. Aggregates of 7-day embryonic liver cells rotated at 28 °C for 24 hours with the supernatant prepared from 7-day embryonic liver cells. ×30. h. Section of an aggregate shown in Fig. 2g. ×300.

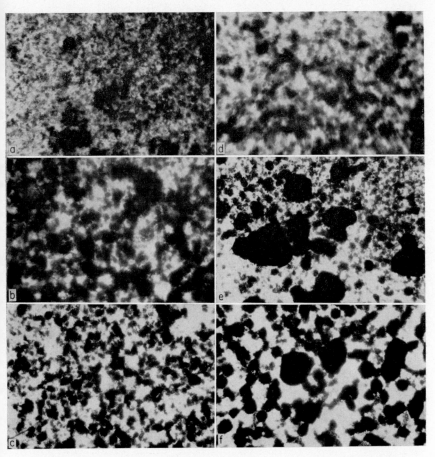

Fig. 3. a–c. Control cultures of 18-day embryonic liver cells in the standard culture medium at 38 °C after one day (a), three days (b), and five days (c). d–f. Cultures of 18-day embryonic liver cells in the supernatant prepared from 7-day embryonic liver cells at 38 °C after one day (d), three days (e), and five days (f). ×30.

on freshly dissociated liver cells from a 7-day embryo, but it had no effect on 18-day embryonic liver cells.

The supernatant prepared from 18-day embryonic liver cells had no effect at 28°C on liver cells from 7-day or 18-day embryos. These facts indicate that the liver cells from older embryos lost their competence to react on the active supernatant and that the supernatant prepared from older embryos lost its activity on the competent cells.

Induction of the Aggregation by the Supernatant

Eighteen-day embryonic liver cells were incubated with the supernatant obtained from 7-day embryonic liver cells on a gyratory shaker at 38°C for several days. Some small cell clusters were observed after one day and they grew well to fairly large aggregates after three days (Fig. 3). Histological examination revealed that in the aggregates some cell arrangement forming the parenchymal structure was observed. After five days in continuous rotation culture, relatively compact aggregates were formed. In control cultures in which 18-day embryonic liver cells were rotated in the standard culture medium, no such aggregates were observed even after five days of rotation culture.

Chemical Composition of the Supernatant

Some chemical compositions of the supernatant obtained from 7-day embryonic liver cells were determined. Cell-free supernatants were prepared from reciprocation culture of liver cells in Tyrode's solution after shaking at 38°C for 2 hr. This supernatant prepared in Tyrode's solution showed to have an aggregation-promoting activity on 7-day embryonic liver cells at 28°C.

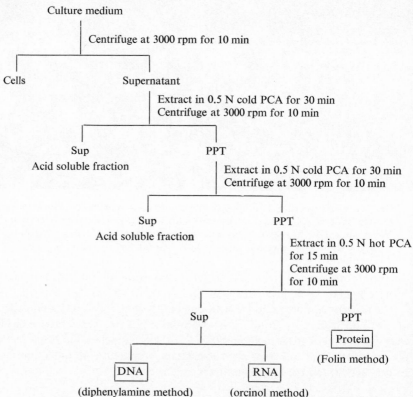

FIG. 4. Procedure showing chemical determination on nucleic acids and protein in the cell-free supernatant prepared from reciprocation cultures of liver cells.

UV absorption spectrum of the supernatant had a peak at 260 mμ, suggesting that the supernatant contains nucleic acids or their derivatives.

Nucleic acids and protein in the supernatant were chemically determined as shown in Fig. 4. After removal of cold 0.5 N perchloric acid soluble fraction, the nucleic acid fraction was extracted at 90°C for 15 min. DNA was determined by a modification of Burton's method[9] with diphenylamine. RNA was determined by Mejbaum's method.[10] Results of chemical determination on nucleic acids and protein in the supernatants from five independent experiments showed that averages 27 μg DNA, 143 μg RNA and 588 μg protein were contained in the supernatant prepared from the culture of 10^6 cells per ml Tyrode's solution. Sugars and other components remained undetermined.

DISCUSSION

Cell-free preparations with aggregation-promoting activity were isolated from living sponge cells[11] and embryonic chick retina cells.[2] The aggregation-promoting preparations from sponge cells had a selective activity on cells of different species, and some quantitative differences in amounts of carbohydrates and in amino acid composition were found in the preparations from different species of sponges.[12] It was found that the aggregation-promoting preparations from embryonic chick retina cells contained protein, hexosamine, and hexuronic acid.[2] In the present experiment the cell-free supernatant with aggregation-promoting activity prepared from 7-day embryonic liver cells contained relatively large amounts of RNA and protein, and a small amount of DNA. This small amount of DNA detected may be derived from degradation of cells during reciprocation culture. Which components of the supernatant are responsible for aggregation-promoting activity requires further study.

It was found that precultivation of dissociated cartilage cells as monolayer reduced progressively the tendency of cells to aggregate, accompanying a decline in the chondrogenic activity of the cells.[13,14] This decrease in aggregability of the monolayer-grown cells may be related to change in the aggregation-promoting activity in the supernatant as prepared in the present experiment and it may contain some mechanism common to the age-accompanying decline in aggregability of liver cells.

SUMMARY

1. Liver cells dissociated from chick embryos at varying ages from 7-day to 20-day were tested for their aggregability in rotation cultures. An age-dependent decline in cohesiveness of progressively older embryonic liver cells was observed.

2. Mixed suspensions of liver cells from younger and older embryos produced aggregates of an average diameter intermediate between aggregates obtained from the cells of each age alone of both embryos.

3. Cell-free supernatant prepared from reciprocation culture of 7-day embryonic liver cells had an activity in enhancing aggregation of 7-day embryonic liver cells at 28°C, but it had no effect on those from 18-day embryonic liver cells. The supernatant from 18-day embryonic liver cells had no effect on both liver cells from 7-day and 18-day embryos.

4. The supernatant prepared from 7-day embryonic liver cells showed an activity of inducing the formation of aggregates from 18-day embryonic liver cells by rotation culture at 38°C for three or more days.

5. Chemical determination of the cell-free supernatant prepared from 7-day embryonic liver cells revealed that the supernatant contained relatively large amounts of RNA and protein with a small amount of DNA.

ACKNOWLEDGEMENTS

The author thanks Miss T. Nagatani for her assistance in much of this work.

REFERENCES

1. A. MOSCONA, *Exptl. Cell Res.*, **22**, 455, 1961.
2. A. MOSCONA, *J. Cell. Comp. Physiol.*, **60**, Suppl. 1, 65, 1962.
3. A. MOSCONA, In *Cells and Tissues in Culture*, vol. 1. Ed. by E. N. Willmer, Academic Press, New York, 1965.
4. A. MOSCONA and M. H. MOSCONA, *Exptl. Cell Res.*, **41**, 697, 1966.
5. A. MOSCONA, In *Growth in Living Systems*. Ed. by M. X. Zarrow, Basic Books, New York, 1961.
6. T. NAGATANI and Y. KURODA, *Japan. J. Genetics*, **40**, 407, 1965.
7. Y. KURODA and T. NAGATANI, *Japan. J. Genetics*, **40**, 402, 1965.
8. J. W. GROVER, *Exptl. Cell Res.*, **24**, 171, 1961.
9. K. BURTON, *Biochem. J.*, **62**, 315, 1956.
10. W. MEJBAUM, *Z. Phys. Chem.*, **258**, 117, 1937.
11. A. MOSCONA, *Proc. Nat. Acad. Sci.*, **49**, 742, 1963.
12. E. MARGOLIASH, J. R. SCHENCK, M. P. HARGIE, S. BUROKAS, W. R. RICHTER, G. H. BARLOW and A. MOSCONA, *Biochem. Biophys. Res. Commun.*, **20**, 383, 1965.
13. Y. KURODA, *Exptl. Cell Res.*, **30**, 446, 1963.
14. Y. KURODA, *Exptl. Cell Res.*, **35**, 337, 1964.

DISCUSSION

M. E. KAIGHN: Your data show a differential response to your supernatant factor between cells isolated from 7-day as opposed to 18-day embryonic liver, i.e. there is a lag period of several days before the older cells are stimulated to aggregate. Do you feel this difference in response reflects the capacity of the cells to grow at the time of their isolation? Have you measured their respective growth rates?

Y. KURODA: Difference in the growth rates between the 7-day and 18-day embryonic liver cells may be a possible explanation for the difference in response of the both cells. In monolayer culture the cells from 7-day embryo showed a higher growth rate than those from 18-day embryo. In rotation culture, however, the difference in the growth rate between the 7-day and 18-day embryonic liver cells has not yet been measured. Another possibility to explain the differential response of both cells is that the 18-day embryonic liver cells may require new synthesis of the aggregation-promoting factor for the formation of aggregates, while the 7-day embryonic liver cells may not require such new synthesis of the factor.

J.D.EBERT: (1) Do all liver cells behave alike in your system?

(2) Do you argue that the "chemical factors" apparently necessary in experimental reaggregation also operate in normal development?

(3) Where in the architecture of the cell surface (considering the fine structure of interacting cells) do you believe the factor(s) to "fit"?

Y.KURODA: (1) In aggregates obtained from the 7-day embryonic liver, parenchymal cells formed a typical arrangement of the parenchymal or sinusoid structure. Endothelial cells were located on the surface of the parenchymal structure. Blood cells were usually adhered to each other and formed blood cell clusters or scattered to fulfill the lumen in the aggregates. Age-dependent aggregation patterns may reflect the aggregation of parenchymal cells of which the majority of liver cell population consists.

(2) It is a suggestive activity of the "chemical factor(s)". The fact obtained in the induction experiment may suggest the operation of the "factor(s)" in normal development. Other experiments on the activity of the "factor(s)" in the development are under investigation.

(3) Although I have no experimental data to show the precise "site" of the "factor(s)", the "factor(s)" contains substances which were proved to be present in the intercellular space.

A. DI MARCO: Have you observed any effect of basically or negatively charged substances on aggregation and surface charge of liver cells?

Y. KURODA: It is well known that some divalent cations are required for the formation of aggregates. However, I have no evidence that such divalent cations affect directly the surface charge which may be related to the aggregation.

ORGAN CULTURE OF THE RAT PANCREAS ANLAGE AS A MODEL FOR THE STUDY OF DIFFERENTIATION: I. TECHNIQUES AND COMPARISON OF *IN UTERO* (EMBRYONIC) AND *IN VITRO* (ORGAN CULTURE) MORPHOGENESIS*

Ismail Parsa, Walton H. Marsh and Patrick J. Fitzgerald

Department of Pathology, State University of New York,
Downstate Medical Center, Brooklyn 3, New York, 11203

THE pancreas anlage removed from the day 13 rat embryo has been shown by Chen to continue morphogenesis and differentiation in culture medium.[1] Grobstein and associates[2] have carried out extensive studies of the development of the mouse pancreas anlage in the *in vitro* organ culture system.

We have found a parallel histogenesis and cytodifferentiation between the acinar cell of the embryonic pancreas kept *in utero* and the acinar cell of the pancreas anlage removed from the day 13 embryo and grown in the *in vitro* organ culture. The *in vitro* model may be of use in studies involving repression and de-repression of acinar and islet cell differentiation.

METHODS AND MATERIALS

Female rats, 200 to 210 g, and male rats, 300 to 400 g (Carworth Farms, CFN), were mated. Copulation was witnessed, immediately after, the rats were separated and the time of copulation was taken as zero time for the estimation of the age of the embryo. A consistent weight gain was considered to be evidence of pregnancy. Each female rat was kept separated from male rats for at least 20 days before remating if pregnancy did not occur after a mating.

The organ culture medium consisted of Eagle's basal media[3] 87%, horse serum (Grand Island Biological Co.) 10% and chick embryo extract (Grand Island Biological Co.) 3%. To each 1000 ml of medium, 50 mg of pyruvate,[4] 30 mg of methionine and 36 mg of serine[5] were added.

* Aided by U.S.P.H.S. grants No. 5 ROI-AM 05556.

In the organ culture a 0.45 micron pore size filter, (Millipore Co.) was cemented by Millipore cement to form the bottom of an autoclavable Plexiglass (General Electric Co.) ring, 5/8 in. outside diameter, 1/2 in. inside diameter. Teflon (DuPont Co.) dishes containing 1 ml of medium were used to house the plexiglass ring with filter in a well. The pancreas anlage was placed on the bottom of the millipore filter in the ring and immersed in medium (Fig. 1).

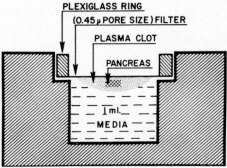

Fig. 1. Culture assembly, schematic diagram.

One ml of chicken plasma and 0.5 ml of chick embryo extract were used to clot the anlage on the millipore filter. Each Teflon block contained 4 to 6 wells. The block was placed in a glass petri dish and the assembly kept in an incubator room at 37.5 °C.

For histologic preparation embryonic pancreas was fixed in a 10 % aqueous solution of formaldehyde (USP), phosphate buffered at pH 7.4, Carnoy's solution (absolute alcohol, chloroform, acetic acid, 6 : 3: 1 v/v), or in glutaraldehyde (Union Carbide), phosphate buffered at pH 7.4. Histologic processing was standard for these fixatives.[6] Histology sections were cut at 2 or 4 microns and were stained with methyl green and thionine,[7] or hematoxylin and eosin.

In Utero Specimen

On each gestation day from day 13 through day 22, ten rats were sacrificed. Under the dissection microscope the dorsal pancreatic primordium was removed from five embryos of each litter and fixed for histologic examination.

In Vitro Organ Culture Specimen

Under ether anesthesia the uterine horn of the 13-day pregnant rat was aseptically removed. The horn was rinsed with warmed Tyrode's solution (37 °C) to eliminate maternal blood and placed in warm horse serum. The embryo was delivered through a longitudinal incision along the uterine horn and transferred to a petri dish containing warmed horse serum. Under the dissecting microscope using transillumination the dorsal rudiment of the pancreas with its mesenchyme was carefully separated from the duodenum and stomach. The rudiment was placed on a warm millipore filter and the excess serum was drained off. Two drops of warmed clotting media were added to the anlage. After formation of the clot the Plexiglass ring with Millipore filter and anlage were seated in the Teflon block with the explant exposed to the medium (Fig. 1). The culture was incubated at $37° \pm 1°$ (C) in an atmosphere of 95% air and as 5% CO_2 saturated with water vapor. The media and atmosphere were changed daily.

At least 30 explants per day for each culture day, from day 1 through day 9 of culture (analogous to days 13 through 22 of *in vivo* embryogenesis), were fixed for histologic processing and light microscopy examination.

RESULTS

In Vivo Embryogenesis

Under the dissecting microscope, the transilluminated dorsal rudiment of the day 13 pancreas anlage was recognizable as a pale yellow–tan duodenal bud adjacent to the stomach. It was surrounded by grey mesenchyme. The smaller ventral pancreas rudiment was located ventrally and caudally.

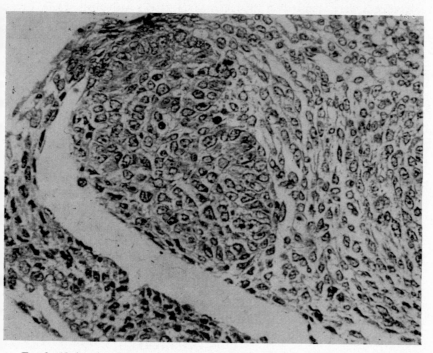

Fig. 2. 13-day dorsal pancreas anlage. Epithelial cells are compact and contain more and heavier staining cytoplasm than the loosely arranged mesenchymal cells. Hematoxylin and eosin. $\times 500$.

Histologic section of the pancreas anlage showed foci of relatively tightly packed epithelial cells with irregular, polyhedral nuclei and a slight amount of cytoplasm, surrounded by more loosely arranged mesenchymal cells with elliptical nuclei and very little cytoplasm (Fig. 2). Mitotic figures were present and more frequent than at later days. The epithelial cells had more abundant, deeper staining cytoplasm, whereas the mesenchymal cells presented very little, pale staining cytoplasm.

Scattered throughout the anlage were clusters of cells with abundant, slightly eosinophilic, vacuolated cytoplasm, spherical nuclei and indistinct cell boundaries. They appeared to be precursors of the islet cells.

On day 14 the epithelial bud was considerably enlarged and its border appeared indented. The lobulation was superficial and on cross section represented a focal outpouching of the epithelium. The lobules consisted of clusters of cells with elliptical to oval nuclei and small amounts of cytoplasm. The nuclear–cytoplasmic ratio was slightly decreased, compared to the previous day, because of the increase of cytoplasm. With the methyl green–thionine stain more cytoplasmic basophilia was discernible than previously. The mesenchyme although compact at the center and in between the lobules, appeared looser at the periphery. A few capillaries containing nucleated red blood cells were seen within the mesenchyme. Cells with eosinophilic vacuolated cytoplasm and spherical nuclei (islet cells?) were present at the periphery of the tubules, along the tubules and singly between individual epithelial cells lining the tube. The dorsal pancreas was still attached to the duodenal wall, although the lumen of its stalk did not communicate with the duodenal lumen.

On day 15 the lobulation of epithelium was more extensive and multiple large lobules were present. In the histologic section they represented well-defined clusters of epithelium about 2 cell layers thick around a small lumen. The nuclei were elliptical to round, basally located, and contained large prominent nucleoli. The epithelial cell cytoplasm was extensive and moderately basophilic. At that time a small lumen was present in the epithelial clusters.

On day 16 the coarse lobules of the previous day had become finely lobulated. Histologically, day 16 was marked by the predominance of a tubular pattern (Fig. 3, A and B). Ramifying smaller tubules were seen. The tubules were lined by 1 to 2 cell layers of epithelium. With the increase of the lumen size, the epithelial cells became somewhat flatter with a decrease in the relative amount of apical cytoplasm. The narrow rim of apical cytoplasm on day 16 contained an eosinophilic substance and there was the suggestion that ill-defined eosinophilic granules were present. The epithelial cell nuclei were mostly round to oval and contained large nucleoli. In the methyl green–thionine stain, increased basal cytoplasmic basophilia was present.

The foci of islet cells became larger and were still contiguous with the acinar cells.

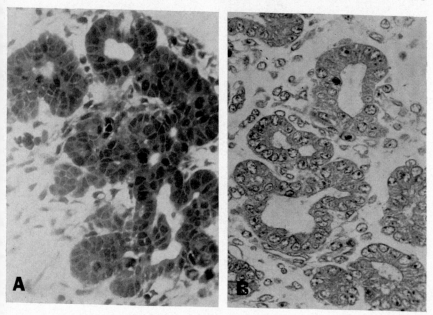

FIG. 3. A. 16-day embryonic pancreas *in utero*. B. Pancreas anlage removed from 13-day embryo and cultured *in vitro* for 3 days. Tubules lined with epithelium are prominent features in both systems. Hematoxylin and eosin. × 500.

During the next 3 days of gestation acinar structures were formed gradually by a focal ballooning out of epithelium from the tubules. On these days the acinar cells had only a narrow rim of apical cytoplasm free of zymogen granules which otherwise were abundant and easily recognizable within the cytoplasm. On day 17 the acinar cell cytoplasmic zymogen granule collection was heavy enough so as to be seen under the dissecting microscope as indistinct, relatively opaque cytoplasm. Islet vascularity became apparent at days 17 and 18. By day 19 acini were well formed (Fig. 4, A and B).

The changes during the last 3 days of gestation consisted mostly of increases in the acinar cell cytoplasm, in its cytoplasmic basophilia and in a greater accumulation of zymogen granules such as shown on day 22 (Fig. 5, A and B).

Roughly similar stages of morphogenesis occurred in the organ culture of the pancreases at the comparable day of gestation *in utero*, i.e. day 2 in culture (day 13 of anlage plus 2 days in culture) was similar in development to the day 15 *in vitro* embryo. At days 19 through 22 there appeared to be more zymogen granules in the cytoplasm of the *in vitro* organ culture acinar cells than in the *in utero* acinar cells. However, the overall growth of the explant was much less than in the *in utero* growth. The average wet weight of the pancreatic rudiment on day 13 was about 11.8 ± 6.1 µg (24 samples), crudely

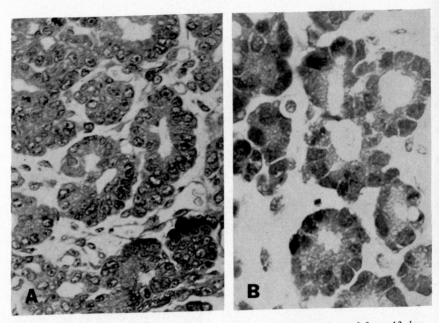

FIG. 4. A. 19-day embryonic pancreas. B. Pancreas anlage removed from 13-day embryo and cultured *in vitro* for 6 days. In both models acini are formed. Zymogen granules can be seen towards the apical portion of the acinar cells. Hematoxylin and eosin. ×500.

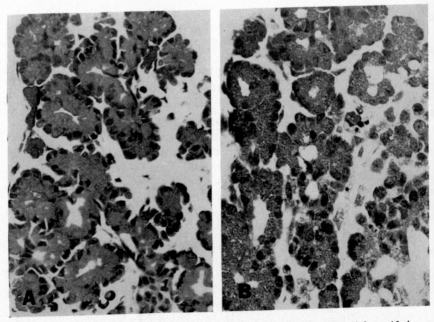

FIG. 5. A. 21-day embryonic pancreas. B. Pancreas anlage removed from 13-day embryo and cultured *in vitro* for 8 days. Cells and acini in both examples resemble the adult pancreas acinar cells. Hematoxylin and eosin. ×320.

approximate because of the varying amount of mesenchyme present. It increased *in vitro* to an average wet weight of about $572 \pm 210 \, \mu g$ on day 22 (24 samples) in organ culture.

DISCUSSION

The similarity of morphogenesis between the *in utero* embryonic development and the *in vitro* organ culture is striking. There was a parallel, almost synchronous appearance in the two systems of cords of epithelial cells, followed by tubule, lobule and acinar development. There was variation from animal to animal on the same day and even some overlap from one day to another, but, in general, the differentiation was remarkably similar. Cytoplasmic basophilia, nucleolar size, nucleolar staining with thionine, and the appearance of cytoplasmic eosinophilic staining substance occurred at about the same time, early in development, in both systems and in both there

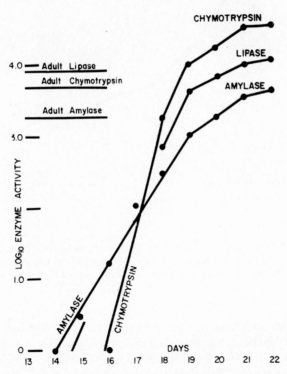

FIG. 6. Plot of pancreatic embryo enzyme activities during embryogenesis. Early lipase activity is an extrapolated one because of the lack of sensitivity of the method.

were recognizable zymogen granules at about day 17. Electron microscopy studies of the *in vitro* cultures of mouse pancreas have shown the presence of ribosomes at day 13, ergastoplasm at days 14 to 15 and zymogen granules at day 16.[8]

Our biochemical studies have shown that pancreatic amylase activity could be detected in the embryo at day 14 and chymotrypsin activity at day 16. Amylase detected on day 13 was of very low activity. Extrapolation of our day 18 findings with lipase suggests that lipase activity appeared at day 15–16 (Fig. 6).[9] Grobstein and colleagues had previously shown that amylase could be detected in the mouse at day 13 of embryogenesis and that there was a parallel increase of amylase production in the *in vivo* pancreas embryo and the *in vitro* organ culture of the day 13 anlage.[10]

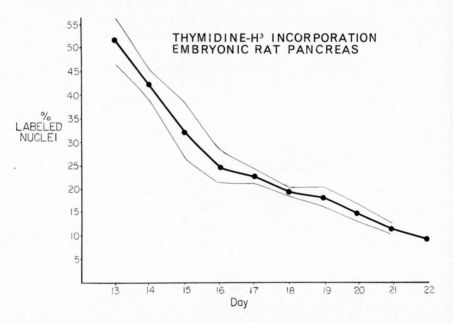

FIG. 7. Embryonic rat pancreas. Plot of the mean of the labeling indices with 95% confidence limits. The labeling index is the number of labeled acinar cell nuclei over the total number of acinar cell nuclei counted. Autoradiography with thymidine-H³.[11,12]

Autoradiographic labeling with thymidine-H³ [11,12] of epithelial cells early during embryogenesis and later labeling of recognizable acinar cell nuclei with thymidine-H³ was studied by us.[13] It was shown that about 50% of epithelial nuclei were labeled at day 13 in the embryo and this dropped to about 10% in the acinar cell at day 22 of embryogenesis (Fig. 7). Labeling of epithelium in the organ culture showed a parallel fall in labeling indices.[14]

Although our interest was primarily in acinar cell morphogenesis, similar parallel development of the islet cells appears to occur in both the *in utero* and *in vitro* organ culture systems. At days 13 or 14 there were noted cells both in the epithelium of the tubule, and alongside of it, which had more eosinophilic cytoplasm than the adjacent epithelium and often had a more spherical, smaller nucleus. These cells appeared to grow as buds along the sides of the tubules and in many sections seemed to be isolated in clusters. In a study of serial sections, however, one could not determine whether they were contiguous or continuous with other structures.

Use of the Model

Since organ culture permits the study of tissue free from the influences of a nervous system and of humoral factors except those present in the medium, or the effects of other organs, the parallel morphogenesis of the *in utero* and in the *in vitro* organ culture of the pancreas suggests that the organ culture may be used as a model for the study of the repression and depression both of acinar[2] and islet cell[15] morphogenesis. Similar days of appearance and parallel increases of the activities of amylase[10] and insulin[15] in the two different systems also indicate additional parameters which may be modified. The parallelism in the changes of acinar nuclear labeling as determined by thymidine-H³ autoradiography[13,14] also indicates that the *in vitro* pattern of DNA synthesis is representative of *in vivo* synthesis and hence might be of use in studying effects of compounds thought to be involved in the regulation of DNA synthesis.[16] It may be possible to add repressor or de-repressor substances to the medium and observe their effects upon the various parameters, morphologic or functional, of differentiation and thereby gain information concerning repression and de-repression as it operates during embryogenesis.

ACKNOWLEDGEMENTS

We are indebted to Dr. H. Clarke Anderson for helpful suggestions.

REFERENCES

1. J. M. CHEN, The cultivation in fluid medium of organized liver, pancreas and other tissue of fetal rats. *Exp. Cell Research*, **7**, 518–529, 1954.
2. C. GROBSTEIN, Cytodifferentiation and its controls. *Science*, **143**, 643–650, 1964.
3. H. EAGLE, Amino acid metabolism in mammalian cell cultures. *Science*, **130**, 432, 1959.
4. H. EAGLE and K. PIEZ, The population dependent requirement by cultured mammalian cells for metabolites which they can synthesize. *J. Exp. Med.*, **116**, 29–42, 1962.
5. R. Z. LOCKART, Jr., and H. EAGLE, Requirements for growth of single human cells. *Science*, **129**, 252–254, 1959.
6. J. F. A. MCMANUS, and R. W. MOWRY, *Staining Methods*, pp. 26–46, New York, P. B. Hoeber, Inc. 1960.
7. A. L. ROQUE, N. A. JAFAREY and P. COULTER, A stain for the histochemical demonstration of nucleic acids. *Exp. and Mol. Path.*, **4**, 266–274, 1965.

8. F. KALLMAN and C. GROBSTEIN, Fine structure of differentiating mouse pancreatic exocrine cells in transfilter culture. *J. Cell Biology*, **20**, 399–413, 1964.

9. W. H. MARSH, I. PARSA and P. J. FITZGERALD, Comparative enzymatic activity of the regenerating adult pancreas and the embryonic pancreas. *Proc. Amer. Assoc. Path. and Bact.* (abstract) *Am. J. Path.* (in press), 1966.

10. W. J. RUTTER, N. K. WESSELLS and C. GROBSTEIN, Control of specific synthesis in the developing pancreas. *Nat. Cancer Institute Monograph* No. 18, 51–61, 1964.

11. P. J. FITZGERALD, M. L. EIDINOFF, J. E. KNOLL and E. B. SIMMEL, Tritium in radioautography. *Science*, **114**, 494–498, 1951.

12. H. J. TAYLOR, P. S. WOODS and W. L. HUGHES, The organization and duplication of chromosomes as revealed by autoradiographic studies using tritium-labeled thymidine. *Proc. Nat. Acad. Sci.*, **43**, 122–128, 1957.

13. I. PARSA, W. MARSH and P. J. FITZGERALD, Nuclear labeling and tissue radioactivity (thymidine-H^3) of the embryonic rat pancreas during embryogenesis. *Fed. Proc.*, **25**, 535, 1966.

14. I. PARSA, W. H. MARSH, L. ROSENSTOCK and P. J. FITZGERALD, DNA synthesis of embryonic rat pancreas acinar cell *in vivo* and in organ culture. *Amer. J. Path.* (Abstract) (in press), 1967.

15. W. J. RUTTER, Ontogeny of specific proteins during pancreatic development. *Proc. Nat. Acad. Sci.* (abstract). *Science*, **150**, 383, 1965.

16. F. JACOB, and J. MONOD, Genetic regulatory mechanisms in the synthesis of protein. *J. Mol. Biology*, **3**, 318–354, 1961.

ESTABLISHMENT
OF POLAR ORGANIZATION
DURING SLIME MOLD DEVELOPMENT

Iкυο Takeuchi*

Department of Biology, Osaka University, Toyonaka, Osaka, Japan

THE fruiting body of the cellular slime mold, *Dictyostelium discoideum*, consists of a cellular stalk supporting a mass of spores. Upon germination each spore releases a small amoeba. These amoebae undergo a vegetative growth phase, during which they feed and multiply. Following this, there is an interphase period in which certain cellular changes occur. This is followed by the aggregation of many amoebae into cell masses, which eventually assume slug shape. The cell mass, which is called a slug, migrates over the substratum and finally rises into the air to form a fruiting body (Fig. 1).

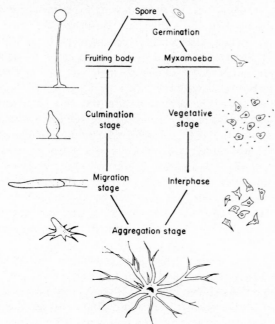

FIG. 1. Life cycle of *Dictyostelium discoideum*.

* Present address: Department of Botany, Kyoto University, Kyoto, Japan.

10a NAM

The slug, which is about 1 mm long, is known to have a definite polarity along its axis. This is shown by its polar nature of the movement and also by the fact that grafting is possible only between the fragments of the same polarity. Polar organization of the cell mass is important in its future differentiation. The anterior part of a slug or any fragment of it always differentiates into stalk cells, while the posterior part differentiates into spore cells (Raper, 1940). Furthermore, when the cells are separated from the cell mass, they differentiate into neither of them. The question to be raised here is how such polar organization is established during the process of cell aggregation and the formation of the cell mass.

Several years ago we made some immunohistochemical studies on this organism, using fluorescein conjugated antispore sera. The results, which will not be presented in detail here, indicate the following (Takeuchi, 1963). The preaggregation cells showed a considerable variation in the amount of combining groups in their cytoplasmic granules capable of reacting with antispore sera, despite the fact that the cells are entirely totipotent in their developmental fate (that is, they could become either stalk cells or spore cells). These various cells are collected randomly into the early aggregation centers. The distribution of these cells in the late centers suggests that during the slug formation the cells are rearranged and sorted out according to their variation.

This possibility is supported by the following experiment concerning the reconstruction of slugs from the dissociated cells (Takeuchi and Sato, 1965). I should mention in passing that all the experiments in this report were conducted with a cell population derived from a single cell. The right group of slugs in Fig. 2 was composed of amoebae which had been labeled with ^{3}H-thymidine during the vegetative stage, while the left group of slugs was composed of unlabeled cells. The posterior third was cut out of many of the labeled slugs, and they were dissociated to free cells in 1 mM solution of EDTA containing 0.15 M NaCl (Takeuchi and Yabuno). After complete disaggregation $CaCl_2$ was added, and the cells were washed in standard salt solution (Bonner, 1947). The anterior fragments of the unlabeled slugs were treated in the similar way. Both of the dissociated cells were mixed together and placed on a non-nutrient agar plate. Some time later they aggregated and formed new slugs. These slugs were fixed in Carnoy and the distribution of the labeled cells within the slug was examined by using the autoradiographic technique.

The number of reduced silver grains in a certain area was counted in four different regions along the axis and plotted in per cent of controls in the anterior-most region. As shown in Fig. 2, more labeled cells were found in the posterior part of the reconstructed slugs. That is to say, the cells which had originally come from the posterior fragments were sorted out to the posterior during the process of slug reconstruction. This means that there is some difference in character between the anterior and posterior cells.

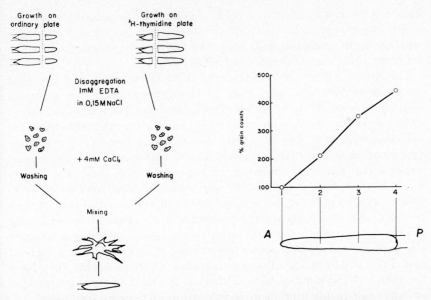

FIG. 2. Slug reconstruction from dissociated cells. Experimental procedures (left) and results (right). The number of reduced silver grains in a certain area was counted in four different regions indicated above and plotted in percent of controls in the anterior-most region. A and P denote the anterior and posterior part of the slug respectively.

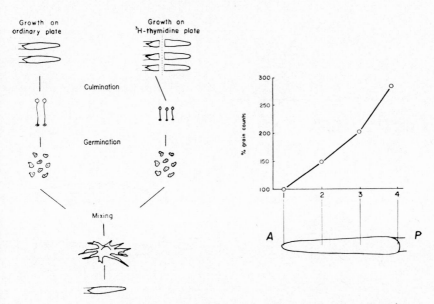

FIG. 3. Slug formation from germinated amoebae without undergoing vegetative growth. For explanation, see text and Fig. 2.

This difference may arise as a result of differentiation occuring in the slug, or else this may represent the cell variation which has already been present among the preaggregation cells. The following experiment shows that the latter is the case (Takeuchi and Sato, 1965).

The posterior fragments were cut out of the labeled slugs and incubated on a non-nutrient agar plate (Fig. 3). Through regulation they construct small but normal fruiting bodies. The spores were taken and allowed to germinate in the complete absence of food bacteria. On the other hand, the unlabeled slugs were allowed to form fruiting bodies without any operation. The amoebae germinated from these spores were mixed with the labeled amoebae and placed on a non-nutrient agar plate. After these cells formed slugs, they were fixed in Carnoy and their autoradiograms were examined as before. As shown in Fig. 3, the slugs contained more labeled cells in their posterior part. When the anterior fragments of the labeled slugs were used in similar experiments, the distribution of the labeled cells in the resulting slugs was just the opposite, more labeled cells being found in their anterior part. These results indicate that a certain kind of variation is present among the spores of the same clone, and that on the basis of this variation the amoebae germinated from them were sorted out when they form slugs.

The question arises as to how and when the cell variation necessary for the sorting-out is produced during the normal course of development. The following experiment indicates that this is produced by cell divisions during the vegetative growth period. Both the anterior and posterior fragments were

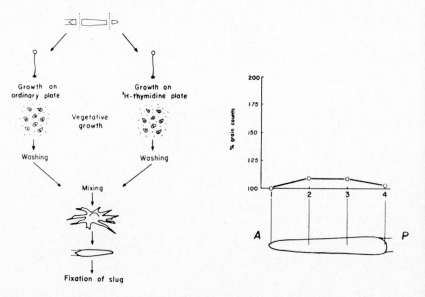

FIG. 4. Slug formation from germinated amoebae after undergoing vegetative growth. For explanation, see text and Fig. 2.

taken from the same slug, and each of them was allowed to construct a fruiting body (Fig. 4). The spores were inoculated in the presence of bacteria so that germinated amoebae might grow and multiply. For labeling, the spores of either fruiting body were grown on a medium containing ^{3}H-thymidine. After a period of growth, both the labeled and unlabeled cells were washed free from bacteria, and after mixing they were placed on a non-nutrient agar plate. The distribution of the labeled cells in the resulting slugs was examined as before. As shown in Fig. 4, the labeled cells were distributed uniformly throughout the slug. This result makes a contrast with the previous experiment of non-growing cells, and indicates that when a cell multiplies its progenies do not occupy the same relative position as the progenitor. Instead, it produces a battery of cells of different possibilities.

All these results support previous immunohistochemical results. By undergoing a number of cell divisions during the vegetative stage, a population of amoebae becomes considerably heterogeneous with respect to their cytoplasms. We do not know whether this variation resulted from a certain kind of unequal cell division occuring during the growth period, or represents differences in time of completing vegetative growth. We have some evidence which seems to rule out the second possibility, but we need more results to be certain. At any rate, during the process of aggregation these various cells are collected randomly into aggregation centers. When a slug is being formed from the center, the cells are sorted out according to certain properties of each amoeba and rearranged in a certain order within the slug.

If the above idea is correct, when the preaggregation cells are divided into two groups according to a certain variation, these groups of cells may be sorted out during the process of slug formation. The following experiment indicates this to be the case. The specific gravity of preaggregation cells was determined by centrifuging the cells in a density gradient of dextrin solution. It was found that the cells showed a considerable variation in their specific gravity, but the average value was estimated at 1.061* in one experiment. Both labeled and unlabeled cells were centrifuged in the dextrin solution of specific gravity 1.061 for 20 min at 150 g (Fig. 5). From the labeled cell population we took sedimented cells, that is the cells of higher specific gravity. Floating, that is lighter cells, were taken from the unlabeled population. After washing, both the cells were mixed and placed on a non-nutrient agar plate. The distribution of the labeled cells in the resulting slugs indicates that the heavier cells tended to be sorted out to the anterior position.

The above result also implies that as a result of such sorting-out, an anteroposterior gradient will be established along the axis of the slug depending on the specific gravity of the cells. A similar result has been obtained with ^{3}H-uridine incorporation (Sato and Takeuchi, 1965). It was found that during slug formation the cells were sorted out according to the activity of ^{3}H-uridine incorporation in the interphase period. This resulted in the produc-

* This value varies depending on culture conditions.

tion of a gradient within the slug by reason of the amount of incorporated uridine. In this way, the presence of cell variation among the preaggregation cells followed by this sorting-out of the cells during slug formation will lead to the establishment of many kinds of gradient along the axis of the slug. The gradient thus established by the sorting-out mechanism would in turn regulate the pattern of cell differentiation which will take place in the cell mass.

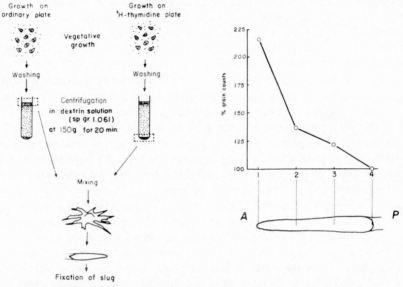

FIG. 5. Slug formation from lighter and heavier preaggregation cells. For explanation, see text and Fig. 2. In this case, the number of reduced silver grains was plotted in per cent of controls in the posterior-most region.

Now, apart from such an intrinsic factor, some extrinsic factors may also stabilize or amplify the gradient established by the sorting-out mechanism. Studies on the respiration of this organism during its development revealed that the slugs had a lower rate of respiration than the cells dissociated from them (Takeuchi and Mine). As indicated in Table 1, the Q_{O_2} of the dissociated cells amounted to about 150% of the Q_{O_2} of the slug. Probably the rate of diffusion of oxygen into the slug limits the rate of respiration of the cell mass. Recently, the rate of respiration of the anterior fragment of the slug was compared with that of the posterior fragment of the same slug by using a microtechnique. Our preliminary results showed that the Q_{O_2} of the anterior fragments was about twice that of the posterior fragments, as shown in Table 1. The anterior and posterior fragments were both dissociated and their rate of respiration was compared. Preliminary results indicate that there is no significant difference in the rate of respiration between the cells dissociated from the anterior and posterior fragments (Table 1).

TABLE 1. RELATIVE RATE OF RESPIRATION OF THE CELL MASS
AND ITS CONSTITUENT CELLS

	Relative rate of respiration %
Slug	100
Dissociated slug cells	153
Anterior slug fragments*	100
Posterior slug fragments*	60
Dissociated anterior cells*	100
Dissociated posterior cells*	105

Rate of respiration was compared between the two in each column.
* Rate of respiration was estimated on the basis of the number of cells.

It was then considered that the difference in the rate of respiration between the anterior and posterior fragments can be explained by the difference in their shape. As indicated by the diagrammatic drawing of the slug in Fig. 1, the anterior part has a higher surface to volume ratio than the rest of the slug, thus receiving a greater supply of oxygen. This in turn would result in a higher rate of respiration in the anterior region. Although we do not know what determines the shape of the slug, these results demonstrate the possibility that some extrinsic factor, such as the local difference in oxygen pressure within the cell mass, amplifies the original gradient established by the sorting-out mechanism.

ACKNOWLEDGEMENTS

The experiments reported in this paper were originally carried out in collaboration with Dr. T. Sato and Miss H. Mine of Osaka University, to whom I should like to express my sincere gratitude. Also I am indebted to Prof. N. Kamiya of Osaka University for his interest and encouragement during the course of this study. This study was supported by Grants-in-Aid from the Ministry of Education of Japan.

REFERENCES

J. T. BONNER (1947), J. Exp. Zool., 106, 1.
K. B. RAPER (1940), J. Elisha Mitchell Sci. Soc., 56, 241.
T. SATO and I. TAKEUCHI (1965), Jap. J. Exp. Morphol., 19, 105 (abstract, in Japanese).
I. TAKEUCHI (1963), Develop. Biol., 8, 1.
I. TAKEUCHI and H. MINE, in preparation.
I. TAKEUCHI and T. SATO (1965), Jap. J. Exp. Morphol., 19, 67 (in Japanese).
I. TAKEUCHI and K. YABUNO, in preparation.

DISCUSSION

E. ZEUTHEN: You have expressed the opinion that the respiration role in the thicker parts of the slug is limited by the diffusion of oxygen. Have you checked this idea by comparing Q_{O_2} of the anterior and posterior fragments at a series of oxygen tensions?

I. TAKEUCHI: No, we have not, but that is one of the things we would like to examine in the near future.

H. TERAYAMA: (1) I wonder if the cells in a posterior part might migrate into an anterior part or vice versa during the movement of the slug.

(2) I wonder what would happen if you repeat the fragmentation experiments. Do you expect the depolarization tendency or not?

I. TAKEUCHI: (1) It was shown by Dr. Raper many years ago that there is little mixing of the cells within the slug during a certain period of migration, except that a small number of cells move forward or backward, as noted by Dr. Bonner. This minor mixing may have something to do with the cell divisions occurring in the cell mass (though they are much less frequent than during the vegetative growth period). At any rate, we could say that the majority of cells retain their relative position within the slug while it migrates.

(2) I do not expect any depolarization tendency. Because, as I have shown, when a cell multiplies, it produces cells of all different tendencies. In other words, even when we take the cells from the posterior (or anterior) fragment, they will produce the same range of variation as observed in the original population after undergoing vegetative growth.

J. D. EBERT: Please describe the pattern of cell division within the slug. Is replication a requirement for differentiation?

I. TAKEUCHI: It has been observed that a wave of cell division starts from the anterior to the posterior part during the slug formation, and there are also some sporadic cell divisions occurring in the slug during the early part of migration. I do not think replication in the slug is required for cell differentiation. Dr. Bonner showed that the construction of a normal fruiting body from a slug fragment was not necessarily accompanied by cell division. In a mutant strain of Dr. Sussman, he reported normal morphogenesis of a cell population with no changes in the cell number under certain conditions.

F. H. KASTEN: A beautiful movie picture of streaming phenomena in slime molds was presented at a recent meeting of the American Society for Cell Biology. Since there are differences in the streaming pattern in different regions of the slime mold, can you tell us to what extent this phenomenon is observed in your material and how it might influence the DNA labeling pattern?

I. TAKEUCHI: I think you are talking about the *Myxomycetes* plasmodium, which is a kind of coenocyte and has no cell boundary in its inside. The cellular slime molds I talked about are entirely different organisms. They have cellular structure at any stage of development and the slugs are composed of a large number of cells. Therefore, such streaming pattern as can be seen in the *Myxomycetes* plasmodium is not observed in the slugs of these organisms.

PHYSIOLOGICAL
AND BIOCHEMICAL STUDIES
ON THE FRUIT BODY FORMATION
IN *BASIDIOMYCETES*

TOMOMICHI YANAGITA, YOKO TSUSUÉ,
FUSAÉ KOGANÉ and SHIZUYO SUTO

Institute of Applied Microbiology, University of Tokyo, Bunkyo-ku, Tokyo

FUNGAL hyphae grow only at their tips.[1] Such an apical dominance is one of the important characteristics of fungal growth. Another noticeable characteristic in fungal development is the formation of spore bearing organs at the rear of hyphal apices. Thus, the functioning site in a fungal hypha is located at close to its apex but not at the apical cell itself. Based on these facts, it may be considered that fungi are the prototypes of higher plants in their mode of growth and differentiation.

In view of the above considerations, we have recently started an investigation on the morphogenesis in the Basidiomycetes, *Coprinus macrorhizus* Rea f. *microsporus* Hongo, strain 708, supplied by Prof. K. Kimura of Okayama University. Fungi of this genus are known to respond to light and low temperature in fruit body formation as do the higher plants.

The dicaryotic strain, A^7B^7/A^8B^8, of this fungus was cultured, unless otherwise specified, on the potato-sucrose agar (pH 6.0) covered with cellophane membrane on the surface. In some experiments, we employed a synthetic medium of the following composition: aspartic acid, 5 g; KH_2PO_4, 5 g; $MgSO_4 \cdot 7H_2O$, 0.5 g; $FeCl_3 \cdot 6H_2O$, 4 mg; niacinamide, 150 μg; thiamine hydrochloride, 150 μg; sucrose, 20 g; agar, 20 g; and water to 1 liter (pH 6.0). It is interesting to note that, if niacinamide and thiamine are removed from the synthetic medium, the fruit bodies finally formed are normal in shape but almost sterile.

As has been shown by some workers, when a fungus is transplanted successively using pieces of mycelia as inoculas, the vigour of the fungus will become less as the result of frequent transplantations, and it will finally lose the capacity to produce fruit bodies. This was also found to be the case in this fungus; during the course of the degeneration, it formed normally shaped but sterile fruit bodies. To avoid such a degenerative change in the present fungal culture, we devised a special method of inoculation. Several pieces of fruit body primordia, in which no formation of spores was yet occurring, were taken from stock cultures, put in a saline solution and homogenized aseptically in a Waring blendor. The homogenate was spread with a bent glass rod or spotted with a pipette on media contained in Petri dishes. This method of inoculation was found to be effective in obtaining fungal cultures which formed normal fruit bodies.

EFFECT OF LIGHT ON FRUIT BODY FORMATION

When the fungus was grown at 30 °C in the light (white fluorescent light of about 500 lux), it produced visibly tiny primordia of less than 1 mm in height after 3 days, and, as seen in Fig. 1 A, it finally completed fruit bodies after 7 days if the temperature was shifted down to 20 °C after 5 days. Following the dissemination of spores at the 7th day, the pilei of the fruit bodies gave rise to a rapid autolysis (Fig. 1 B). By contrast, when the fungus was grown in complete darkness with the same temperature shift as mentioned above, malformed fruit bodies consisting of long stipes with very tiny undeveloped pilei on the tops were formed (Fig. 1 C). The further cultivation of the malformed bodies in the dark resulted in the growth of hairy hyphae from them, suggesting the occurrence of dedifferentiation.

The timing and the duration of illumination effective for the fruit body formation were determined by changing the light and dark regimes during the course of cultivation. The results showed that the illumination period from 3 to 5 days was indispensable for the formation of the normal fruit bodies. Moreover, two successive intermittent illuminations for 12 hr each in this period were also found to be effective. In practice, the fungus was cultured in the light throughout the developmental cycle.

The most effective wave length of light was found to lie in a region from 440 to 485 mμ as determined by using various combinations of light filters. It is worth referring to the fact that the sporulation of some fungi is induced by near-ultraviolet[2] or blue[3] light.

The light intensity causing the morphogenesis appeared to be very low.

EFFECT OF TEMPERATURE ON FRUIT BODY FORMATION

As already described by earlier workers,[4] in some species of fungi, different intensities of sporulation were induced by changes in temperature and an abundant sporulation was obtained by a large change in temperature.

The fungus was inoculated as a spot in the center of a Petri dish and cultured at 30 °C for 3 days, the fungus then forming a round hyphal colony of about 4 cm in diameter. If the whole culture was then incubated at 10 °C for 24 hr, no appreciable further growth of the colony was noticed during this period. When the colony was again cultured at 30 °C in the light, a num-

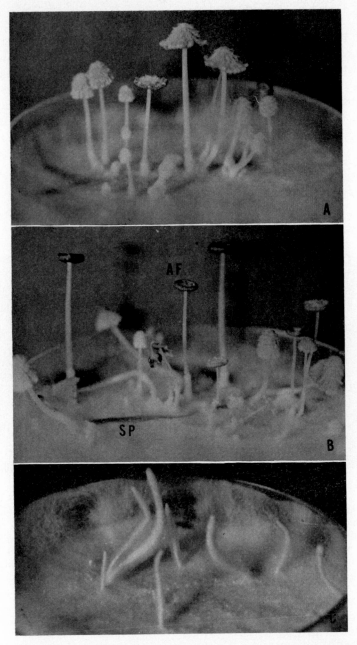

FIG. 1. Photographs showing developmental stages of fruit body formation under different culture conditions. A. Normal fruit bodies formed in the light. B. Autolyzed fruit bodies (AF) showing black spore prints (SP) on the mycelial mat. C. Malformed fruit bodies ("dark stipe") were formed in the complete darkness. The photograph was taken long after the formation of the "dark stipe". See the formation of hairy hyphae on the surface of stipes.

ber of tiny primordia became visible the next day forming a ring pattern. Figure 2A shows the so-called fairy ring thus formed on the plate. The fact is worth noticing that the location of the ring was about 6–8 mm inside the margin of the colony at the time of the cold treatment. This fact indicates that the induction of fruit body formation by the low temperature occurred in the hyphae at sites rear to the apices. Moreover, it was revealed that,

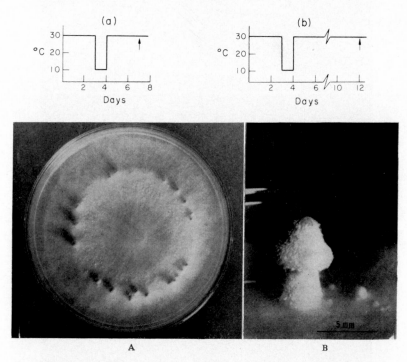

FIG. 2. Photographs showing the effect of cold treatment on the fruit body formation. A. An inoculum was spotted on the centre and cultured in the light according to the temperature programme as shown in the diagram. Photograph taken at the 7th day (indicated by an arrow) after inoculation shows the so-called fairy ring in fruit body formation. B. Fungus was cultured in the light according to the temperature programme as shown in the diagram. Photograph taken at the 12th day (indicated by an arrow) shows a dwarf and sterile fruit bodies.

when the inoculum was spread over the medium, the yield and the synchrony of formations of mature fruit bodies per plate were improved by the cold treatment at 10°C.

Another characteristic effect of low temperature on the fruit body formation was found in the maturation of fruit bodies. In the next experiments, the culture temperature was programmed as follows: (1) cultured at 30°C for 3 days, 10°C from 3 to 4 days, and back to 30°C, followed by a temporary shift to 20°C for more than 6 hours on the 7th day, and (2) same temperature programme as above, but with or without temporary treatment at 20°C for shorter than 2 hr. In the case of the former programme, normal fruit bodies were formed about 24 hr after the end of the cold treatment at 20°C, whereas in the latter, the dwarf and sterile fruit bodies as shown in Fig. 2B were formed about 5 days after the cold treatment. Thus, the sporulation and the growth of stipes seemed to be dependent on treatment at 20°C for several hours.

In recapitulation, the effect of low temperature on the fruit body formation was exhibited in two stages of morphogenesis, namely, in the induction of primordia formation and in the maturation of fruit bodies.

MORPHOLOGICAL CHANGE

The histological appearance of the fungal culture during the early stages of fruit body formation was examined. Buller[5] reported that the initial step of fruit body formation in *Coprinus* is the appearance of a cluster of several irregular hyphae originating from a single cell unit on a hypha. Such rudiments, however, have not yet been observed in our test organism. Figure 3 shows the cross sections of rudiments of the more advanced stages. In earlier stages of primordia formation, the central portion of the rudiment, from which the fruit body was being derived, was strongly basophilic (Figs. 3A and B). Cells in the centre of the rudiments became shortened, some of them became ovoid or round, forming a cluster in the early stage (Fig. 3B). Then a compact band of longer cell-forms began to appear showing the formation of the stipe rudiment, thus showing the formation of the pileus rudiment (Fig. 3C). The formation of gills then became apparent when the size of the primordia was nearly 1 mm in height (Fig. 3D).

During the course of fruit body formation, basidia formation on the surface of the gills became apparent at about the 5th day and black spores were formed on the basidia at the stage of elongation of the stipes. Then, the opening of the pilei followed and the dissemination of spores ensued. After the spore dispersal, the pilei rapidly autolyzed. The final stage of fruit body maturation from the beginning of stipe elongation to the autolysis was finished within about 9 hr. It is an important fact that when the fruit bodies formed were sterile, as influenced by some culture conditions, autolytic process never occurred.

CHANGES IN CELLULAR COMPONENTS

Figure 4A indicates the course of fresh and dry weight changes during the fruit body formation. These weights increased rapidly in the final stage of fruit body formation. When the dry weights of the stipes and the pilei were measured separately, the increase in the pilei preceded that in the stipes. Fresh weight measurement on the pilei and the stipes revealed that the water

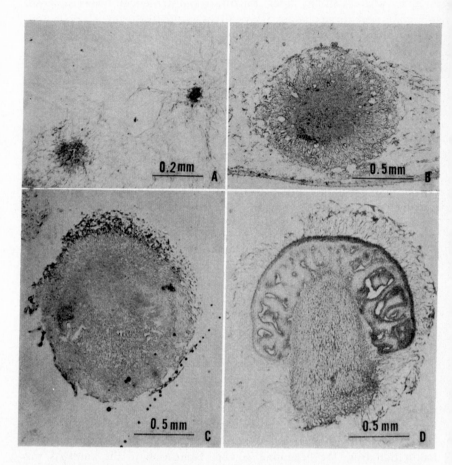

FIG. 3. Photomicrographs showing developmental stages of fruit body primordia. Paraffin-embedded specimens were sectioned and stained with methyl green-pyronin.

content in the stipes increased markedly during the stage of stipe elongation. This suggests that the elongation of stipes is highly dependent on the turgor in stipe cells and that the active dispersal of spores occurring at the culminated stage of stipe elongation is dependent on the active water transfer into the gills from the stipes. The latter was indeed suggested by Wells[6] based on observations on the fine structure of a spore-bearing sterigma in *Schizophyllum commune*.

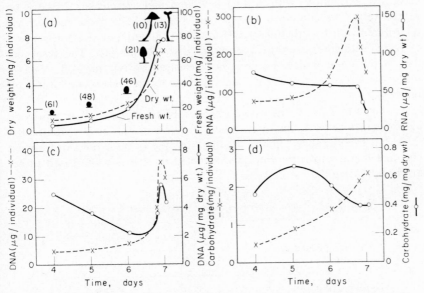

FIG. 4. Changes in amount of cellular components during the development of fruit bodies. Culture temperature: 0–5 days, 30°C; after 5 days, 20°C. Figures in parentheses and silhouettes shown in (a) represent numbers of individuals pooled for analyses and shapes of developing fruit bodies, respectively, taken at desired times. Aliquots of lyophilized samples were subjected to measurement of various cellular components. For the analysis of nucleic acids, cellular components were fractionated by the Schmidt–Thannhauser–Schneider method.[7] (a) Changes in dry weight and fresh weight of developing fruit bodies. (b) Change in RNA content in developing fruit bodies. RNA was measured by the method of Dische and Borenfreund.[8] (c) Change in DNA content in developing fruit bodies. DNA was measured by the indole reaction.[9] (d) Change in carbohydrate content in developing fruit bodies. Carbohydrate was measured by the anthrone reagent.[10]

In Figs. 4b–d are shown the typical patterns of changes in cellular components during the fruit body formation. The increase in amount of RNA per individual preceded that of DNA and the latter increased rather abruptly in the period of active sporulation. Contents of the nucleic acids per dry weight were kept almost constant (RNA) or decreased gradually (DNA) before the final maturation of fruit bodies. This might probably be influenced by the increase of other major cellular materials, such as carbohydrates.

CHANGES IN STEROLS

It has generally been known that fungi contain a considerable amount of sterols. Recent investigations of Lampen *et al.*[11] have shown that sterols in fungi are localized mainly in the cell membranes. However, the physiological significance of sterols in fungal cells has not yet been clarified.

The fungal samples taken at intervals were lyophilized and subjected to successive extractions with boiling methanol and acetone for 2 and 8 hr, respectively. The combined extract was evaporated and the residue was hydrolyzed with 1 N ethanolic KOH for 2 hr. The unsaponified fraction was extracted with ether from the hydrolyzate dissolved in water. The sterol content in the unsaponified fraction was measured by two methods; one by ultraviolet-absorption at 282 mμ and the other by the Liebermann–Burchard reaction.[12]

As shown in Table 1, the values obtained by these methods were in good agreement. In general, the total content of sterols in fruit bodies was smaller than that in vegetative hyphae. In the pilei, the content remained almost unchanged prior to their autolysis, while in the stipes it increased rather markedly during the maturation.

TABLE 1. STEROL CONTENTS IN THE UNSAPONIFIED FRACTION

Development stage		Percentage dry weight	
		UV-Absorption†	L–B Reaction‡
Hyphae (3rd day)		0.55	0.56
Pilei	Primordia	0.47	0.48
	Matured	0.46	0.48
	Autolyzed	0.34	0.33
Stipes	Primordia	0.39	0.41
	Matured	0.58	0.58
	Autolyzed	0.59	0.62

† UV-absorption at 282 mμ taking ergosterol as standard.
‡ Liebermann–Burchard reaction.[12]

Qualitative analysis of sterols was carried out by various methods. Gas-liquid partition chromatography was performed in an apparatus of Barber-Colman Model 10 (glass tube column, 4 mm $\times$ 1.5 m; 100–120 mesh chromosorb W coated with 1.5% XE 60; carrier gas, 2 kg/cm^2 argon; working temperature, 213°C; reference standard, cholesterol). Mass-spectrometry was kindly carried out by Dr. A. Tatematsu of Maijo University and Dr. S. Okuda of this Institute. Ultraviolet-absorption spectra and the Liebermann–Burchard reaction were also used for the identification of sterols.

First, hyphae grown under various conditions were compared for their sterol patterns. The monocaryotic hyphae (A^7B^7) and the dicaryotic hyphae (A^7B^7/A^8B^8) showed essentially the same pattern as shown in Fig. 5A. The dicaryotic hyphae grown (1) by shaking in yeast extract-glucose medium, (2) by shaking in potato-sucrose medium, and (3) statically on potato-sucrose agar also showed the similar sterol pattern. These observations suggest that the sterol pattern of vegetative hyphae was not influenced greatly by culture conditions.

In the gas-chromatographic pattern shown in Fig. 5A, two major components, I and II, were identified as ergosterol and 22-dihydroergosterol, respectively. There is, however, a possibility that the latter is 7-dehydro-campesterol, which is the 24-epimer of 22-dihydroergosterol. A minor component which appeared in between these two major components (relative retention time: 1.50) has not yet been identified. The finding was characteristic in that the amount of ergosterol was far less than that of dihydroergosterol in many hyphal samples.

Figure 5B shows the sterol patterns of fruit bodies taken at different developmental stages. The unsaponified fractions were extracted separately from the pilei and the stipes of these samples. In general, the relative amount of ergosterol increased in the fruit bodies. This was indeed more evident in pilei than in stipes. In the former, the reverse situation in the ratio ergosterol to dihydroergosterol was found as compared with that of hyphae.

In the course of maturation of fruit bodies, the sterol pattern of pilei remained almost unaltered, whereas that of stipes showed a marked change. When the fruit body became matured, a distinct peak appeared at the relative retention time of 1.75. Unfortunately, however, the chemical nature of the third component has not yet been identified. Taking into consideration the above finding that the sterol content in pilei did not change, and in the stipes it did increase, the augmented fraction in stipes may correspond to the amount of the additional third component.

As already mentioned, we have found several culture conditions under which sterile fruit bodies can be obtained. The cultivation of this fungus in the complete darkness caused the formation of malformed bodies consisting only of stipes. These will be called hereafter the "dark stipes". The cultivation of primordia at 30°C without temporary incubation at 20°C resulted in the formation of dwarf and sterile fruit bodies, which will be named the "dwarfs". Normally shaped but sterile fruit bodies, which will be referred to as "steriles", were formed when grown on the synthetic medium devoid of niacinamide and thiamine. Employing these abnormal fruit body samples, sterol patterns by gas-chromatography were obtained as shown in Fig. 5C.

It is thus clear that in these sterile samples no detectable amount of the third component was present. From these observations, one is tempted to speculate that the third component may possibly be related to the sporulation phenomenon in this fungus. It is also of great interest that such a differ-

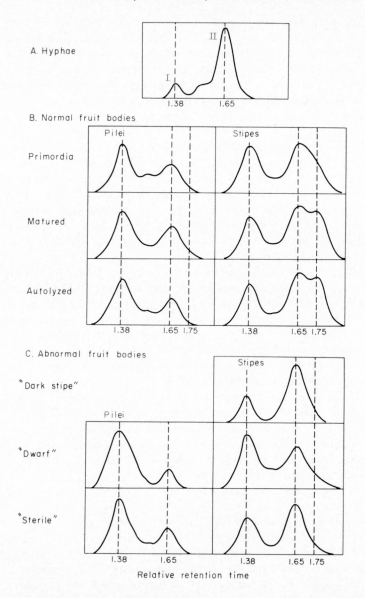

FIG. 5. Patterns of gas-liquid chromatography of the unsaponified fraction of the fungus at various developmental stages. Relative retention time is expressed taking cholesterol as reference standard. For gas-liquid chromatography see text. A. Sterol pattern of dicaryotic hyphae grown on the potato-sucrose agar. B. Sterol patterns of normal fruit bodies at different developmental stages. Primordia employed for the present experiment were about 5 mm in average height. C. Sterol patterns of malformed fruit bodies of matured stage. For the formation of malformed fruit bodies see text.

ence in sterol pattern between the normal and sterile fruit bodies was observed in the stipe but not in the pileus where sporulation actually occurs. By contrast, there is an alternative possibility with regard to the physiological significance of the third sterol component. As already mentioned, when fruit bodies are fertile, their pilei autolyze after the dispersal of spores, whereas when they are sterile, they never autolyze. In other words, the sporulation and the autolysis of the pileus are intimately correlated with each other. Upon consideration of this fact, it may also be possible to infer that the third sterol component contained only in the stipes is related to the autolytic phenomenon.

ACKNOWLEDGEMENTS

The authors are highly indebted to Dr. S. Okuda of this institute and to Dr. N. Ikekawa of Institute of Physical and Chemical Research for their valuable suggestion in the study of sterols. The authors' thanks are also due to Dr. A. Tatematsu of Meijo University who kindly performed mass-spectrometry of our sterol samples and to Prof. K. Kimura of Okayama University who kindly gave us the present fungal strains.

REFERENCES

1. M. O. REINHARDT, *Jahrb. Wiss. Botan.*, **23**, 479, 1892.
2. C. M. LEACH and E. J. TRIONE, *Plant Physiol.*, **40**, 808, 1965.
3. P. B. MARCH, E. E. TAYLOR and L. M. BASSLER, *Plant Disease Rept.*, Suppl. No. 261, 1959.
4. V. W. COCHRANE, *Physiology of Fungi*, p. 362. John Wiley & Sons, Inc., New York, 1958.
5. A. H. R. BULLER, *Researches on Fungi*, vol. IV, p. 166. Reprinted by Hafner Pub. Co., New York, 1958.
6. K. WELLS, *Mycologia*, **57**, 236, 1965.
7. E. VOLKIN and W. E. COHN, In *Methods of Biochemical Analysis*, vol. I, p. 287. Ed. by D. Glick, Interscience Pub., New York, 1954.
8. Z. DISCHE and E. BORENFREUND, *Biochim. et Biophys. Acta*, **23**, 639, 1957.
9. G. CERIOTTI, *J. Biol. Chem.*, **214**, 59, 1953.
10. J. H. ROE, *J. Biol. Chem.*, **212**, 335, 1955.
11. J. O. LAMPEN, P. M. ARNOW, Z. BOROWSKA and A. I. LASKIN, *J. Bacteriol.*, **84**, 1152, 1962.
12. R. SCHOENHEIMER and W. M. SPERRY, *J. Biol. Chem.*, **106**, 745, 1934.

ON THE TRANSFORMATION
AND RECOVERY
OF THE CROWN-GALL TUMOR CELL*†

ARMIN C. BRAUN

The Rockefeller University, New York

I SHOULD like to consider at this time some of our studies that deal with a neoplastic disease of plants commonly known as crown gall. This disease is initiated by a tumor-inducing principle (TIP) that is elaborated by the specific bacterium *Agrobacterium tumefaciens* (Smith and Town, Conn.). The TIP can transform competent plant cells into tumor cells in short periods of time. Once the cellular transformation has been accomplished, the continued abnormal and autonomous proliferation of the host cells becomes an automatic process that is entirely independent of the inciting bacterium.[1,2] The transformation of a normal plant cell to a tumor cell is not, however, a one-step process but takes place gradually and progressively, leading in a 3–4-day period to a rapidly growing, fully autonomous tumor cell type. Cells transformed in shorter periods of time represent lower grades of neoplastic change.[3] The degree of cellular transformation achieved reflects the period in the normal wound-healing cycle in which the cellular alteration occurs. It is just before active cell division begins in the normal wound-healing cycle that cells are transformed into tumor cells of the most rapidly growing type.[4]

If sterile tumor tissues are isolated from either primary or secondary tumors and planted on a simple chemically defined culture medium consisting of mineral salts, sucrose, and three vitamins, the tissues grow profusely and indefinitely. This medium does not support the continued growth of normal cells of the type from which the tumor cells were derived. This is, therefore, a selective medium that permits us to distinguish, at a tissue culture level, between normal plant cells and crown-gall tumor cells.

The question that obviously arises is why should tumor cells grow profusely and indefinitely on that simple culture medium while normal cells do

* Travel grant No. GB-5524 from the National Science Foundation is gratefully acknowledged.

† The investigations reported here were supported in part by a research grant (PHS Grant No. CA-06346) from the National Cancer Institute, Public Health Service, and a research grant (NSF Grant No. GB-197) from the National Science Foundation.

not? In order to answer that question we must know something about the factors that regulate normal cell growth and division. Growth in all plants and animals results either from the enlargement of cells or from the combined processes of cell enlargement and cell division. These fundamental growth processes are controlled in plants by the quantitative interaction of two growth-regulating substances, the auxins and the cytokinins. The auxins are concerned with cell enlargement while the cytokinins act synergistically with the auxins to trigger cell division.[5] The response of plant cells to these hormones can best be illustrated with the use of certain mature cell types that have lost, as a result of their maturation, the capacity to synthesize those substances. If, for example, a fragment of mature tobacco pith parenchyma tissue is isolated and planted on a simple chemically defined culture medium, the cells of that tissue will not enlarge, and will not divide but will remain quiescent. If, now, an auxin is added to the culture medium at a concentration of 1 p.p.m., the cells will enlarge greatly in size but they will not divide. If, in addition to an auxin, a cytokinin such as kinetin (6-furfurylaminopurine) is added to the culture medium at a concentration of 0.5 p.p.m., an abundant growth accompanied by profuse cell division results. The cytokinin without an auxin is completely ineffective in promoting either cell enlargement or cell division. It is clear, therefore, that two substances, which act synergistically, are essential to regulate the fundamental growth processes of cell enlargement and cell division in higher plant species.

Since tobacco pith parenchyma cells have lost, as a result of their maturation, the capacity to synthesize the two essential growth-regulating hormones, it was of interest to learn how such cell types would respond if transformed into tumor cells. If, for example, only the auxin system is activated as a result of the transformation process, then the altered cells should enlarge greatly in size but they should not divide. If, on the other hand, the cytokinin system is activated without a corresponding activation of the auxin system, neoplastic growth should not result. Only if both growth substance-synthesizing systems are activated simultaneously will growth accompanied by cell division and, hence, tumor formation result. The results of those studies showed that when healing pith parenchyma cells were transformed, typical crown-gall tumors developed.[6] This simple experiment demonstrates, then, that although normal pith parenchyma cells could not and did not synthesize either auxins or cytokinins, following their transformation to tumor cells both growth-regulating hormones were produced in greater than regulatory amounts. If this were not true, continued growth accompanied by cell division and, hence, tumor formation would not have occurred in that test system. That those substances are, in fact, synthesized by actively growing tumor tissue was demonstrated by isolation and chemical characterization of the biologically active substances. The main auxin component, as might be expected, was found to be β-indoleacetic acid. The cytokinin, on the other hand, turned out to be an interesting compound.[7] It has a nicotinamide ring, a glucose sugar moiety, sulfur in the form of sulfate or sulfonate, and one or more methyl groups. That compound is biologically active at a concentration of less than 150 µg per liter.

The persistent activation in the plant tumor cell of the auxin and cytokinin synthesizing systems would in itself appear adequate to account for the continued abnormal and autonomous proliferation of the crown-gall tumor cell. Further studies demonstrated, however, that that was by no means the entire explanation.[3]

It was indicated earlier in this discussion that the transformation of a normal plant cell to a fully autonomous crown-gall tumor cell is not a one-step process but takes place gradually and progressively over a 3–4-day period. When sterile tumor tissues showing varying grades of neoplastic change were isolated and planted on a simple chemically defined culture medium, such tissues retained indefinitely in culture their characteristic growth patterns. Cells transformed during a 34-hr period grew very slowly in the host and in culture, while those altered during a 50-hour period grew at a moderately fast rate on a basic culture medium. Very rapidly growing tumors were initiated when host cells in the region of a wound were exposed to the tumorigenic agent for 72 or more hours. Since cell strains that showed varying grades of tumorous change were very stable in culture, they were admirably suited for a study of the factors required for rapid autonomous growth. For such studies the rapidly growing, fully autonomous tumor cell types were used as the standard. Such cells can synthesize in optimal or near optimal amounts all of the growth factors required for their continued rapid proliferation from mineral salts, sucrose, and three vitamins present in the basic culture medium. The moderately fast growing tumor cells transformed during a 50-hr period required that the basic culture medium be supplemented with an auxin, glutamine and myo-inositol if those cells were to achieve a growth rate comparable to that of the fully transformed tumor cells grown on the basic culture medium. These were the minimum nutritional requirements to achieve that rate of growth. That these were the actual metabolites involved was demonstrated experimentally.

The very slowly growing tumor cells transformed during a 34-hr period required, in addition to an auxin, glutamine and inositol, asparagine or aspartic acid as well as a purine and pyrimidine to achieve a growth rate comparable to that of the fully transformed cell grown on the basic medium. The difference between the three types of tumor cells appears, then, to be quantitative since all three types can grow continuously although at different rates on the basic culture medium.

Normal cells of the type from which the tumor cells were derived do not grow at all on the basic culture medium. Thus, the difference between the normal cells and the three types of tumor cells is qualitative. The normal cells can be forced into rapid growth by supplementing the basic culture medium with the same metabolites required for the rapid growth of the most slowly growing type of tumor cell. In addition, however, the normal cells possess an absolute exogenous requirement for cytokinin, the hormone that triggers cell division. The normal cells cannot synthesize that substance; all three types of tumor cells have learned, as a result of the transformation process, to produce it. The persistent activation of the cytokinin system in the tumor cell appears to represent a very important distinction between normal and tumor cell types since it is the key metabolite that keeps the tumor cells dividing persistently.

An experiment such as that described above gives us a great deal of information concerning the workings of this system. It demonstrates that as a result of the transition from a normal cell to a fully autonomous tumor cell, a series of quite distinct but well-defined biosynthetic systems, which represent the entire area of metabolism concerned with cell growth and division, become progressively and persistently activated. The degree of activation of those systems determines the rate of growth of a crown-gall tumor cell. It is clear, moreover, that autonomy, in this instance, finds its explanation in terms of cellular nutrition. The tumor cells have learned, as a result of their transformation, to synthesize all of the essential growth factors that their normal counterparts require but cannot make for cell growth and division.

Finally, this study demonstrates that as the result of the transition from a normal cell to a fully autonomous tumor cell, a profound and persistent reorientation in the pattern of synthesis occurs. The switch in metabolism from that found in a resting cell to that present in a persistently dividing cell is triggered by irritation accompanying a wound. It may be permanently fixed in the plant tumor cell not only by the TIP in crown gall but by other diverse agents as well. That pattern of synthesis is maintained in the plant tumor cell by virtue of the fact that the two hormones that regulate cell growth and division in plants are continually synthesized by such cell types. The other metabolites shown to be produced by tumor cells are required for the synthesis of the nucleic acids, the mitotic and enzymatic proteins and, in the case of inositol, the membrane system of the cell. These metabolites are essential to permit the pattern of synthesis concerned with cell growth and division to be expressed.

The question that now arises is how the diverse biosynthetic systems shown to be functional in the plant tumor cell become persistently activated. It is clear that some very fundamental cellular mechanism must be involved in the simultaneous or perhaps sequential activation of those systems. Although the entire answer to that problem is not as yet in hand, some progress appears to have been made. Our studies indicate that the regulation of essential biosynthetic systems in normal and tumor cells appears, in part at least, to reflect differences in the properties of the membrane systems of the two cell types.[8,9] It has been possible to effectively replace all three of the organic requirements for rapid growth of the moderately fast growing tumor cell and to replace five and in part six of the seven organic requirements for the rapid growth of normal cells by simply raising the concentration of four salts, KCl, $NaNO_3$, NaH_2PO_4 and $(NH_4)_2SO_4$ in the basic culture medium. It has not as yet been possible to activate the system concerned with the synthesis of cytokinin with ions.

Time course studies using an anion, a cation, and a charged organic molecule have shown that the net uptake of all three is significantly greater in tumor cells than in normal cells when the two cell types are growing at essentially the same rate.[10] The evidence suggests, moreover, that active transport mechanisms are not involved, at least not for the solutes studied. These studies indicate that tumor cells take up ions very efficiently from dilute salt solutions while their normal counterparts do not. This, then, represents a fundamental difference between a normal plant cell and a crown-gall tumor cell since it permits the activation by ions of a large segment of metabolism concerned specifically with cell growth and division.

The finding that there are differences in the properties of the membrane systems in normal and tumor cells has led to a study of the phospholipid patterns of the two cell types. Striking differences in these patterns have been found in preliminary studies. The normal cells contain between four and six times as much phospholipid as do the fully autonomous tumor cells when both cell types grow at essentially equal rates. A choline-containing phospholipid represents by far the major phospholipid component of the normal cell but constitutes only a minor part of the total phospholipid in the tumor cell. The major component of the tumor cell is one that absorbs at 240 mμ. A knowledge of the composition of the phospholipids of the two cell types could be useful because it offers the possibility of specifically altering the properties of the membrane systems and thus, in effect, the regulation of biosynthetic metabolism in such cell types.

It is a generally accepted belief that heritable changes of the type that we have been considering are of a permanent and irreversible type. This would certainly appear to be true of the typical crown-gall tumor cell since many such cell types have been kept under constant observation in the laboratory for more than a decade without showing the slightest indication that they have become any less tumorous than when first isolated.

We attempted in our studies to distinguish between somatic mutation at the nuclear gene level involving either the deletion or permanent rearrangement of genetic information, on the one hand, and epigenetic changes which are concerned with alterations in the expression but not with the integrity of the genetic potentialities that are normally present in a cell.

For studies of this type plants offer several very distinct advantages as experimental test objects. The first of these is that somatic cells of certain plant species may remain totipotent throughout the life of the plant. The second advantage derives from the unique manner in which dicotyledonous plant species grow. Primary growth in such species results from the rapid division, subsequent elongation, and finally the differentiation of the meristematic cells at the extreme apex of a shoot or root. By combining the properties of totipotency and unique growth characteristics, it was possible to regularly achieve the recovery of the crown-gall tumor cell.[11]

The typical crown-gall tumor cell possesses a pronounced capacity for proliferation, a very limited capacity for differentiation, and such tumor tissue lacks entirely the ability to organize structures. If, however, totipotential cells of certain plant species are transformed to a moderate degree but not fully, then one obtains a morphologically very different type of tumorous growth. The new growth, a teratoma, is composed in part of a chaotic assembly of tissues and organs that show varying grades of morphological development. Sterile tissue fragments isolated from the abnormal but organized structures and planted on a basic culture medium grow profusely and indefinitely, as do the fully transformed tumor cells. They differ from the latter in that the teratoma tissues retain indefinitely in culture a pronounced capacity to organize abnormal leaves and buds. That such teratoma tissue is composed entirely of tumor cells and not of a mixture of normal and tumor cells was demonstrated unequivocally by isolating a number of clones of single-cell origin.[11] Such single-cell clones grown on the basic medium behaved in every respect as did the tissues from which they were derived. Since the teratoma tissues of single-cell origin possessed a capacity to organize tumor buds, they were admirably suited and were used for studies dealing with the nature of the heritable cellular change underlying the tumorous state in crown gall.

In those studies it was hypothesized that if tumor shoots derived from the abnormal tumor buds present in the teratoma tissue could be forced into rapid but organized growth, a recovery might occur if the primary cellular change leading to autonomy was of an epigenetic type but not if it involved somatic mutation at the nuclear gene level. The results of those studies showed that if tumor shoots derived from tumor buds were forced into rapid but organized growth as a result of a series of graftings to healthy plants, they gradually recovered and ultimately became normal in every respect. Recovery was complete. The results of those studies make somatic mutation at the nuclear gene level appear highly unlikely since mutations are not generally considered to be lost as a result of rapid and organized growth. Since the nuclei of the normal and tumor cells appear to be genetically equivalent, the results suggest rather that we are dealing in this instance merely with a change in the expression of the genetic information that is present in a cell. The biosynthetic systems concerned with cell growth and division are gradually and progressively activated during the transformation process and those systems are again gradually repressed during recovery.

The question that now arises is whether a reversal of the neoplastic state is something that is unique to plant tumor systems or whether it has broader biological implications. There have been within the past few years a number of examples of this type, some of which are very well documented, that have come from the animal field. We need only to recall the studies of Seilern-

Aspang and Kratochwil[12,13] on the European newt, the work of Mizell[14] with the Lucké tumor implants into the tail of the tadpole of the frog, and that of Kleinsmith and Pierce[15] on the teratocarcinoma of the mouse. A significant number of other examples, some of which are not so well documented as those cited above, could be listed as demonstrating that a reversal of the neoplastic state can, in fact, occur in tumor systems of the most diverse type. This, in turn, would suggest that in those instances in which a reversal of the neoplastic state is achieved the nuclei of the tumor cell and its normal counterpart are genetically equivalent and thus no change in the integrity of the genetic information present in a cell has occurred. In many instances in which a reversal of the tumorous state has been accomplished experimentally, multipotential tumor cells were involved. This raises the question as to why multipotential cells may recover from the tumorous state while most differentiated cell types do not? The answer to that question would appear to rest in the fact that multipotential cells are endowed with broad morphogenetic or regenerative capabilities. Such cell types can very effectively remodel metabolic patterns by primary gene action. The apparent irreversibility of the tumorous state in the great majority of instances may simply reflect an inability of most differentiated cell types to undergo intracellular regenerations of the type that are characteristic of the multipotential cell. It would appear that if we are really to understand the tumor problem we shall have to learn, as the multipotential cell has learned, how to switch the pattern of synthesis in a cell from one that makes it grow as a tumor cell to one that will restore it to its normal or at least benign state. It would thus appear that a long step forward toward accomplishing that end may possibly be achieved by learning how the genes of a cell are selectively activated and repressed.

REFERENCES

1. A. C. BRAUN, *Am. J. Botany*, **34**, 234, 1947.
2. A. C. BRAUN, *Phytopathology*, **41**, 963, 1951.
3. A. C. BRAUN, *Proc. Nat. Acad. Sci., U.S.*, **44**, 344, 1958.
4. A. C. BRAUN and R. J. MANDLE, *Growth*, **12**, 255, 1948.
5. J. R. JABLONSKI and F. SKOOG, *Physiol. Plantarum*, **7**, 16, 1954.
6. A. C. BRAUN, *Cancer Research*, **16**, 53, 1956.
7. H. N. WOOD, *Colloques Internat. Centre Nat. Recherche Scient.*, No. 123, Gif-sur-Yvette, Régulateurs Naturels de la Croissance Végétale, 97. Paris, 1964.
8. H. N. WOOD and A. C. BRAUN, *Proc. Nat. Acad. Sci., U.S.*, **47**, 1907, 1961.
9. A. C. BRAUN and H. N. WOOD, *Proc. Nat. Acad. Sci., U.S.*, **48**, 1776, 1962.
10. H. N. WOOD and A. C. BRAUN, *Proc. Nat. Acad. Sci., U.S.*, **54**, 1532, 1965.
11. A. C. BRAUN, *Proc. Nat. Acad. Sci., U.S.*, **45**, 932, 1959.
12. F. SEILERN-Aspang and K. KRATOCHWIL, *J. Embryol. Exptl. Morphol.*, **10**, 337, 1962.
13. F. SEILERN-ASPANG and K. KRATOCHWIL, *Wiener klin. Wochenschr.*, **75**, 337, 1963.
14. M. MIZELL, *Am. Zoologist*, **5**, 215, 1965.
15. L. J. KLEINSMITH and G. B. PIERCE, Jr., *Cancer Research*, **24**, 1544, 1964.

DISCUSSION

H. ENDO: You have mentioned that the grafted tumor grew expansively, but such a tumor has no metastatic activity?

A. C. BRAUN: True metastases do not and cannot occur in plants. Plant cells are cemented tightly together and with few exceptions do not and cannot move. The secondary tumors that are found in certain plant species are not due to the movement of tumor cells from the primary growth to other parts of the plant.

R. KINOSITA: Could you explain what sort of changes pertain to the expression of the genetic potentiality?

A. C. BRAUN: Our studies have shown that during the transition from a normal plant cell to a fully autonomous tumor cell, progressive changes in the permeability of the membrane systems occur. The tumor cells take up ions and other solutes very efficiently from dilute solutions; the normal cells do not. Increased permeability of the tumor cell membranes represents a fundamental difference between a normal cell and a plant tumor cell, since it permits the activation by ions of five and in part six of the seven essential biosynthetic systems shown to be persistently activated, as a result of transformation, in the plant tumor cell. Only the biosynthetic system concerned with the synthesis of the cell division-triggering hormone cannot be activated by ions. This, then, would appear to be one type of change that influences gene expression either directly or indirectly.

IV. CONTROL OF CELL GROWTH, CELL TRANSFORMATION AND CANCER INDUCTION BY VIRUS

MOLECULAR ASPECTS OF THE RELEASE OF TUMOR CELLS FROM THE CONTROL OF PROLIFERATION

Hiroshi Terayama, Haruki Otsuka, Keiko Sakuma and Hideo Yamagata

Department of Biophysics and Biochemistry, Faculty of Science,
University of Tokyo, Tokyo, Japan

The proliferation of the normal tissue cells in adult animals is considered to be regulated by some homeostatic mechanism or mechanisms, while malignant tumor cells appear to have a tendency to uncontrolled or unlimited growth. Prior to mitosis as well as cell division, replication of DNA takes place in cells. Most of the cells in many resting tissues of adult animals are in the G1 phase instead of the G2 phase, suggesting that the cells are regulated at a stage prior to the DNA synthesis (Fig. 4).

In regenerating tissues, DNA synthesis as well as the activities of several key enzymes involved in DNA synthesis starts to increase after a certain lag, reaches a maximum level and then returns to a normal low level again in a due course.[1]

Among various metabolic activities related with the DNA synthesis, enzymatic activities involved in the *de novo* as well as the salvage synthesis of thymidine deoxynucleotides are considered to be the most important with respect to the regulatory mechanism (Fig. 1).

Thymidine kinase, thymidylate kinases, thymidylate synthetase, and deoxycytidylate deaminase are known to be induced in the regenerating liver at 16 hr after partial hepatectomy, reaching a maximum level at 24–30 hr. The activities of these key enzymes are quite high in the proliferating tissues as well as in the tumors, while they are low in the normal, resting tissues.[2−6]

Biochemical attempts aimed at disclosing any abnormality in the feedback inhibition of these key enzymes from the tumor cells have so far accumulated rather pessimistic results. It seems, therefore, to be reasonable to look for abnormality of the tumor cells in the regulatory mechanism of the activity of the operon responsible for proliferation, or in some other mechanisms.

Physiological as well as biochemical activities of animal cells, including the growth of cells, are usually regulated by hormones. Hemingway[7,8] reported that mitotic activity of regenerating rat-liver was higher in adrenalectomized

327

rats. Administration of cortisone counteracted the effect of adrenalectomy. Recently, Mochizuki[9] found a diurnal rhythm in the mitotic activity of the regenerating rat-liver. This rhythm, however, was shown to disappear in adrenalectomized rats. These papers suggest that adrenal hormones, particularly cortisone, may inhibit the mitosis in rat-livers. On the other hand, Bullough[10-13] observed that mitotic activity of mouse epidermal cells was inhibited by adrenalin with the co-operation of a tissue-specific protein which was referred to as *chalone* by him. He suggested that the adrenalin-chalone complex interferes with the growth of the cells in the G2 phase as well as in the M phase.

Since these studies were carried out mainly by measuring the mitotic indices, it is rather difficult to see exactly where and how these hormones may exert their actions.

In this paper I wish to present two series of experiments that we consider important with respect to the altered growth control mechanism in tumor cells.

The first series of experiments are concerned with the effect of adrenal hormones, especially of adrenalin, upon the induction of the key enzymes in regenerating rat-livers.

TABLE 1. EFFECT OF ADRENALIN INJECTIONS UPON ^{14}C-TdR INCORPORATION INTO DNA IN LIVER, REGENERATING LIVER AND AZO DYE-INDUCED PRIMARY HEPATOMA (SAKUMA AND TERAYAMA)

Tissue	Treatment	(Number of rats)	cpm/mg DNA
Normal liver	No injection	(5)	117 ± 23
Normal liver	Sham	(4)	103 ± 30
Normal liver	Adrenalin[a]	(5)	76 ± 4
24 hr reg. liver	Sham	(5)	993 ± 50
24 hr reg. liver	Adrenalin[b]	(5)	217 ± 113
Azo dye hepatoma	Sham		2725
Azo dye hepatoma	Adrenalin[c]		2080
Tumor-bearing liver	Sham		783
Tumor-bearing liver	Adrenalin[c]		816

Adrenalin was injected (s.c.) every hour for 24 hr.
(a) 25 µg, (b) 15 µg, (c) 20 µg.

In these experiments, adrenalin (15–25 µg) was injected every hour subcutaneously during a period from partial hepatectomy to sacrifice at 24 hr. As shown in Table 1, injections of adrenalin reduced DNA synthesis as measured by 90 min incorporation of injected (i.p.) ^{14}C-thymidine into DNA. On the other hand, DNA synthesis in 3'-methyl-4-dimethylaminoazobenzene induced primary hepatoma cells was not affected to any considerable extent even when the tumor-bearing rats had been injected with adrenalin for 24 hr.

Next, experiments were carried out to see if the inhibitory effect of adrenalin upon DNA synthesis may or may not be due to inhibition in the induction of key enzymes, which, under the ordinary conditions starts at 12–16 hr after partial hepatectomy and is almost completed by 24 hr. In Group I,

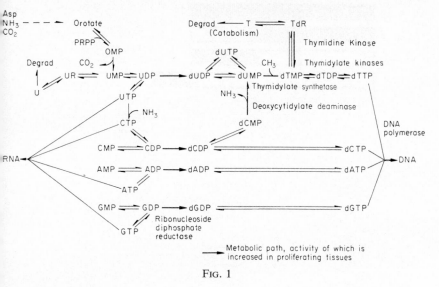

FIG. 1

adrenalin (20 μg) was injected subcutaneously every hour until sacrifice at 26 hr after partial hepatectomy. In Group II, adrenalin was injected only for the first 16 hr, and in Group III, it was injected only for the last 10 hr. The rats in Group IV were treated with the adrenalin injections from 24 hr after partial hepatectomy. Thymidine kinase from the liver of rats killed at 26 hr in Group I, II and III, and at 32 hr in Group IV, was assayed by the method of Bollum and Potter[2].

Results given in Table 2 show that adrenalin injections retard the induction of thymidine kinase when it is given within 24 hr after partial hepatectomy as in Groups I, II and III. However, the effect of adrenalin appears to become less pronounced when given after the induction has been almost accomplished (Group IV). In Group V, rats which had previously been transplanted with ascites hepatoma cells were treated with adrenalin in a similar way for 15 hr. In this case, however, thymidine kinase activity in the tumor cells was apparently not affected at all by the adrenalin treatment.

These observations clearly indicate that adrenalin may be involved, either directly or indirectly, in the repression of the so-called proliferation operon in rat-liver nuclei and that hepatoma cells are rather insensitive, or resistant, to the repressive effect of adrenalin.

Next, we wish to present briefly the effect of cortisone on the regenerating rat-liver. Five milligrams of cortisone were injected intramusculatly daily for

TABLE 2. EFFECT OF ADRENALIN UPON THYMIDINE KINASE IN REGENERATING LIVER
AND ASCITES HEPATOMA (SAKUMA AND TERAYAMA)

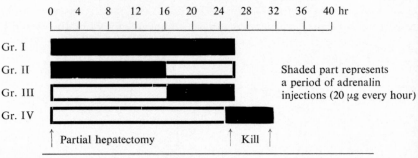

	Group	Treatment	Thymidine kinase activity (% P-tion)
Regenerating liver	I	Sham	27.7 (10)[a]
		Adrenalin	12.1 (10)
	II	Sham	27.3 (8)
		Adrenalin	8.7 (9)
	III	Sham	31.0 (5)
		Adrenalin	13.0 (8)
	IV	Sham	22.4 (9)
		Adrenalin	17.2 (9)
AH-414	V	Sham (15 hr)	30 ± 10 (3)
		Adrenalin (15 hr)	30 ± 1 (3)

five successive days prior to partial hepatectomy. As demonstrated by
Fig. 2, initiation of DNA synthesis as well as induction of thymidine kinase
can be almost completely prevented by cortisone pretreatment. But a single
dose of cortisone given at the time of partial hepatectomy does not affect the
course of regeneration. The observation that unphysiologically large doses of
cortisone are needed to bring about any serious effect upon regeneration of
rat-liver suggests that the effect of cortisone is more pharmacological
than physiological. At the moment, the authors consider that this effect of
cortisone may be ascribed to a decrease in the number of lymphocytes in experimental animals. A role of lymphocytes in the tissue regeneration has
been cited by many workers.[14]

Thus far we have presented a short report on the effect of adrenal hormones upon the DNA synthesis in the regenerating rat-liver, and now I want to
proceed to the second topic.

Several papers[15-17] have shown that cell saps, or aqueous extracts from
the liver, inhibit the proliferation of tissue-cultured cells. Since we have been
interested in DNA synthesis and its regulation, we investigated several tissue
extracts to see how they influence DNA synthesis in ascites hepatoma cells

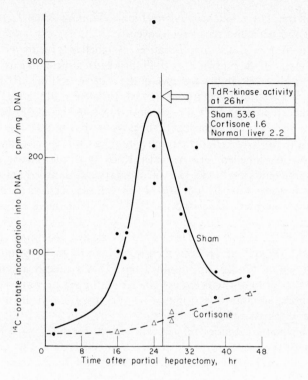

FIG. 2

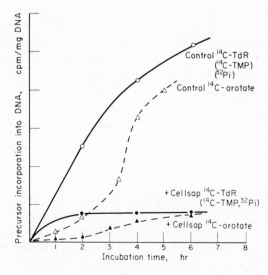

FIG. 3

suspended in the Hanks' balanced salt medium containing serum, lactalbumin hydrolysate and penicillin plus streptomycin.

As shown in Fig. 3, the incorporation of labelled precursors such as ^{14}C-thymidine, ^{14}C-thymidylate or ^{32}Pi proceeds almost linearly within several hours. However, in the presence of the rat-liver cell sap (105,000 × g, 1 hr), the incorporation was inhibited almost completely after a lag of about one hour. Incorporation of ^{14}C-orotate into DNA in the ascites hepatoma cells exhibited slightly different kinetic behaviour, but in this case also the addition of the rat-liver cell sap strongly inhibited the incorporation. The incorporation of labelled orotate into RNA was not interfered with markedly by the addition of cell sap. Incorporation of labelled amino acids into nuclear histones was also not affected by cell sap. Thus it appears that rat-liver cell sap inhibits rather specifically DNA synthesis in ascites hepatoma cells.

The effect of the cell sap was reversible for ascites hepatoma cells preincubated with the cell sap for 3 hr, then washed and transferred into a new medium containing no cell sap, resumed the DNA synthesis as rapidly as before. Viability of tumor cells did not change during the incubation as tested by the nigrosine staining. Electronmicroscopic observations did not reveal any damage in the subcellular structures.

Several enzymatic activities related with DNA synthesis have been investigated with the ascites cells preincubated in cell sap for 3 hr and compared with cells incubated in the absence of the cell sap. As shown in Table 3,

TABLE 3. KEY ENZYME ACTIVITIES IN AH 414 CELLS INCUBATED FOR 3 HR
IN THE PRESENCE OR ABSENCE OF CELL SAP (OTSUKA AND TERAYAMA)

Key enzymes	Enzyme activity of AH 414 cells after incubation	
	with call sap	without cell sap
Thymidine kinase[a]	32.5	43.6
Deoxycytidylate deaminase[b]	$0.91 \times 10_4$	1.20×10^4
Cell-free TMP incorp. into DNA[c]	2.46×10^3	2.83×10^3

(a) Percent of ^{14}C-TdR phosphorylation.
(b) Counts of deaminated product(s) from ^{14}C-dCMP.
(c) Counts of DNA after incubation with ^{14}C-TMP.

only a slight decrease in the activity of key enzymes such as thymidine kinase, deoxycytidylate deaminase or the activity to incorporate ^{14}C-thymidylate into DNA in the cell-free system is observed with the cells incubated in the presence of the cell sap as compared with the cells incubated in the absence of the cell sap. Thus it seems likely that the effect of the cell sap is not due to the loss of the key enzymes from the cells.

The active principle in the cell sap was found to be effective only upon the DNA synthesis in the cells, since it failed to shows any effect upon the DNA synthesis in the cell-free system regardless of the strandedness of the DNA used as a primer, as demonstrated in Table 4.

TABLE 4. EFFECTS of PARTIALLY PURIFIED ACTIVE PRINCIPLE UPON CELL-FREE DNA-SYNTHESIS

Expt. No.	Reaction system	^{14}C-TMP incorporat. (cpm/mg DNA/90 min)
I	Complete (heated DNA as primer)	2830
	+ partially purified active principle	3020
	− DNA	86
II	Complete (native DNA as primer)	2009
	+ partially purified active principle	1844
	− DNA	86

The activity to inhibit the DNA synthesis in the tumor cells was found in the $105,000 \times g$ supernatant from the rat-liver homogenate as well as from rat-liver nuclear sonicates. By adjusting the cell sap at pH 4.5, the activity was shown to be in the supernatant part instead of in the precipitates. The supernatant fluid was neutralized and subjected to the differential precipitations using ammonium sulfate. The activity was mainly found in a fraction precipitated between 0.4 and 0.6 saturations. Active precipitates by the ammonium sulfate treatment were dialysed against tris buffer, pH 7.6, then subjected to chromatography on DEAE-cellulose column. The activity was recovered as a preliminary eluate, which corresponds to a passing-through fraction.

Electrophoretic mobility experiments also suggest that the active principle in the cell sap may be a near neutral or slightly basic protein. Activity of the cell sap was destroyed by heating at 100°C for 15 min.

The distribution of inhibitory principle among cell saps from various normal tissues and tumors has been investigated. Activity was detected in almost every normal tissue without regard to the proliferative status of the tissues. It was found in the cell saps from rat-liver, 24 hr regenerating rat-liver, rat-kidney, rat-spleen, rat-brain, rabbit-lymph nodes, etc.

However, the cell saps from various kinds of malignant tumors did not demonstrate the presence of the active principle. We have investigated 3′-Me-DAB induced primary rat-hepatomas, rat ascites hepatomas such as AH-225, AH-414 and AH-601, rat Rhodamine sarcomas, mouse ascites leukemia SN-36 and mouse Ehrlich carcinomas, and we found that all these cell saps from tumor cells are completely devoid of the active principle. However, as exceptions so far, we found Friend virus leukemia (mouse spleen) and Morris minimal deviation hepatomas 7793 and 7795 showed the presence of the inhibitory principle in the cell sap. These tumors are, however, somewhat different from the other malignant tumors: one is a virus-dependent tumor and another is a type of minimal deviation hepatoma, which are shown to hold the capacity to bind with the carcinogenic aminoazo dyes as well as some other differentiated functions of the normal liver.

It sounds strange to report that the active principle occurs in the highly proliferating normal tissues such as 24 hr regenerating rat-liver and -lymph nodes. Two possible explanations may be presented.

Firstly, in the normal tissue cells, the active principle may be present in the other phases except the S phase and it is somehow deleted in the S-phase. It is known that the percentage of the cells in the S phase does not exceed 30% of total liver cells even at the highest proliferative status.[18] Secondly, the active principle may exist throughout the whole life-cycle, but somehow may be inactivated upon entering the S phase by some anti-principle which may be produced at the time of entering the S-phase and counteracts the inhibitory activity of the active principle. The nature of this anti-principle has not yet been fully clarified, but this hypothesis seems to be interesting when we consider that cells need a synthesis of specific RNA and proteins upon entering the S phase as reported by E. W. Taylor.[19] This hypothesis may also provide us with a reasonable explanation of our finding that the active principle purified partially from the cell sap as described above does inhibit the DNA synthesis in the ascites tumor cells but does not inhibit the DNA synthesis in the regenerating rat-liver slices as shown in Table 5. If there is an anti-principle accumulated in the regenerating rat liver cells in the S phase, the active principle added externally may not be able to exert its action upon the cells. Otherwise, we have to assume that the active principle has access only to the tumor cells but not to the regenerating liver cells.

TABLE 5. EFFECT OF PARTIALLY PURIFIED ACTIVE PRINCIPLE UPON DNA SYNTHESIS IN REGENERATING RAT-LIVER SLICES

Expt. No.	Reaction system	^{14}C-TdR incorporat. (cpm/mg DNA/2 hr)
I	Control	8500
	+ Partially purified active principle	7160
II	Control	2300
	+ Partially purified active principle	2360

Contrary to the normal cells, it seems that most malignant tumor cells have lost the ability to produce the inhibitory principle throughout their whole life-cycle. Upon entering the S phase, the tumor cells do not have to produce the anti-principle because they do not have the inhibitory principle from the beginning, and therefore, their DNA synthesis may be inhibited by the externally added active principle from the normal rat liver (see Fig. 4).

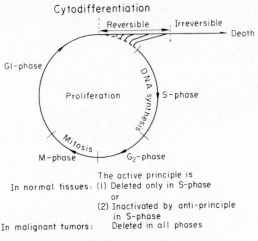

FIG. 4

At the end of my presentation, I wish to add one comment, which seems to be quite interesting with respect to the carcinogenic mechanism. Our active principle found in the cell sap from normal tissues appears in many respects to resemble a protein fraction of the cell sap, which was reported by Sorof et al.,[20,21] as a major component responsible for the binding of carcinogens, and was referred to as slow h_2 fraction. Quite recently, Freed and Sorof[22] reported that the slow h_2 fraction inhibited the proliferation of several tissue-cultured cells when added in the culture medium. The biochemical nature as well as the distribution of active principle, appear to resemble quite closely the slow h_2 protein.

ACKNOWLEDGEMENTS

This work was supported by grant CA-05246-05, from the National Cancer Institute, N.I.H., U.S.A., and by a Grant-in-Aid for Scientific Research from the Ministry of Education, Japan.

REFERENCES

1. N. L. R. BUCHER, International Rev. Cytology, **15**, 245, 1963.
2. F. J. BOLLUM and V. R. POTTER, Cancer Res., **19**, 561, 1959.
3. S. M. WEISSMAN, R. M. S. SMELLIE and J. PANE, Bioch. Biophys. Acta, **45**, 101, 1960.
4. H. H. HIATT and T. B. BOJARSKI, Bioch. Biophys. Res. Communs., **2**, 35, 1960.
5. N. FAUSTO and J. L. VAN LANCKER, J. Biol. Chem., **240**, 1247, 1965.
6. G. F. MALEY and F. MALEY, J. Biol. Chem., **236**, 1906, 1961.
7. J. T. HEMINGWAY and D. B. CATER, Nature, **181**, 1065, 1958.

8. J.T.HEMINGWAY, *Experientia*, **15**, 189, 1959.
9. M.MOCHIZUKI, *J. Fac. of Sci., Univ. of Tokyo*, Sec. IV, **10**, 471, 1965.
10. W.S.BULLOUGH, *Proc. Roy. Soc.*, **B154**, 540, 1961.
11. W.S.BULLOUGH, *Nature*, **199**, 859, 1963.
12. W.S.BULLOUGH, *Exptl. Cell Res.*, **35**, 629, 1964.
13. W.S.BULLOUGH, *Exptl. Cell Res.*, **36**, 193, 1964.
14. B.J.BRYANT, *J. Cell Biol.*, **18**, 515, 1963.
15. H.KATSUTA, T.TAKAOKA, M.MORI, M. YASUKAWA, S.SAITO and S.SUZUKI, *Japan J. Exper. Med.*, **27**, 443, 1957.
16. S.SUZUKI, *Japan J. Exper. Med.*, **29**, 341, 1959.
17. M.HORI and T.UKITA, *J. Biol. Chem.*, **51**, 322, 1962.
18. J.W.GRISHAM, *Cancer Res.*, **22**, 842, 1962.
19. E.W.TAYLOR, *Exper. Cell Res.*, **40**, 316, 1965.
20. S.SOROF, P.COHEN, E.C.MILLER and J.A.MILLER, *Cancer Res.*, **11**, 383, 1951.
21. S.SOROF, E.M.YOUNG, M.M.McCUE and P.L.FETTERMAN, *Cancer Res.*, **23**, 864, 1963.
22. J.J.FREED and S.SOROF, *Bioch. Biophys. Res. Communs.*, **22**, 1, 1966.

DISCUSSION

F.H.KASTEN: Your interesting differences between regenerating rat-liver and tumor tissue in response to adrenalin need to be considered in relation to cytologic differences between these tissues. In regenerating liver, there are definite polyploid classes present in addition to DNA synthesis whereas azo dye-induced hepatomas have their DNA classes obliterated with only a stem line remaining. My question is, do you know what effect adrenalin has on the histology of these tissues and particularly on polyploidy?

H.TERAYAMA: Since we have been using rats weighing about 150 g, both diploid and tetraploid cells should exist in the liver. At the moment I do not know whether adrenalin effect can be selective to only one of these. We might be able to see it if we use very young rats, the livers of which are known to be consisted only of diploid cells. We have not checked the polyploidy problem as you mentioned. It is certainly a very interesting problem and I would like to do it in the future probably by autoradiographic methods, combined with microspectrophotometric techniques.

GROWTH POTENCY
OF INDIVIDUAL HUMAN DIPLOID CELLS
IN VITRO

MASA-ATSU YAMADA

Department of Pathology, National Institute of Health, Tokyo

IN 1961, Hayflick and Moorhead described a culture technique standardized for obtaining human diploid cell strains.[1] The cell strains thus obtained are characterized by their limited life spans, as compared with aneuploid cell lines multiplying indefinitely. The growth of diploid cells can be maintained actively for 40 to 50 transfers, then ceases rapidly. According to Hayflick, the actively growing stage, which continues for about 40 transfers from the primary culture, is referred to as phase II and the stage of declining growth terminating in cell death as phase III. Phase I, or the primary culture, ends with the formation of the first confluent sheet.

After their report had appeared, the results were confirmed[2-5] and efforts were directed towards elucidation of mechanisms operating in this limitation in proliferating capacity, or the differences between the characters of the cells in phase II and those in phase III.[6-8] Hayflick and Moorhead[1] showed that at the end of phase II, cell degeneration begins to take place and mitotic activity to lessen preliminary to the appearance of phase III. During the period of phase II, diploid chromosomal constitution was well retained except for few cases of aneuploidy. Aneuploid changes, however, appeared predominantly following a long period of the apparent chromosome stability, the earliest change being coincident with the beginning of phase III.[6] Macieira-Coelho and others[7] found marked heterogeneity in the length of growth cycle at the later transfer levels, as compared with those in phase II. This limitation in proliferative capacity of human diploid cell strains does not seem to be related to any essential lack of metabolites or to the presence of toxic materials in the medium. Also, no infection by latent viruses or mycoplasmas was demonstrated to be responsible for the cell degeneration. Finally, Hayflick[8] concluded that the time at which human diploid cell strains cease to divide was not essentially a function of the number of transfers but rather of the number of cell doublings and that this limitation would reflect an intrinsic character of diploid cells connected with the aging process.

In any investigations thus far reported, no precise observations have been made of the shift in growth mode during phase II. In these years we have studied the growth potency of individual cells in the diploid population through transfers and demonstrated that even in phase II, the cells deflect to non-dividing state randomly but at a rather constant rate, and the tendency rapidly increases at the later stage, phase III.[9-10] In this paper, I shall mainly discuss the growth potential of the individual cells at the early actively growing stage, or phase II.

Shift in Growth Pattern in Serial Transfers of Human Diploid Cells

At first, we followed Hayflick's experiments using almost the same technique with the human embryonic lung tissues and confirmed a limited life span of 40 to 50 transfers. We then slightly modified the cultural conditions: first we used Eagle's minimal essential medium instead of the basal medium, though the 10% serum content was the same. Second, he made subcultures by a split ratio of 1:2, i.e. one culture divided into 2 bottles when the cell sheet became confluent, usually every 3 to 4 days. But in our case, the seed size was almost constant at the level of 1×10^6 cells/bottle with 12 ml of the medium. In so doing, the growth pattern may be delineated more clearly. Subcultures were made every 3 to 4 days. In our diploid cell strains more than 95% of the metaphase cells were found to be diploid up to the 40th transfer generation.

Figure 1 shows the shift in growth pattern in serial transfers of NIHT 2 cells, one of our diploid cell strains thus maintained. Each curve is made connecting cell numbers at planting and harvest. In this simple way it can be seen that the growth potency of the population is going to decrease even in early transfers such as from the 10th to 20th, or from the 20th to 30th. Since the figure roughly represents the change in the growth pattern during trans-

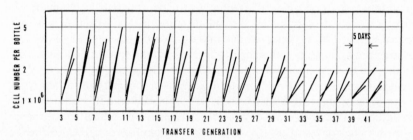

FIG. 1. Shift in growth pattern of NIHT 2 cells in serial transfers. Data were obtained from a routine transfer of a human diploid cell strain, NIHT 2, of a seed size of about 1×10^6 in 12 ml of Eagle's minimal essential medium supplemented with 10% calf serum. At each transfer, which were usually done every 3 to 4 days, a culture was divided into 2 to 5 bottles according to the harvest size. After 30 transfers, where the harvest size used to be less than 2×10^6, two cultures, for example, were pooled and divided into 3 bottles for keeping the seed size at more than 1×10^6. Each curve in the figure is depicted by connecting the seed size with the harvest size. The interval between two successive longitudinal lines represents 5 days. For a simplification, two curves at every other transfer generation are shown in Fig. 2.

fers, we closely examined the growth curve of the cell strain at various transfer levels. Figure 2 outlines a typical experiment in which estimation was
made at the same time, with the same batch of medium. The cells at different
transfer levels were taken from the frozen stock subcultivated 2 or 3 times

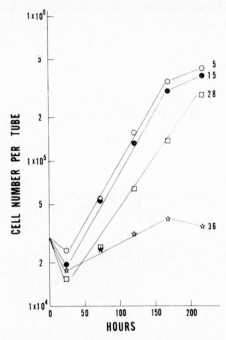

FIG. 2. Growth curves of NIHT 2 cells at four transfer levels. NIHT 2 cells at the
5, 15, 28, and 36th generation of routine transfers were seeded initially with 3×10^4
cells per tube in 1.5 ml of the medium. Cell counts were made at 2-day intervals
from 24 hr after seeding. Three separate tube cultures were used for each point on
the curve. Figures on the far right indicate transfer generations.

after thawing. As seen in Fig. 2, unexpectedly, the growth rate of cells in the
logarithmic phase is actually similar among the 5th, 15th, and 28th generations. However, initial decrease in cell number, or time lag, is found to increase as the transfer proceeds. At the 36th generation, the growth rate obviously decreases.

Figure 3 demonstrates summarized data on changes in the growth rate, expressed as doubling time in the logarithmic phase of diploid cell strains at the
various transfer levels. The data include diploid cell strains, NIHT 2, 4, and
5 in our laboratory as well as WI 38 established by Hayflick. It can be seen
from the figure, that the doubling time of the cells before 30 transfers remains virtually constant and the mean being 33.2 ± 4.5 hr. After the 30th
generation, the doubling time is shown to increase up to 100 hr or more, resulting in the cessation of cell multiplication.

The decreasing tendency of the growth rate during early serial transfers (Fig. 1) may reflect elongation of time lag, but not an actual increase in doubling time in the logarithmic phase. In other words, the results suggest that non-dividing cells gradually accumulate by serial transfers and that some fractions of their number are lost from the population by the procedure of subculture.

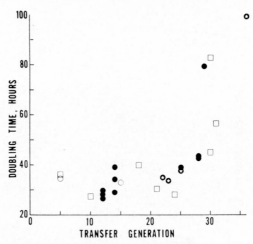

FIG. 3. Doubling time of diploid cell strains at various transfers. At the various transfer levels, growth curves were estimated as described in Fig. 2. Doubling time was calculated from the regression line in the logarithmic phase of the individual curves. The data include four human diploid cell strains: WI 38 (asterisk in solid circle), NIHT 2 (solid circle), NIHT 4 (open circle), and NIHT 5 (open square). All the strains show the same tendency to changes in doubling time according to the transfer levels.

Variation in Growth Potency among a Diploid Cell Population to Form Colonies

Figure 4 shows a Petri dish of diploid cells on the 14th day of cultivation in a CO_2 cabinet. One hundred cells were plated with 5 ml of Eagle's minimal essential medium with 20% calf serum. The batch of sera had been checked in advance and selected by the standard of nearly 100% of plating efficiency with HeLa S3 cells. By this precaution, high plating efficiency was to be expected in human diploid cell strains and was found as far as our examinations extended.

The colony of diploid cells shows a typical fibroblastic one with a rough edge and hairly curled arrangement. The colony size is subject to wide variation in marked contrast to the similarity with HeLa S3 cells. This had been demonstrated already by Puck and others in their papers[11-13] which made use of their elegant cloning techniques for animal cells in culture. We took several clonal lines of our diploid cell strains and found a wide variation of colony size even in the clonal populations. This observation suggests that a variation in colony size may depend upon an intrinsic cell property and not altogether on nutritional inadequacy.

We estimated the plating efficiency of the cells at various transfer levels from the number of colonies larger than a given size. In order to obtain the validity of such a comparison, we employed the same batch of serum throughout the experiment. As shown in Fig. 5, more than 50% of cells form large colonies at the 8th generation, and the efficiency decreases as the transfer proceeds. After the 20th generation, the figure becomes less than 10%. As a result, it can be said, we now have evidence that some fractions of the population lose the potential to form large colonies when subcultured, even at early transfer levels.

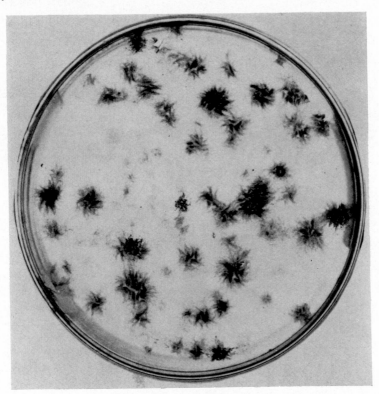

FIG. 4. Colonies developing from mono-dispersed NIHT 5 cells. One hundred NIHT 5 cells at the 11th transfer generation in routine transfers were plated in a 60 mm Petri dish with 5 ml of Eagle's minimal essential medium supplemented with 20% calf serum and incubated for 2 weeks in a CO_2 cabinet. Acetic alcohol fixation and 0.1% crystal violet stain. Colonies display typical fibroblastic features with rough edge and characteristic hairy whorl and reveal a wide variation in size. The variation in colony size still held true for plating of a clonal population isolated from a large colony.

A consideration of such characteristics of the cell strains has readily led to the following questions: does the progeny of a cell lose the capacity to divide randomly or systematically? How long do the non-dividing cells persist? The next experiment will reply to these questions.

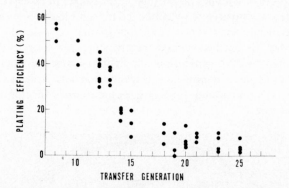

FIG. 5. Shift in plating efficiency of NIHT 2 cells during transfers. At various transfer levels, single-cell culture was performed with a small number of NIHT 2 cells as described in Fig. 4. The same batch of calf serum was employed throughout the experiment. At the 8th generation or later, all the colonies revealed fibroblastic morphology. As seen in Fig. 4, there appeared a wide variation in colony size and the plating efficiency was tentatively defined here as the percentage of number of colonies larger than 2 mm in diameter 2 weeks after plating, to the number of cells plated.

Clonal Growth Curve of Human Diploid Cells

We examined the growth potential of individual cells in the process of colony formation (Fig. 6). For the observation of clonal growth, we employed the window technique devised by Marcus[14] in which a metal plate with small round windows is applied under a Petri dish with double-faced scotch tape. When a suitable number of single cells is seeded to obtain one or two cells per window, the multiplication of individual cells can be observed at intervals through the windows by an inverted microscope. In the case of human diploid fibroblasts, windows 6 mm in diameter, were required for one week observation, since fibroblasts move more vigorously than HeLa S3 cells. The figure on the right in Fig. 6 shows clonal growth of 30 diploid cells at the 10th generation and the one on the left that of HeLa S3 under the same culture conditions.

Twenty-four hours after HeLa S3 cells were plated, two fifths of the total individual cells become 2 cells, and almost all the cells more than 2 cells by 48 hr. As an exception, 2 individual cells do not divide by 72 hr, but multiply thereafter. In fact, all the cells plated and attached to the glass were observed to divide at least once during the observation period. The curves showing exponential growth approximate with each other after temporary incubation. We calculated the mean growth curve for the 30 cells and found that it takes a time lag of 10 hr and a generation time of 22 hr. These figures coincide well with those in mass culture.

As compared with HeLa S3, the individual growth curve of human diploid cells was shown to scatter characteristically. The fastest curve corresponds to that of HeLa S3 and some fractions persist as one cell and do not divide at all by 140 hr, or through the whole period of observation. This suggests that life spans of the non-dividing cells are much longer than the generation time of dividing cells. In any case no cells divide by 24 hr after being seeded. From these curves with a wide variation, we tentatively estimated the mean generation time to be 27 hr. However, because these populations actually contain non-dividing cells which have appeared randomly at the stages seen in Fig. 6, it is likely that the figure represents the mean doubling

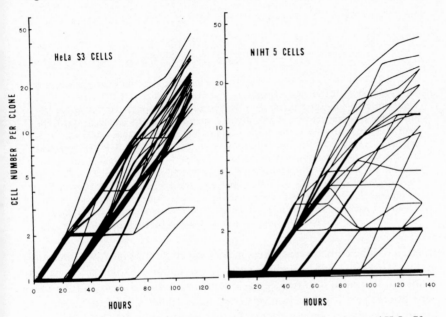

FIG. 6. Clonal growth curve of HeLa S3 and NIHT 5 cells. Two thousand HeLa S3 cells were plated into a 60 mm Petri dish with 5 ml of Eagle's minimal essential medium supplemented with 20% calf serum and incubated in a CO_2 cabinet. On the next day, they were applied to a metal plate with 16 windows, 1.2 mm in diameter, under the dish with double-faced scotch tape. About 30 clones which appeared solely in the center of windows were selected to be observed. The number of cells forming each clone was counted once a day and plotted against the incubation period in the left figure. In the case of NIHT 5 cells, the motility is higher than that of HeLa S3 and cells forming a clone dwelt sparsely. For the above reason, 6 mm windows were employed and only 200 NIHT 5 cells (10th generation) were plated into a dish with 5 ml of the medium with the result that 1 or 2 clones were found in a window. For about 30 clones individual growth curves are shown in the right figure.

time instead of the generation time. In case of HeLa S3 cells, on the contrary, both figures are in good accordance with each other.

Generation Time of Human Diploid Cells

With the above idea as a guide, we determined the true generation time of human diploid cells by means of autoradiography with ³H-thymidine (Fig. 7). The material was a NIHT 5 cell strain at the 13th generation. The

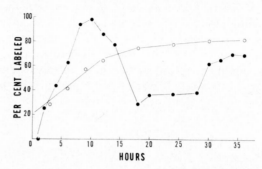

FIG. 7. Distribution of labeled mitosis and interphase after pulse- and continuous labeling with ³H-thymidine, respectively. For pulse-labeling, NIHT 5 cells (13th generation) growing exponentially onto coverslips were exposed to 0.1 μC/ml of ³H-thymidine (specific activity of 2 C/mM) for 30 min. After the labeling period, the cells were washed twice with culture medium and fresh medium was added supplemented with 2 × 10⁻⁵ M non-radioactive thymidine. The coverslips were removed at intervals thereafter, fixed in acetic acid-ethanol (1:3) and processed autoradiographically with NR-M 2 emulsion (Konishiroku Photo Industrial Co., Tokyo). Per cent labeled mitosis is plotted against the period of chasing (solid circle). For continuous labeling, the cells were exposed to 0.02 μC/ml of ³H-thymidine. The coverslips were removed at intervals, fixed, and processed autoradiographically. Per cent labeled cells were scored in this experiment (open circle).

fraction of labeled mitoses, expressed by solid circles in Fig. 7, was measured at intervals after pulse-labeling with ³H-thymidine. From the curve, we estimated the mean duration of each compartment of a cycle and the mean generation time. The duration of G1 phase is 10 hr; S, 9 hr; G2, 5 hr; and the total generation time is 24 hr. These figures are a little bit longer than previous estimates of 18 hr by Defendi and Manson[15] or by Macieira-Coelho and others,[7] but still obviously shorter than the doubling time of 33 hr we have presented in Fig. 3.

Considering that only a fraction of cells capable of division participated in calculating the generation time by the pulse-labeling method, the discrepancy of the values between generation time and doubling time suggests that some cells in a diploid population would transfer into a non-dividing state during a growth cycle in logarithmic phase, even when the cells are at the earlier transfer levels or in phase II. The fact was, furthermore, demonstrated by the data on the fraction of labeled cells after continuous labeling with

^{3}H-thymidine (indicated by open circles in Fig. 7). The figures do not reach 100% after an incubation period equivalent to one generation time. We have also proved that the level of per cent labeled cells after 24 hr of labeling decreases as the transfer proceeds.

Analysis of DNA Synthesizing Cells at Clonal Level

The relation between growth potency and capacity of DNA synthesis are reviewed in Table 1. A small number of NIHT 5 cells at the 16th generation were plated onto large coverslips inserted in dishes and incubated for 7 days in a CO_2 cabinet. During the period, colonies appeared of a variety of sizes, from one cell alone to colonies consisting of more than 100 cells. At the time, the cells were incubated with ^{3}H-thymidine for 48 hr corresponding to 2 generation times. Then, coverslips were removed, fixed, and processed autoradiographically. Fractions of DNA synthesizing cells were estimated according to colony size. Column 1 in the table indicates the groups classified by logarithmic scale of cell number per clone based on 2, and the ranges of standard deviation. Column 2 shows the number of clones which participated.

TABLE 1. FREQUENCY OF DNA SYNTHESIZING CELLS PER CLONE

Two hundred NIHT 5 cells at the 16th generation were plated with 5 ml of Eagle's minimal essential medium supplemented with 20% calf serum into 60 mm Petri dishes which contain a coverslip, 32 × 24 mm in size. Seven days after incubation, the cells were incubated with 0.05 μC/ml of ^{3}H-thymidine for 48 hours. The coverslips were then removed, fixed, and processed autoradiographically.

No. of Cells per Clone	No. of Clones	No. of Cells Labeled / Total No. of Cells	Per Cent Labeled
1	44	5/44	11.4
2	13	1/26	3.9
4 (3 - 6)	17	17/70	24.3
8 (7 - 12)	12	24/106	22.6
16 (13 - 24)	12	67/200	33.5
32 (25 - 48)	8	175/243	72.0
64 (49 - 96)	13	627/951	65.9
128 (97 -)	8	914/1177	80.8

Column 3 indicates the fraction of labeled cells and column 4 their percentage. As seen in the table the percentage of labeled cells is really low with small clones consisting of 1 or 2 cells. It can be mentioned that the bigger the colony size, the higher was the percentage. Furthermore, even in large colonies consisting of more than 32 cells, one can still see 20 to 30% of non-labeled cells. The fraction of cells has a generation time longer than twice the mean. These would actually be considered as non-dividing cells.

G1 Block in Diploid Cell Population

All the data I have mentioned here have been concerned with evidence for the presence of non-dividing cells in a young population of human diploid cell strains. The frequency of appearance of such cells seems to be random but at a fairly constant rate as a whole during the early transfers. The accumulation of the non-dividing cells seems to reflect a decrease in growth rate or, more precisely a delay in time lag through transfers. Now we shall have to ascertain in which part of the cycle the non-dividing cells are located.

We demonstrated that DNA content per cell measured by the Schmidt-Thanhauser method decreases when the culture enters a stationary phase at a given transfer level or proceeds to the later transfer levels. In the 17th transfer generation, for example, the DNA content per cell was measured as 1.63×10^{-11} g, 1.40, and 1.18 at the 3rd, 5th, and 8th day of cultivation, respectively. By microspectrophotometry, we have also been able to demonstrate an accumulation of G1-phase cells at the stationary phase. The data entirely conflict with those of Macieira-Coelho and others,[16] who have reported that in the stationary phase, or at the later transfer levels, the cells were blocked in the G2 period. Analyzing their data carefully, the interpretation of G2 block would not be valid. Cells were first labeled with ^{3}H-thymidine for 24 hr nearly at the stationary phase. During the period, the labeled cells would be distributed around the whole cycle. Experiments were then made for another 24 hr and it was shown that no labeled cells passed through mitosis during the period. Their results seem to indicate that cells may be blocked in any part of the cycle not necessarily in G2 phase. On this matter, I discuss details elsewhere.[17]

In summary, we have shown here the transition of a growth compartment in a population of diploid cells to non-dividing ones, even at phase II which has so far been considered simply as a luxuriently growing stage. After Hayflick's report, attention has mainly been concentrated on the transition of phase II to phase III as a whole population. When the culture of human diploid cell strains enters phase III, drastic changes take place, followed by total degeneration of the culture. If the limited life span is an expression of aging or senescence at the cellular level as stated by Hayflick,[1,8] any sign of aging would be likely to be found not only at phase III, but also at the phase II even to a slight degree. Our finding would be the case. Growth of individual diploid cells seems to be controlled by halting at the cross roads of multiplication or non-division of a cycle in a similar way to the somatic cells in the animal body. As cited by Konigsberg,[18] a common pattern in many segregated stem cell populations *in vivo*, such as those represented by the basal layers of stratified squamous epithelium, is that subsequent specialization in structure and function is associated with emigration out of the proliferative zone. Cytodifferentiation, therefore, may be a response to the new cellular environment of the migrating cells. However, if we consider the situation in which both daughter cells resulting from a single stem cell are located in the same environment, as in the case of hemopoietic tissues, such stem cells *per se* would be the sites of action for control mechanisms.

Till and others[19] discovered this kind of control mechanism *in vivo* using colonies of the bone marrow cells developing in the spleen of X-irradiated mice, and showed that a "birth-and-death" process may be operative during the growth of spleen colonies. According to their model of proliferation during the growth of spleen colonies, the two processes, "birth" and "death", occur at random in the population. This implies that individual cells are not closely regulated. However, it appears possible that their studies of the progeny of single cells may display a random feature of hemopoietic function, while a study of large populations of cells reveals the ordinary behavior of the whole system. The growth pattern of human diploid cell strains described here would be considered as an *in vitro* model system for analyzing this kind of growth control.

REFERENCES

1. L.HAYFLICK and P.S. MOORHEAD, *Exptl. Cell Res.*, **25**, 585, 1961.
2. In a symposium held at Opatija, Yugoslavia, 1963, several papers dealt with the cessation of the growth in human diploid cell strains (see Proceedings: Human Diploid Cell strains, Blasnikova Tiskarna, Ljubljana, Yugoslavia, 1963).
3. G.J.TODARO, S.R. WOLMAN and H.GREEN, *J. Cellular Comp. Physiol.*, **62**, 257, 1963.
4. M.C. YOSHIDA and S.MAKINO, *Japan J. Human Genetics*, **8**, 39, 1963.
5. C.P.MILES, *Cancer Res.*, **24**, 1070, 1964.
6. E.SAKSELA and P.S.MOORHEAD, *Proc. Nat. Acad. Sci.*, **50**, 390, 1963.
7. A.MACIEIRA-COELHO, J.PONTÉN and L.PHILIPSON, *Exptl. Cell Res.*, **43**, 20, 1966.
8. L.HAYFLICK, *Exptl. Cell Res.*, **37**, 614, 1965.
9. M.YAMADA, *Kagaku*, **34**, 295, 1964.
10. M.YAMADA, *Japan J. Med. Sci. Biol.*, in press.
11. T.T.PUCK, P.I.MARCUS and S.J.CIECIURA, *J. Exptl. Med.*, **103**, 273, 1956.
12. T.T.PUCK, S.J.CIECIURA and H.W.FISHER, *J. Exptl. Med.*, **106**, 145, 1957.
13. T.T.PUCK, S.J.CIECIURA and A.ROBINSON, *J. Exptl. Med.*, **108**, 945, 1958.
14. P.I.MARCUS and T.T.PUCK, *Virology*, **6**, 405, 1958.
15. V.DEFENDI and L.A.MANSON, *Nature*, **198**, 359, 1963.
16. A.MACIEIRA-COELHO, J.PONTÉN and L.PHILIPSON, *Exptl. Cell Res.*, **42**, 673, 1966.
17. M.YAMADA, H.OHNO and N.NAKAZAWA, data to be published.
18. I.R.KONIGSBERG and S.D.HAUSCHKA, In *Reproduction: Molecular, Subcellular and Cellular*. Ed. by M.Locke, Academic Press, New York, 1965.
19. J.E.TILL, E.A.McCULLOCH and L.SIMINOVITCH, *Proc. Nat. Acad. Sci.*, **51**, 29, 1964.

DISCUSSION

M.E.KAIGHN: Did you subclone the large clones and determine the growth properties of the resultant population of individual subclones?

M.YAMADA: Clonal cell populations showed also wide variation in colony size as wild strains.

STUDIES ON THE NUCLEIC ACID
METABOLISM OF LEUKEMIC CELLS

Gyoichi Wakisaka, Toru Nakamura, Shigeru Shirakawa
Yataro Yoshida, Akio Todo, Akiko Kano
and Hiroyoshi Sawada

The First Division, Department of Internal Medicine, Faculty of Medicine,
Kyoto University, Kyoto, Japan

As is well known, nucleic acids play an important role in the control of cell proliferation and cell maturation. However, only a few studies have been reported on the quantitative aspects of nucleic acid metabolism and the kinetics of proliferation and maturation of human leukemic cells.[1-11] In this study the nucleic acid metabolism of human leukemic leukocytes was investigated employing chemical assay of the incorporation of radioactive precursors into nucleic acids of the cells *in vitro*, autoradiography of leukemic cells using thymidine-^{3}H, and microspectrophotometric determination of DNA content of the cell nucleus after Feulgen staining.

NUCLEIC ACID METABOLISM OF LEUKEMIC CELLS
AS OBSERVED FROM THE INCORPORATION
OF ADENINE-^{14}C INTO RNA AND DNA

Leukocytes were isolated from the blood of patients with leukemia and normal controls according to the method described by Skoog and Beck,[12] and the leukocytes thus obtained were suspended in Tyrode's solution in a concentration of 10^8 cells/ml. To 0.5 ml of this leukocyte suspension were added 0.5 ml of the autologous serum, 0.5 ml of Tyrode's solution and nucleic acid precursors labeled with ^{14}C or ^{3}H, and the mixture was incubated at 37°C for 3 hr. Then, the nucleic acids were extracted from leukocytes by the method of Hecht and Potter,[13] and the radioactivity in each fraction was measured with a gas flow counter. The amount of nucleic acid was determined by the Beckman's UV spectrophotometer at 260 mμ. For the extraction of nucleic acid bases, Bendich's method[14] was used.

In peripheral leukocytes of normal controls adenine-^{14}C was highly incorporated into RNA. On the contrary the incorporation of adenine-^{14}C into DNA was negligible. In peripheral leukocytes obtained from patients with leukemia the specific activity of RNA was within the normal range in most cases, while the specific activity of DNA was increased above normal in all cases of leukemia except for one case of chronic lymphatic leukemia (Fig. 1). The increase of specific activity of DNA was especially marked in

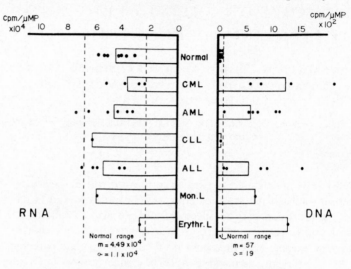

FIG. 1. Adenine-^{14}C incorporation into nucleic acids of leukemic leukocytes.

chronic myeloid leukemia. The ratio of specific activity of DNA to the percentage of immature cells was lower in acute leukemia than in chronic myeloid leukemia, suggesting that the average DNA synthesis of immature cells is lower in the former than in the latter (Fig. 2). There was no correlation between the percentage of immature cells and specific activity of RNA

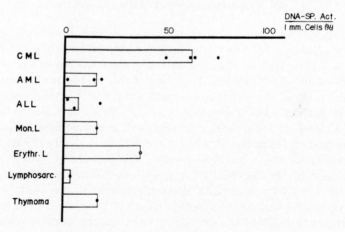

FIG. 2. Ratio of DNA-specific activity to immature cells.

and DNA. However, when only the cases of chronic myeloid leukemia were taken into consideration, there was a close relationship between the percentage of immature cells and the specific activity of DNA. The correlation coefficient was 0.96, which was statistically significant at the level of 1%. Therefore, it might be assumed that there is no individual difference in the average DNA synthesis of immature cells in chronic myeloid leukemia.

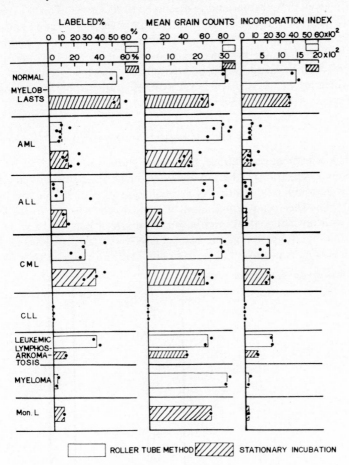

FIG. 3. Thymidine-^{3}H uptake by leukemic blast cells (autoradiography).

In order to investigate the difference of the specific activity of DNA according to the types of leukemia at cellular levels, autoradiographic studies using thymidine-^{3}H were performed *in vitro* on bone marrow cells obtained from patients with leukemia, and the percent of labeled cells and grain counts were observed. The total grain count per 100 blast cells which was obtained by multiplying the mean grain counts by percent of labeled cells was as follows:

on average, 1158 in chronic myeloid leukemia (4 cases), 352 in acute myeloid leukemia (7 cases), 157 in acute lymphatic leukemia (2 cases) and 378 in monocytic leukemia. These values were parallel to the specific activity of DNA, which was determined chemically by the incorporation experiment of adenine-^{14}C into DNA.

There was a marked difference in the percent of labeled cells between chronic myeloid leukemia and acute myeloid leukemia. This was very low in acute leukemia as compared with chronic myeloid leukemia and normal controls, while there was not much difference in the mean grain counts (Fig. 3). Therefore, the low incorporation rate of adenine-^{14}C into DNA in acute myeloid leukemia might be attributed to the low percentage of blast cells which are at the stage of DNA synthesis rather than the prolongation of the stage of DNA synthesis.

KINETICS OF LEUKEMIC CELLS

As has been mentioned above, the *in vitro* studies on thymidine-^{3}H uptake of leukemic cells showed that, in acute leukemia, the percent of labeled cells was low as compared with that of normal controls and chronic myeloid leukemia, while the mean grain count of labeled cells was not lower as compared with normal controls and basophilic erythroblasts on the same smear. These findings suggest that the rate of DNA synthesis of labeled cells in acute leukemia is not lowered as compared with normal controls.

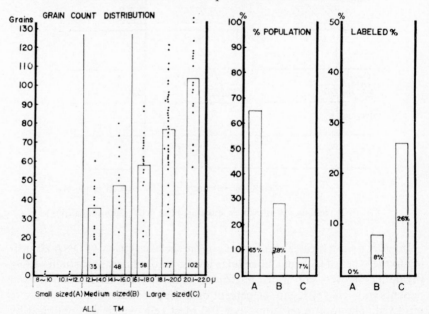

FIG. 4. Thymidine-^{3}H uptake by leukemic blast cells in relation to cell size.

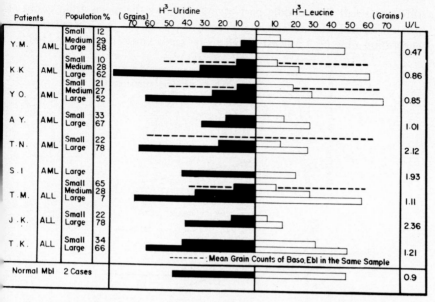

FIG. 5. Uridine-³H and leucine-³H uptake by leukemic cells (mean grain counts and uptake ratio).

In order to further analyze the DNA synthesis of leukemic cells, they were divided into three groups according to the size of the cells, namely large, middle-sized and small cells, and the uptake of thymidine-³H was observed *in vitro*. Most of the small cells, which represented the major component of the leukemic cell population, were not labeled, while the large cells were labeled in a high percentage and the percent of labeled cells in the middle-sized cell group was in between (Fig. 4). The uptake of uridine-³H and leucine-³H was also high in proportion to the size of the cells, being comparable to that of basophilic erythroblasts in large cells, while the uptake of uridine-³H and leucine-³H of small cells was low (Fig. 5). It may be postulated from these findings that the leukemic cells are composed of two compartments, namely dividing and non-dividing compartments, and the cells of the dividing compartments exhibit pretty rapid proliferative activity.

Further observations were made on the *in vivo* uptake of thymidine-³H by leukemic cells in patients with chronic myeloid leukemia and acute leukemia. Figure 6 shows the *in vivo* labeling of mitotic figures of bone marrow cells in the course of time in a patient with chronic myeloid leukemia injected intravenously with 5 mc of thymidine-³H.

The percentage of labeled mitotic figures was observed in the course of time, and from these results the length of G_2 period was estimated to be 3 hr, and that of S period from 17 to 20 hr. The second peak of percent labeling could not be observed, because the number of bone marrow puncture was small. The middle part of Fig. 6 shows the mean grain count as plotted against time on semilogarithmic paper. From this observation the generation time (T_G) was estimated to be approximately 4.5 days (108 hr).

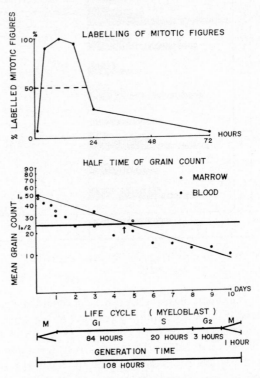

FIG. 6. *In vivo* labeling with thymidine-^{3}H in chronic myeloid leukemia. (C.T. 29, M.)

The lower part of Fig. 6 summarizes the life cycle of myeloblasts in this patient; $T_G = 108$ hr, $G_1 = 84$ hr, $S = 20$ hr, $G_2 = 3$ hr, and $M = 1$ hr (the value of M was estimated from the literature).

Figure 7 shows the percentage of labeled cells at various stages of maturation by time in days after the intravenous injection of thymidine-^{3}H in the above-mentioned patient with chronic myeloid leukemia. It is interesting to observe that in initial labeling the percentage of labeled cells was higher in bone marrow cells (21 %) than in peripheral blood cells (12 %), and that the percentage of labeled cells was highest in promyelocytes followed by that of myelocytes and myeloblasts in this order. Similar findings were obtained *in vitro* in another patient with chronic myeloid leukemia. The percentage of labeled myeloblasts increased with the lapse of time, reaching its maximum at 24 hr, both in the bone marrow and peripheral blood. These findings suggest that the percentage of labeled myeloblasts increases by division of labeled myeloblasts and that a large part of the labeled myeloblasts enter the peripheral blood without going into a maturation pool. The percentage of labeled myeloblasts decreased after 24 hr and demonstrated again a slight increase at 6 days. Also, from these results the cellular cycle of myeloblasts was estimated to be from 4 to 5 days.

As to promyelocytes, the initial labeling index was high, but it decreased rapidly with the lapse of time, indicating that the labeled promyelocytes matured into myelocytes after their division. The time course of labeled myelocytes corresponded well to this finding. Metamyelocytes, staff cells (band forms) and polymorphonuclears showed no initial labeling. They were labeled at a later stage in the order of maturation. In this case, the differential count of bone marrow cells was: 3.6% myeloblasts, 3.0% promyelocytes, 15.5% myelocytes, 23.6% metamyelocytes and 54.7% neutrophil staff cells and polymorphonuclears. It is interesting to note that the percentage of labeled myeloblasts was low as compared with that of normal controls. This finding suggests that some of the myeloblasts are undergoing homoplastic division and remain in the non-dividing compartment, and some of the myeloblasts mature into promyelocytes in response to demand.

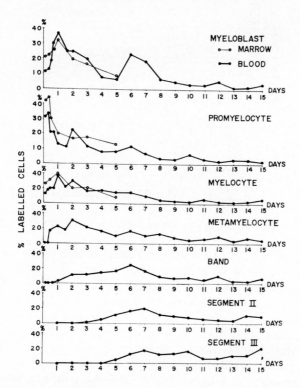

FIG. 7. Time sequence of labeled cells in chronic myeloid leukemia. (C.T. 29, M.)

The upper part of Fig. 8 shows the time course of labeled mitotic figures in the bone marrow of a patient with acute myeloid leukemia after a single intravenous injection of 5 mc of thymidine-^{3}H. From this observation, the length of the S period was estimated to be 20 hr. There was no second peak in the labeling index. The middle part of Fig. 8 illustrates the time course of mean grain count plotted on semilogarithmic paper.

From the half time of the mean grain count, the generation time (T_G) was calculated to be approximately 3.5 days (84 hr). The lower part of Fig. 8

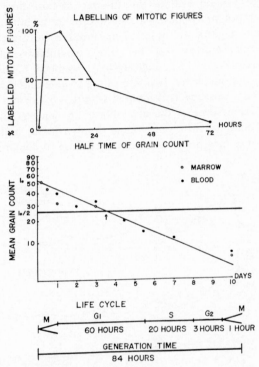

FIG. 8. *In vivo* labeling with thymidine-^{3}H in acute leukemia. (T.M. 29, F.)

summarizes the life cycle of leukemic blast cells in this case; $T_G = 84$ hr, $G_1 = 60$ hr, $S = 20$ hr, $G_2 = 3$ hr, and $M = 1$ hr.

Figure 9 shows the time course of labeling index of leukemic blast cells in the bone marrow and peripheral blood. Black circles show the labeling index of *in vivo* labeling, and white circles that of *in vivo* and *in vitro* double labeling. In bone marrow the initial labeling index was 2.5% and it increased until 24 hr, indicating that the initially labeled blast cells may have increased in number by division. In the *in vivo* and *in vitro* double labeling, the labeling index at 24 hr was twice as high as that in the *in vivo* labeling, probably due to labeling of blast cells which entered the phase of DNA synthesis at this time.

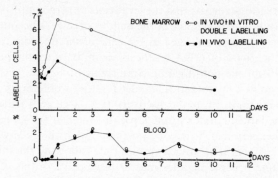

Fig. 9. Time sequence of labeled cells in acute leukemia. (T.M. 29, F.)

In the peripheral blood there was no initial labeling either *in vivo* or *in vitro*. After 10 hr there appeared a small number of labeled cells in the peripheral blood and the labeling index increased from 24 to 72 hr. All of the labeled cells were of middle size or small size. These findings suggest that the cells, which had completed DNA synthesis and undergone mitosis in the bone marrow, appeared in the peripheral blood. Namely, in this case, the blast cells in the peripheral blood were considered to belong to the non-dividing compartment. However, there remains a possibility that these blast cells in the peripheral blood may settle in some tissues or organs and re-enter a phase of DNA synthesis there. The labeling index of blast cells in the peripheral blood in the *in vivo* and *in vitro* double labeling experiment was similar to that in the *in vivo* labeling experiment.

Figure 10 gives the time course of labeling index in blast cells of different sizes. The high initial labeling index of large blast cells decreased with the passage of time, whereas the labeling index of small blast cells, which were not labeled initially, increased with the lapse of time. The labeling index of middle-sized blast cells was lower than that of large cells at the initial stage, and it remained at nearly the same level during 24 hr.

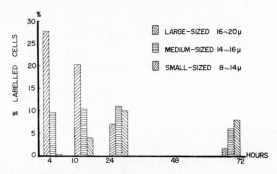

Fig. 10. Time sequence of labeling index in different blast cell size in acute leukemia.
(T.M. 29, F.)

These observations suggest that the large cells undergo mitosis and become middle-sized cells, and that some of the middle-sized cells become small cells, while the others become enlarged as the cells of the dividing compartment, and re-enter the phase of DNA synthesis again, or enter the peripheral circulation as cells of the non-dividing compartment.

Figure 11 shows a hypothetical model for the proliferation of leukemic cells in acute leukemia. From the discovery that large cells which showed a decrease in the labeling index in the lapse of time had a high labeling index in the *in vitro* experiment, it may be postulated that the cells which entered a phase of DNA synthesis after a certain time following the initial labeling had by that time become middle-sized or large cells. In this case the grain counts of leukemic blast cells at the initial labeling were almost similar to those of basophilic erythroblasts, which were found on the same smear.

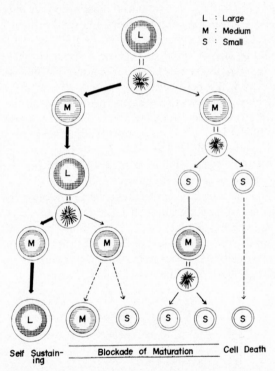

Fig. 11. Kinetic model for cell proliferation in leukemia.

From these results it seems likely that, although the generation time of leukemic blast cells in acute leukemia is prolonged as compared with that of normal myeloblasts, it is not so long as expected from the low labeling index in the *in vitro* experiment, and that the cells in the dividing compartment have nearly the same length and activity of the phase of DNA synthesis as normal cells, although the length of G_1 period is prolonged. Because the absolute number of blast cells is increased and they are undergoing homoplastic division, it is not surprising that the leukemic blast cells can proliferate rapidly in spite of their prolonged generation time.

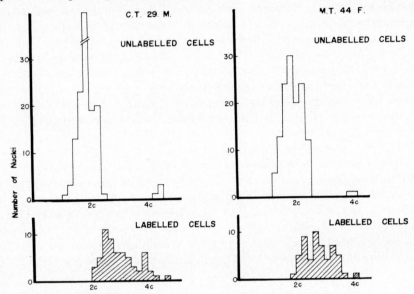

FIG. 12. DNA histogram of labeled and unlabeled cells (chronic myeloid leukemia).

In this case S/T_G was $20/84 = 0.24$ (24%), while the observed percent of initial labeling was 2.4%. From these results it may be concluded that in acute leukemia, the cells of the non-dividing compartment make up a large proportion of the blast cells. On the other hand, in chronic myeloid leukemia, there was an agreement between the calculated labeling index and the observed one, indicating that there is no non-dividing compartment in chronic myeloid leukemia.

In order to learn the correlation between the DNA content of leukemic blast cells and their DNA synthesis, simultaneous observation of DNA content and the uptake of thymidine-^{3}H was made on the same cells by employing the technique of microspectrophotometry and autoradiography. Figure 12 gives the distribution pattern of DNA content of labeled and unlabeled cells in the bone marrow of two patients with chronic myeloid leukemia. As shown in the upper part of this figure, most of the unlabeled cells had a DNA content of 2c value, while only a small number of unlabeled cells had one of 4c value. On the contrary, all of the labeled cells had DNA content between 2c and 4c values, indicating that all the cells in the phase of DNA synthesis, as judged from their DNA content, incorporated thymidine-^{3}H within 1 hr by *in vitro* experiment with thymidine-^{3}H (1 µc/ml, exposure 4 weeks).

Figure 13 shows the pattern of DNA synthesis of leukemic cells in bone marrow in two patients with chronic myeloid leukemia. The lower part shows the relationship between mean grain counts and DNA content. Although there was a wide variation in the grain counts of cells with the same DNA content, the mean grain count was highest in the cells with DNA content in the middle of 2c and 4c, that is in the cells at the middle of DNA synthesis. The middle part of this figure illustrates the mean grain counts of cells as plotted against their DNA content, indicating that the grain counts reach the maximum at the middle of S period and decrease gradually thereafter. The upper part shows the schematic pattern of the rate of DNA synthesis of leukemic cells, as constructed from the curve in the middle part of the figure, assuming that the grain counts represent the rate or speed of DNA synthesis in a given time.

Figure 14 shows the pattern of DNA synthesis of leukemic cells at various stages of maturation in a case of chronic myeloid leukemia. As shown in the left part, the percent of cells in the stage of DNA synthesis (namely the cells with DNA content between 2c and 4c values) was highest in promyelocytes (39.6%), and was followed by that of myelocytes (23.6%). In myeloblasts the percent of cells in the stage of DNA synthesis was low (20%), and most of the cells had DNA contents of about 2c value, indicating that most of the cells were at the stage of G_1. The hatched column shows the percent of cells

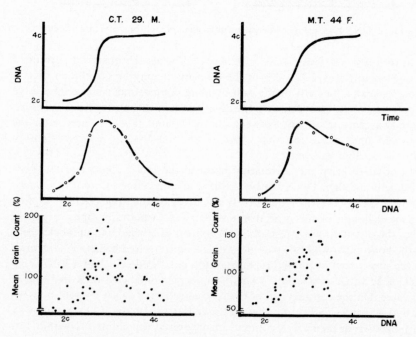

FIG. 13. Pattern of DNA synthesis of leukemic cells (chronic myeloid leukemia).

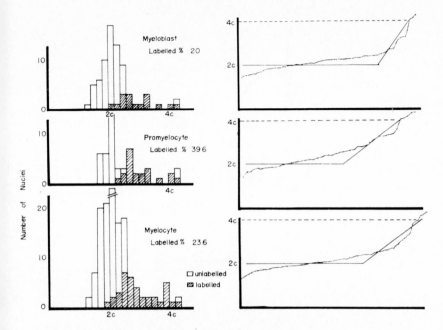

FIG. 14. DNA histogram and life cycle pattern of leukemic cells in chronic myeloid leukemia. (C.T. 29, M.)

labeled by thymidine-^{3}H. Most of the cells at the stage of DNA synthesis were labeled by thymidine-^{3}H. The right part of Fig. 14 presents the distribution of DNA content of cells at various stages of maturation. The DNA content of individual cells was plotted at equal intervals, beginning with the lowest value from the left. From this figure, it is possible to estimate the ratio of cells at the stage of G_1 and S and the life cycle of cells in a given population. As shown in the middle part the number of cells at the stage of S period was largest in promyelocytes, followed by that of myelocytes. In myeloblasts, on the other hand, there were more cells at the stage of G_1 as compared with promyelocytes and myelocytes.

Figure 15 shows the distribution of DNA content of labeled and unlabeled leukemic blast cells in a patient with acute lymphoblastic leukemia. Most of the cells with DNA content of 2c and 4c values were not labeled, while the cells with DNA content between 2c and 4c values were all labeled. In other cases of acute leukemia with low labeling index, some of the cells with DNA content between 2c and 4c values were not labeled, indicating that there may be an arrest of DNA synthesis in some of the cells in acute leukemia (Fig. 16). Anyhow, in acute leukemia most of the cells had the DNA content of 2c value, and only a small number of cells had the DNA content between 2c and 4c values, indicating prolongation of G_1 period.

12a NAM

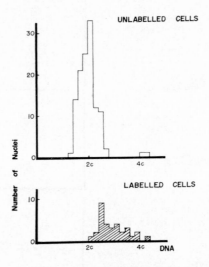

FIG. 15. DNA histogram of labeled and unlabeled cells (acute lymphoblastic leukemia, T.K. 18, F.).

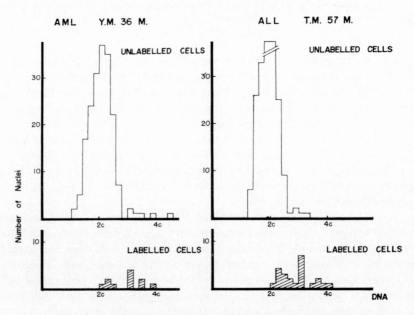

FIG. 16. DNA histogram of labeled and unlabelled cells in acute leukemia.

Figure 17 shows the DNA content of leukemic blast cells according to the cell size. As shown in the left part, most of the small cells (nuclear diameter < 7.5 μ) had DNA content of 2c value. In middle-sized cells (nuclear diameter 7.5–10.0 μ), a small number of cells had the DNA content between 2c and 4c values, while among large cells (nuclear diameter > 10 μ) most of the cells had a DNA content between 2c and 4c values. In the three cell groups, a majority of the cells with DNA content of 2c value were not

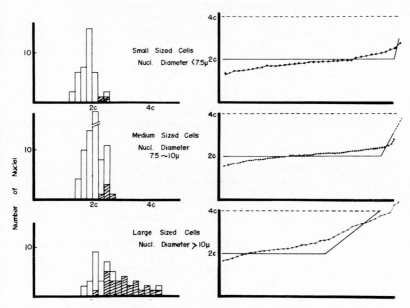

Fig. 17. DNA histogram and life cycle pattern of blast cells of different size (acute lymphoblastic leukemia, T.K. 18, F.).

labeled, while most of the cells with DNA content between 2c and 4c values were labeled. The right part of Fig. 17 presents the distribution of DNA content of leukemic blast cells according to cell size in the same case of acute lymphoblastic leukemia. In small cells, most of them were at the stage of G_1, while in large cells there were more cells with DNA content between 2c and 4c values than in middle-sized and small cells. These results suggest a prolongation of G_1 period in small cells.

METABOLISM OF RNA IN LEUKEMIC CELLS

Leukemic cells were collected by the dextran method from the peripheral blood of patients with acute leukemia before treatment or at the stage of acute exacerbation. The cells were suspended in Tris-Hanks solution and incubated with ^{32}P at 37 °C for 1 to 2 hr. Then the cells were suspended in the extraction medium, and DNA and RNA were extracted using phenol and ethanol as described in Fig. 18. The precipitated nucleic acids were subjected to MAK (methylated albumin Kieselgur) column chromatography, and s-RNA, DNA and r-RNA were separated by changing the NaCl concentration from 0.2 M to 1.4 M, using 0.05 M phosphate buffer of pH 6.7 and 5×10^{-4} M EDTA.

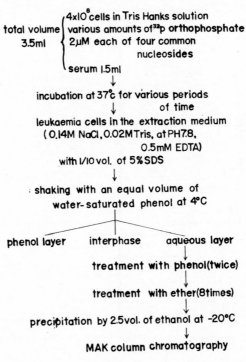

FIG. 18. Methods of incubation of leukemic cells and isolation of nucleic acids.

On each fraction, optical density was measured at 260 mμ and the radioactivity was counted in a gas-flow counter. For time course study, the extraction of nucleic acids was performed according to the method of Hecht and Potter,[13] and the separation of nucleus was made by Schneider's method using 0.25 M sucrose.

In the Tris-Hanks solution the specific activity of DNA and RNA increased linearly with the lapse of time, indicating that the leukemic cells maintain their activity for 3 hr in our incubation system. In a case of acute lymphoblastic leukemia, after 60 min incubation, the peak of ^{32}P radioactivity did not coincide with that of optical density. There appeared peak 1 (P_1) between DNA and r-RNA, and peak 2 (P_2) after r-RNA (Fig. 19). After 120 min incubation, there appeared ^{32}P radioactivity at s-RNA, DNA and r-RNA, but peak 1 and peak 2 were still present. Similar results were obtained likewise in a case of acute myeloblastic leukemia. Peak 2 and probably peak 1 are rapidly labeled RNA with rapid turnover, which is different from s-RNA and r-RNA, and they are mostly nuclear RNA.

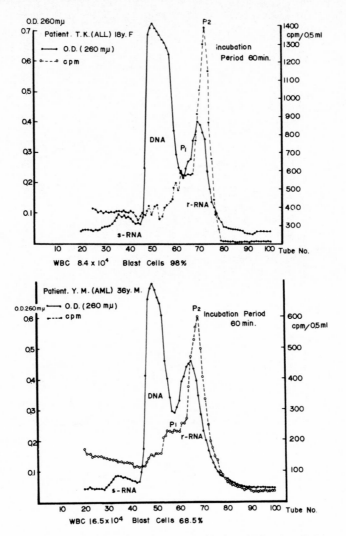

FIG. 19. MAK column chromatogram of nucleic acids from leukemic cells in acute leukemia.

In a case of chronic myeloid leukemia, the peak of rapidly labeled RNA was not observed after 60 min incubation (Fig. 20). At the time when this patient developed an acute blastic crisis, there appeared peak 1 and peak 2 as well as the peaks at s-RNA, DNA and r-RNA after 60 min incubation. At

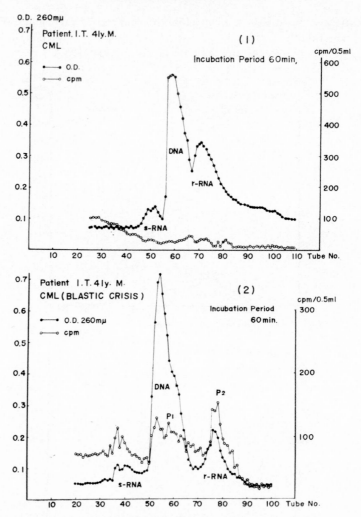

FIG. 20. MAK column chromatogram of leukemic cells in chronic myeloid leukemia
before and after blastic crisis.

the time of remission of the blastic crisis following treatment with steroid hormone and 6-MP, the peaks of rapidly labeled RNA were not observed
(Fig. 21). Table 1 summarizes the percent of the radioactivity of peak 1 and
peak 2 versus total radioactivity, and the specific activity of peak 1 and peak 2
in relation to total nucleic acid after 60 min incubation. In cases where the
number of blast cells was large, the percent of peak 1 and peak 2 was high,
and the specific activity of peak 1 and peak 2 was highest in a case of acute
lymphoblastic leukemia, followed by that of acute myeloblastic leukemia,
blastic crisis of chronic myeloid leukemia and monocytic leukemia.

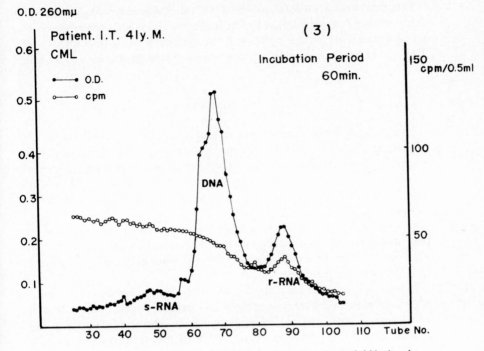

FIG. 21. MAK column chromatogram of leukemic cells in chronic myeloid leukemia after remission of blastic crisis.

TABLE 1. Comparison of P_1 and P_2 Peaks in Human Leukemic Cells in Various Types of Leukemia

CASE	AGE SEX	DIAGNOSIS	WBC (x10⁴)	BLAST CELL %	% OF TOTAL COUNT		S.A. IN RELATION TO TOTAL N.A.(cpm/μM-P)	
					P₁	P₂	P₁	P₂
T. K.	18 F.	ALL	8.4	98.0	19.0	42.0	2800	6050
Y. M.	36 M.	AML	16.5	68.5	11.2	59.5	885	2800
K. N.	46 F.	Mo.L	3.9	89.0	17.0	22.2	330	464
I. T.	41 M.	CML	35.4	2.0 (21.0)	—	—	—	—
I. T.	41 M.	Blastic Crisis CML	18.1	32.0 (86.0)	18.5	17.2	880	820

From these results it was confirmed that two kinds of rapidly labeled RNA are present in leukemic cells of acute leukemia and it may be presumed that immature cells are producing these RNA more intensively than mature cells. The nature and significance of these rapidly labeled RNAs await further investigation.

SUMMARY

Nucleic acid metabolism of human leukemic leukocytes was investigated employing chemical assay of incorporation of radioactive precursors into nucleic acids in the cells *in vitro*, autoradiography of leukemic cells using thymidine-^{3}H, and microspectrophotometric determination of DNA content of the cell nucleus after Feulgen staining. The results obtained were as follows:

1. Specific activity of DNA of acute leukemia leukocytes was lower than that of chronic myeloid leukemia when the leukocytes were incubated with adenine-^{14}C. The difference was particularly remarkable in specific activity of DNA per immature cells.

2. Autoradiographic examination with thymidine-^{3}H revealed lowering of percent labeled immature cells of the bone marrow of acute myeloid leukemia as compared with chronic myeloid leukemia. However, the mean grain counts showed little difference between the both types of leukemia.

These findings suggest that the rate of DNA synthesis of labeled cells in acute leukemia is not lowered as compared with chronic myeloid leukemia.

3. When leukemic cells in acute leukemia were divided into three groups according to their size, most of the small cells, which represented the major component of the leukemic cell population were not labeled, while the large cells were labeled in a high percentage, and the percent of labeled cells in the middle-sized cells was in between. The uptake of uridine-^{3}H and leucine-^{3}H was also high in proportion to the size of the cells. In view of these observations it was suggested that the leukemic cells are composed of two compartments, namely dividing and non-dividing compartment, and the cells of the dividing compartment have pretty rapid proliferative activity.

4. From the observation of the *in vivo* uptake of thymidine-^{3}H, the life cycle of leukemic cells was estimated as follows: in myeloblasts in a case of chronic myeloid leukemia $T_G = 108$ hr, $G_1 = 84$ hr, $S = 20$ hr, $G_2 = 3$ hr, and in myeloblasts in a case of acute myeloid leukemia $T_G = 84$ hr, $G_1 = 60$ hr, $S = 20$ hr and $G_2 = 3$ hr.

5. From the *in vivo* study of labeling index of blast cells in acute leukemia, it was found that the high initial labeling index of large blast cells decreased with the passage of time, while the labeling index of small blast cells, which were not labeled initially, increased with the lapse of time. The labeling index of middle-sized cells was lower than that of large cells at the initial stage, and it remained nearly at the same level during 24 hr.

6. Simultaneous observation of the DNA content and the uptake of thymidine-^{3}H in leukemic cells showed that, in chronic myeloid leukemia, most of the unlabeled cells had DNA content of 2c value, while only a small number of unlabeled cells had DNA content of 4c value. On the contrary, all of the labeled cells had DNA content between 2c and 4c values. In a case of acute lymphoblastic leukemia most of the cells with DNA content of 2c and 4c values were not labeled, while the cells with DNA content between 2c and 4c values were all labeled. In acute leukemia most of the cells had the DNA content of 2c value, and only a small number of cells had the DNA content between 2c and 4c values, indicating a prolongation of G_1 period.

7. Based on the kinetic studies of leukemic cells in acute leukemia, a hypothetical model for the proliferation of leukemic cells was presented.

8. It was confirmed that two kinds of rapidly labeled RNA are present in leukemic cells of acute leukemia, while in chronic myeloid leukemia these rapidly labeled RNAs were not found.

REFERENCES

1. T. NAKAMURA and S. SHIRAKAWA, *Acta haemat. Jap.*, **27**, 705, 1964.
2. G. WAKISAKA, H. UCHINO, S. NAKAMURA, S. SHIRAKAWA, A. ADACHI, M. SAKURAI and K. MIYAMOTO, *Acta haemat.*, **31**, 214, 1964.
3. E. P. CRONKITE, *Kinetics of proliferation of normal and leukemic leukocytes. Methodological approaches to the study of leukemias*, p. 51. Ed. by V. Defendi, The Wistar Institute Press, Phil., 1965.
4. A. M. MAUER, B. C. LAMPKIN and E. F. SAUNDERS, In *Proc. XI. Congress of the International Society of Hematology*, p. 42, Sydney, Australia, 1966.
5. K. OTA, *Acta haemat. Jap.*, **27**, 693, 1964.
6. B. D. CLARKSON, K. OTA, T. OHKITA and A. O'CONNOR, *Proc. Am. A. Cancer Res.*, **5**, 12, 1964.
7. C. G. CRADDOCK and G. S. NAKAI, *J. Clin. Invest.*, **41**, 360, 1962.
8. F. GAVOSTO, A. PILERI, C. BACHI, L. PEGORASO, *Nature*, **203**, 92, 1964.
9. A. M. MAUER and V. FISCHER, *Nature*, **193**, 1085, 1962.
10. S. A. KILLMANN, E. P. CRONKITE, J. S. ROBERTSON, T. M. FLIEDNER and V. P. BOND, *Lab. Invest.*, **12**, 671, 1963.
11. T. KUROYANAGI and M. SAITO, *Tohoku J. Exp. Med.*, **80**, 168, 1962.
12. W. A. SKOOG and W. S. BECK, *Blood*, **11**, 436, 1956.
13. L. I. HECHT and V. R. POTTER, *Cancer Res.*, **16**, 988, 1956.
14. A. BENDICH, *Methods in Enzymology*, vol. 3, p. 715. Academic Press, New York, 1957.

DISCUSSION

I. YAMANE: When we checked the viability of leukemic cells obtained from the patients employing the dye exclusion method with the use of trypan blue, the majority of the examined leukemic cells were dead.

G. WAKISAKA: Thank you very much for your comment. I think your observation is in agreement with our hypothetical model of the kinetics of leukemic cells. In acute leukemia the proportion of cells in the non-dividing compartment is high and these cells may be aged cells, which finally die without attaining maturity.

B.THORELL: Did you try other sera than from these patients in your *in vitro* experiments? Because in some instances of leukemia, the patients' own serum have been shown to contain "growth-inhibition" factor or factors.

G.WAKISAKA: We added autologous serum to the culture medium. We did not add normal control serum to the culture medium, but I think this experiment may be worthwhile trying.

PURIFICATION OF "DIVISION PROTEIN"
OF *TETRAHYMENA PYRIFORMIS*

YOSHIO WATANABE, MARIKO IKEDA* and SHINICHI TAMURA†

Department of Pathology, National Institute of Health, Tokyo

THE ciliated protozoan, *Tetrahymena pyriformis*, was one of the first cells to be synchronized in a high degree by a technique employing a series of heat-shocks.[3] This organism has also been extensively studied with respect to its nutritional requirements, response to chemical and physical agents and synthesis of macromolecular components in relation to its cell life cycle.

One of the most interesting findings is that temperature-synchronized *Tetrahymena* cells fail to enter cell division after the completion of DNA synthesis, unless several prerequisites for division are satisfied in the so-called recovery period—from the end of the heat-treatment (EHT) to a synchronous division.

With regard to the specific protein synthesis in this period, from the studies on delays in division and morphogenesis induced by short exposures to amino acid analogues in synchronized cells, Dr. Zeuthen (present in this Symposium) and his group suggested that the protein indispensable for the oncoming division, designated as "division protein", is to be synthesized in the recovery period.[1,2,7] Furthermore, in this Symposium, one must have known and appreciated Dr. Zeuthen's beautiful work on the close relation between this protein and the fiber-formation in oral apparatus directly involved in cell division of the ciliated protozoan.[8]

From different angles, we started to isolate division protein from the bulk of proteins of *Tetrahymena* in order to explore its chemical nature and biological function. At first, we prepared water-soluble and 0.6 M KCl-soluble protein fractions from the several stages during the recovery period. We separated the two protein fractions further into sub-fractions by the use of a DEAE-cellulose column, and obtained three sub-fractions from the KCl-soluble proteins and ten sub-fractions from the water-soluble proteins. For diagnosing the division protein postulated by Dr. Zeuthen,[7] we set up five reasonable criteria and scrutinized the thirteen sub-fractions with respect to their qualifications for this rationale. As far as the sub-fraction qualified for

* Department of Biology, Tokyo Metropolitan University, Tokyo.
† Department of Zoology, Tokyo University of Education, Tokyo.

including division protein is concerned, peak 7 of the water-soluble protein satisfies all five conditions, whereas none of the others fulfill more than 2.[5,6] However, much more thorough studies on peak 7 are required for the identification of division protein.

First, we mention experimental results concerning the further purification procedure of division protein.

Classification of Water-soluble Proteins by the Gradient Extraction Method with Ammonium Sulfate (AS)

Samples of 100 ml were taken at three selected stages during the recovery period. The cells were washed three times by centrifugation with cold inorganic medium and once with cold distilled water; the packed cells were immediately homogenized after addition of nine times this volume of distilled water, and then centrifuged 15,000 rpm for 15 min. The supernatant after dialysis was used as containing water-soluble proteins. Before gradient elution was started, all water-soluble proteins were precipitated with 100% saturation of AS and inserted into the upper compartment of the column for this method. A 3.0×10.0 cm column with side-arm was used in which the contents can be stirred under increasing pressure. For an accurate elution of dissolved protein at a certain concentration of AS, Geon 427 (powder of polyvinyl chloride) was employed for the packing material of the column. The packing material is partitioned into two compartments by inserting a disk-shaped glass filter at the middle. After the entire contents of the column were equilibrated with 100% saturation of AS and precipitated proteins were charged, only the upper compartment was stirred by a glass rod with propellers, followed by eluting with a linear falling gradient from 100% down to 0% saturation of AS.

Figure 1 represents the effluent patterns, on the gradient extraction with AS, of the water-soluble proteins taken from synchronized cells at 0, 45 and 70 min after EHT, respectively. Among these three stages, 45 min is considered to be the stage synthesizing division protein actively and the cells at 70 min correspond to the early phase of synchronous division. As shown in Fig. 1, the bulk of the water-soluble proteins in each stage precipitates within the range between 75% and 20% saturation of AS. However, the amounts of proteins precipitable with 52.5–42.5% saturation of AS at 0 min are larger than those at 45 or 70 min after EHT. On the other hand, amounts of proteins ranging of both 42.5–32.5% and 32.5–20% saturation of AS at 0 min are smaller than those of the corresponding proteins from the other two stages.

In pulse-labeling experiments (10-min exposure: -10–0, 35–45 and 60–70 min after EHT) with 1 μc/ml of ^{3}H-phenylalanine, the incorporation patterns at the three stages were seemingly similar to the respective effluent patterns (by optical density at 280 mμ) as illustrated in Fig. 1. This held true for incorporation of ^{3}H-methionine.

From these, while it seems difficult to select the fraction containing division protein, water-soluble proteins can tentatively be divided into six groups containing the proteins precipitable with 100–75%, 75–52.5%, 52.5–42.5%, 42.5–32.5%, 32.5–20% and 20–0% saturation of AS, respectively. Accordingly, in the next step, each of the six classified protein fractions is further chromatographed on DEAE-cellulose column as follows.

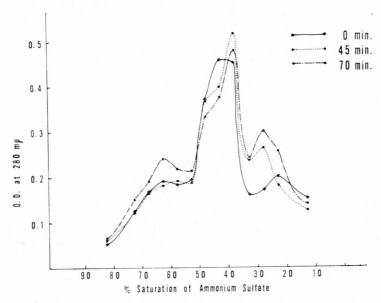

Fig. 1. Effluent patterns, using the linear falling gradient extraction method with ammonium sulfate, of water-soluble proteins of synchronized *Tetrahymena* cells at 0, 45 and 70 min after the end of the heat-treatment, respectively. Synchronization of cell division was practised by application of cyclic temperature-treatment (eight alternate temperature shifts between 34 °C and 26 °C for 30 min each). Synchronous division occurs at 75 min after the end of the heat-treatment, the maximum division index being 85%.

Fractionation of Water-soluble Proteins by Combination of the Extraction Method with AS and DEAE-cellulose Column Chromatography

Synchronized cells were exposed to 5 µc/ml of ^{3}H-methionine for 10 min at three different stages (−10–0, 40–50 and 60–70 min after EHT) in the inorganic medium. After labeling, water-soluble proteins of each stage were prepared and precipitated with AS saturation. The precipitated total water-soluble proteins were, at first, dissolved in 6 ml of 75% saturation of AS, subsequently the remaining precipitate was dissolved, in turn, in each 6 ml of 52.5%, 42.5%, 32.5%, 20% and 0% saturation of AS. Afterwards, each protein solution was dialyzed against cold distilled water and then equilibrated to the starting buffer for chromatography.

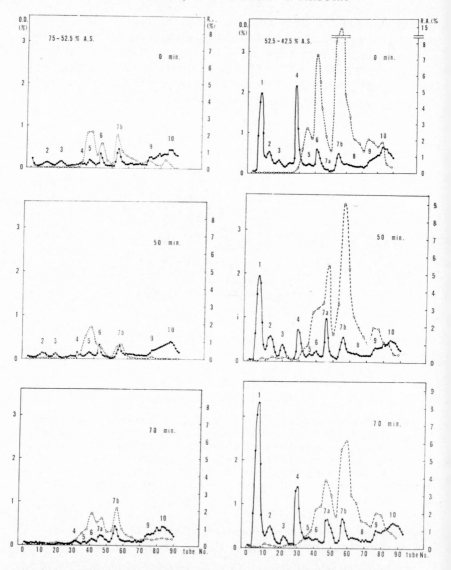

FIG. 2. Fractionation of water-soluble proteins of synchronized cells by the combination of extraction method with ammonium sulfate and DEAE-cellulose column chromatography. The upper, middle and lower rows of diagrams represent chromatographic patterns of proteins taken from the synchronized cells at 0, 50 and 70 min after the end of the heat-treatment, respectively. The first to fourth columns of diagrams show effluent patterns of proteins precipitated with 75–52.5%, 52.5–42.5%, 42.5–32.5% and 32.5–20% saturation of ammonium sulfate, respectively. Each diagram shows the effluent patterns by spectrophotometry (solid line with solid circles) and by radioactivity assay (broken line with open circles). For incorporation of radioactivity, 10-min labeling with $5\,\mu Ci/ml$ of ^3H-methionine was carried out before

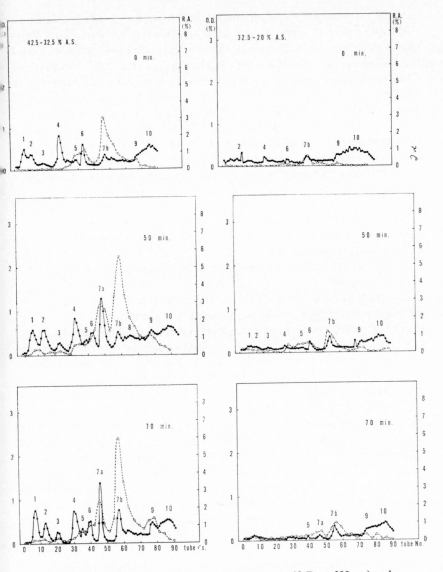

sampling in inorganic medium. Total protein amounts (O.D. at 280 mμ) and cpm (R.A.) of each stage are taken as 100 and individual readings are shown in percentages of the total, for the sake of convenience. The proteins charged on column were eluted with four different phosphate buffers: 80 ml of 0.005 M Na$_2$HPO$_4$, pH 8.8; 100 ml of 0.02 M phosphate buffer in 0.05 M NaCl, pH 7.5; 80 ml of 0.1 M phosphate buffer in 0.1 M NaCl, pH 6.7; 100 ml of 0.18 M NaH$_2$PO$_4$ in 0.18 M NaCl, pH 4.5; in this order. Under the conditions, the most important fraction—peak 7—previously reported (refs. 5 and 6) can be separated into two peaks, peak 7a and b. Ninety tubes were collected, each having 4 ml of effluent. Separated peaks are numbered in each diagram.

Small chromatographic diagrams in Fig. 2 represent the chromatographic patterns of protein amounts and of ^{3}H-methionine incorporation. These twelve diagrams are composed of the four main protein fractions divided by AS taken from the cells at 0, 50 and 70 min after EHT, respectively. Chromatographic patterns from the respective proteins precipitable with 100–75% and 20–0% saturation of AS are omitted from the figure, since no appreciable peak was found by spectrophotometry or by radioactivity assay.

In Fig. 2, the total protein amounts and total radioactivities of the water-soluble proteins of the three stages are each adjusted to the same value, that is 100, for convenience of interpretation. Such manipulation is needed for pulse-labeling experiments in the inorganic medium, since starvation effects are obviously superimposed upon the synthetic abilities peculiar to the stages.[4] Contents of water-soluble proteins in each stage being approximately equal, amounts of a certain sub-fraction would be allowable for comparison with respective stages. In Fig. 2, it is worth noting that "peak 7a" of the two protein fractions (i.e. 52.5–42.5% and 42.5–32.5% AS) is pronouncedly increased both in amount and in incorporation at 50 min after EHT in comparison with those of 0 min.

Returning to the notion of division protein, small protein sub-fraction containing division protein should be increased in amounts from 0 up to 50 min after EHT. Furthermore, by using the pulse-labeling technique, incorporation of radioactive amino acids into the sub-fraction around 45 min (40–50 min) after EHT should be much more pronounced than that of the other two stages (−10–0 or 60–70 min after EHT). On these bases, nearly sixty protein sub-fractions in each stage were tested with respect to their candidacy for the fraction including division protein. As in Fig. 3, we recalculated the percentages of protein amounts and radioactivity in each sub-fraction from the results of Fig. 2 and investigated the change in percentages of them in respective stages. As far as protein amounts are concerned (the upper diagram), peak 7a of both 52.5–42.5% AS and 42.5–32.5% AS and peak 9 of 42.5–32.5% AS can be selected as important candidates for the sub-fraction containing division protein. Furthermore, with regard to ^{3}H-methionine incorporation (the lower diagram), peak 7a of both 52.5–42.5% AS and 42.5–32.5% AS remain as highly probable sub-fractions for containing division protein.

For convenience of illustration, balances of optical density (50 min minus 0 min), radioactivity (50 min minus 0 min) and radioactivity (50 min minus 70 min) in each protein sub-fraction are enumerated in Table 1 in order of largeness. Higher values in each column of the Table must be accountable for further possible fractions including division protein. From these, it is most likely that division protein is present in peak 7a of the water-soluble proteins precipitable with 52.5–32.5% saturation of AS.

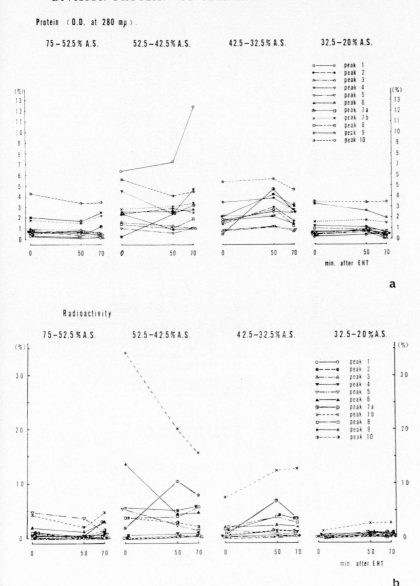

FIG. 3. Changes in percentage of protein amounts (the upper diagram) and of radioactivity (the lower diagram) of each protein sub-fraction against those of the total water-soluble proteins of respective stages. Recalculated from the results of Fig. 2. Materials shown by 75–52.5% AS, 52.5–42.5% AS, 42.5–32.5% AS and 32.5–20% AS represent the protein fractions precipitable within the mentioned saturation of ammonium sulfate. Abscissae represent time after the end of the heat-treatment. Individual readings in ordinate and peak numbers refer to those in Fig. 2.

TABLE 1. BALANCES OF O.D.(50′–0′), R.A.(50′–0′) and R.A.(50′–70′) IN EACH PEAK, ENUMERATED IN ORDER OF LARGENESS.

O.D.(50′)–O.D.(0′)		R.A.(50′)–R.A.(0′)		R.A.(50′)–R.A.(70′)	
Values (%)	Peak No.	Values (%)	Peak No.	Values (%)	Peak No.
4.21	42.5–32.5% AS—peak 7a	8.45	52.5–42.5% AS—peak 7a	3.27	42.5–32.5% AS—peak 7a
2.17	52.5–42.5% AS—peak 7a	6.15	42.5–32.5% AS—peak 7a	2.51	52.5–42.5% AS—peak 7a
2.15	42.5–32.5% AS—peak 9	2.75	42.5–32.5% AS—peak 9	1.11	42.5–32.5% AS—peak 8
1.44	42.5–32.5% AS—peak 2	1.97	42.5–32.5% AS—peak 8	0.66	42.5–32.5% AS—peak 9
1.21	42.5–32.5% AS—peak 8	0.92	42.5–32.5% AS—peak 4	0.66	32.5–20.0% AS—peak 6
1.18	52.5–42.5% AS—peak 1	0.89	32.5–20.0% AS—peak 5	0.43	32.5–20.0% AS—peak 5
0.90	42.5–32.5% AS—peak 1	0.87	32.5–20.0% AS—peak 6	0.28	42.5–32.5% AS—peak 6
0.50	52.5–42.5% AS—peak 2	0.81	42.5–32.5% AS—peak 10	0.28	52.5–42.5% AS—peak 3
0.46	42.5–32.5% AS—peak 6	0.56	32.5–20.0% AS—peak 4	0.24	32.5–20.0% AS—peak 3
0.46	42.5–32.5% AS—peak 3	0.38	42.5–32.5% AS—peak 6	0.22	32.5–20.0% AS—peak 4

Purity of the Sub-fraction, Peak 7a

"Peak 7a" thus obtained is recognized as a single band on electrophoresis with cellulose-acetate membrane (Fig. 4). This band corresponds to one of twelve to fifteen components found in the proteins precipitable with 52.5–42.5% saturation of AS, shown as a control for comparison with each other. Moreover, according to the densitometrical analysis of the membrane, it seems that this small protein sub-fraction would mainly be composed of single protein component from the nature of the curve.

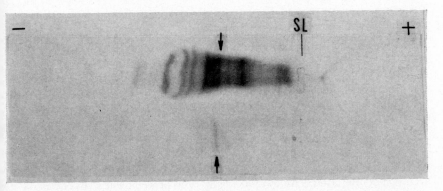

FIG. 4. Electrophoresis of peak 7a taken from the water-soluble proteins precipitable with 52.5–32.5% saturation of ammonium sulfate. A sheet of 6 × 17.5 cm cellulose-acetate membrane (Commercial name: Separax, Jōkō Sangyo Co.) was used. Buffer used here is composed of 9.2 g of tris-aminomethane, 9.35 g of boric acid and 2.34 g of sodium chloride in 1000 ml of distilled water (final pH 8.0). Ionic strength of this buffer corresponds to 0.05 M potassium chloride ($\mu = 0.05$). Electrophoresis was carried out with a constant amperage, 0.06 mA/cm, for 4 hr in the cold room. The upper sample is the protein fraction precipitable with 52.5–42.5% saturation of ammonium sulfate, taken from the cells at 45 min after the end of the heat-treatment (illustrated as a control). The lower sample is the chromatographically separated peak 7a of the water-soluble proteins precipitable with 52.5–32.5% saturation of ammonium sulfate. Arrows indicate the matching band of the two samples. SL: Starting line.

Under these circumstances, we are now inclined to think that isolation of division protein would virtually be succeeded. The chemical nature and biological function of the division protein and localization of the protein in the process of cell division are interesting subjects for our research in the future.

REFERENCES

1. J. FRANKEL, *Compt. Rend. Lab. Carlsberg*, **33**, 1, 1962.
2. L. RASMUSSEN and E. ZEUTHEN, *Compt. Rend. Lab. Carlsberg*, **32**, 333, 1962.
3. O. SCHERBAUM and E. ZEUTHEN, *Exptl. Cell Res.*, **6**, 221, 1954.
4. Y. WATANABE and M. IKEDA, *Exptl. Cell Res.*, **38**, 432, 1965.
5. Y. WATANABE and M. IKEDA, *Ibid.*, **39**, 443, 1965.

6. Y. WATANABE and M. IKEDA, *Ibid.*, **39**, 464, 1965.
7. E. ZEUTHEN, *Biological Structure and Function*, vol. II. Academic Press, London, 1961.
8. E. ZEUTHEN and N. E. WILLIAMS, *Second Intern. Symposium for Cellular Chemistry*, held at Ohtsu, 1966.

DISCUSSION

E. ZEUTHEN: I want to congratulate you on the progress already made in your difficult search for proteins directly involved in cell division. Dr. Williams and I in this symposium would expect that synthesis of division protein(s) relating to the first division is continuous for the first 55 min after the heat-treatments, but we accept that it could be fastest at the times when you search for synthesis of a division protein if for no other reason than because the cell grows fastest around these times. These proteins would be likely to include microtubular precursors not too different from spindle precursors in other cells.

I have two questions: (1) Did you attempt to make protein fibers from your more interesting fractions?

(2) Where in your published chromatograms shall we search for the water insoluble proteins which enter the developing structures of the oral primordium?

Y. WATANABE: Thank you for your pertinent comments. First, I would like to touch on our criteria for diagnosing division protein. The five criteria have been tentatively selected only for convenience because the range of selection is limited to testable cases. Of course, I have felt that to know the accurate time of synthesis of division protein is very important for setting up the criteria. However, it is possible to say at least that the peak 7 we previously reported would have a close connection with cell division, since the fraction fails to increase by addition of 30 mM sodium fluoride which is sufficient to suppress synchronous division completely and the formation of protein in the fraction is most seriously affected by heat-shock. As to the fiber formation *in vitro* from the protein sub-fraction of peak 7a, I have not yet attempted this. With regard to your last question, I think that obviously much more thorough studies are required for making sure of the relation between the division protein that we purified and the water-insoluble microtubular proteins which you found.

S. SPIEGELMAN: Have you done a pulse-chase experiment to determine whether peak 7 turns over during the division cycle?

Also, would you comment on the possibility that peak 7 contains a precursor to spindle-fiber protein?

Y. WATANABE: With regard to the first question of Dr. Spiegelman, pertaining to the turnover of peak 7, I have not yet done a pulse-chase experiment. However, I would like to do it in the future, since it would be very important for discovering the role of division protein.

As to the second point, from several indirect evidences we have had, I am enticed to speculate that the protein in peak 7a and some kinds of proteins in 0.6 M KCl-soluble protein fraction would participate in the remodelling of the structure (including oral apparatus) necessary for the cell division.

T. YANAGITA: Sylvén et al. (Exp. Cell Res., 16, 75, 1959) reported that, in a synchronously dividing yeast culture, protease and dipeptidase activities became higher during (or shortly before) the cell division. Recently Nishi et al. of our laboratory showed that this is also the case in synchronously growing E. coli. (J. Gen. Appl. Microbiol., 11, 321, 1965; 12, 293, 1966). Is there any possibility that your division protein is related to such enzyme(s) involved in the hydrolysis of proteinacious substances in cells?

Y.WATANABE: I am interested in the activation of proteolytic enzymes involved in cell division. However, I cannot answer your question because I have not yet studied it in our system.

H.NAORA: Is there any similarity to the protein isolated by Stevenson and Inoué from sea urchin?

Y.WATANABE: As you know, a microtubular protein obtained by Drs. Kane, Stephens and Inoué is 0.6 M KCl-soluble, while our division protein is water-soluble. In Tetrahymena, besides the division protein, we have worked on the nature of 0.6 M KCl-soluble proteins. The KCl-soluble proteins of Tetrahymena can take place so-called "superprecipitation" by addition of ATP in vitro like contractile proteins in the muscle, whereas some properties (e.g., in terms of visocity) are quite different from myosin or actomyosin.

All we can say at the present is that, from the preliminary experimental result, the S value of our 0.6 M KCl-soluble protein fraction seems to be considerably lower than the 22S of the proteins prepared by Drs. Kane, Stephens and Inoué.

SOME BIOLOGICAL ACTIVITY
OF BASIC NUCLEAR PROTEINS
IN RELATION TO CANCER

T. Ishikawa, S. Odashima, S. Fukuda and T. Suyama

In the first part of this paper we have dealt with the immunological analysis of the highly tumor-specific components of the nuclear substances, such as acidic nuclear globulin, deoxyribonucleic acid, nuclear sap and basic proteins extracted from the tumor cells. In the second part we have discussed the biological significances of each of these factors, particularly the regulatory effect of the cancerous DNA on polyoma virus multiplication in the cultured mouse embryonic cells and the significance of the basic proteins; that is, the effects of histones on the mitochondrial functions.

The cancerous materials used in our experiments were: DAB-induced hepatoma, AH66F and AH127 ascites tumor cells from rats, human gastric cancer and the non-cancerous ones were: normal rat liver and human gastric ulcer materials which were extripated operatively.

Cell nuclei have been isolated from these cancerous and non-cancerous materials by Hogeboom-Schneider's method, while Chauveau's method[1] was particularly needed in the case of histones, and we have examined by an ultramicroscope, contaminations of extranuclear components which should be removed as much as possible.

Particular Antigenic Components of the Cancerous Acidic Nuclear Globulin and Nuclear Sap

We have at first prepared acidic nuclear globulin from the isolated nuclei of the cancerous and non-cancerous materials by the modified procedure of Busch[2] and of Patel *et al.*[3] respectively and then immunized rabbits with these components, Freund's complete adjuvant method being applied. Antisera against the acidic nuclear globulin of tumor cells, taking up one example of AH66F, which is abbreviated as anti/AH66F·NP$_{50}$, reacted with purified test antigens of acidic nuclear globulin from AH66F and normal rat liver, respectively and it resulted in producing some precipitating lines by Ouchterlony's agar diffusion method and by Grabar's immunoelectrophoretic technique. One of these precipitating lines could be retained in the case using AH66F, by the use of anti/AH66F·NP$_{50}$ absorbed with acidic nuclear globulin of normal rat liver, which is also abbreviated as anti/AH66F·NP$_{50}$ minus normal rat liver NRL·NP$_{50}$, in order to remove cross-reacting components. This remaining component can be indeed regarded as highly AH66F tumor-specific. The specific factor was heat-stable at 60°C, acid-stable at pH 5.4, was located at the β-globulin position electrophoretically and could be separated chromatographically. Figure 1 illustrates this specific factor being eluted chromatographically on a DEAE-cellulose column.

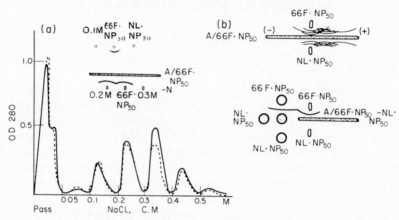

FIG. 1A. Immune-electrophoretic patterns obtained with acidic nuclear globulin developed with rabbit antiserum against acidic nuclear globulin of cancer, absorbed with acidic nuclear globulin of normal rat liver. Anti/AH66F·NP$_{50}$ reacted with AH66F·NP$_{50}$, producing several precipitating lines, immunoelectrophoretically. The both being compared with each other, there remained one out of these lines in the case of AH66F by absorption test using *anti/AH66F·NP$_{50}$-NL·NP$_{50}$*; that is, antiserum against acidic nuclear globulin of AH66F, absorbed with acidic nuclear globulin of normal rat liver. This antigenicity could be regarded as highly AH66F tumor-specific and could be isolated chromatographically. Figure 1B shows a step-wise elution diagram of NP$_{50}$ on DEAE cellulose with 0.02 M acetate buffer. 100 mg of acidic nuclear globulin was applied to a 1.5 × 25 cm column: effluent collected in 5 ml fractions collected. E indicates light absorbance at 280 mμ. Highly tumor-specific factor mentioned above could be found in the effluent with 0.2 M Tris-buffer, but neither in the other fractions of cancer materials nor in every fraction of the non-cancerous material. Acidic nuclear globulin is abbreviated as NP$_{50}$, nuclear sap as SP$_{50}$, acidic nuclear globulin of AH66F as AH66F·NP$_{50}$, that of normal rat liver as NL·NP$_{50}$ and antiserum against nuclear globulin of AH66F, absorbed with acidic nuclear globulin as anti/AH66F·NP$_{50}$-NL·NP$_{50}$.

Figure 3 E shows an immune-fluorescent picture of AH66F, which was stained by fluorescein-labeled antibody against AH66F·NP$_{50}$ absorbed with NRL·NP$_{50}$. The fluorescence is strongly positive in the nuclear sap of AH66F tumor cells themselves, but almost negative in that of normal rat liver, and weakly positive, even, in that of heterologous tumor cells, AH127. It is accordingly assumed that it may be to some extent strain-specific.

By a similar method, we have also ascertained highly tumor-specific factors of acidic nuclear globulin from AH127 which were different from those of AH66F, immunologically. Through these results, it is supposed that each of the tumor cells may contain its own specific factor.

Another Type of Acidic Material in Nucleus, that is DNA

Concerning the antigenic specificity of the cancerous DNA, there are several reports, some authors being affirmative, some negative.[4-14] When the homogenates of the isolated nuclei were treated with fluorocarbon (tri-chloro-trifluoro-ethane), the protein fraction associated with DNP and RNP were gradually eliminated and almost completely eliminated after the 12-time treatments, and we could obtain almost pure DNA-RNA mixture sample. We have investigated carefully the antigenic specificity, of these 12-time fluorocarbon-treated DNAs and in spite of the repeated experiments, we could not detect any immunological differences between normal rat liver and AH tumor cells by the immune-diffusion technique and even by means of the complement-fixation test. The specific properties of cancerous DNA could be demonstrated by the biological tests, for example, examining the regulatory effect of cancerous DNA on the polyoma multiplication which is discussed later.

Basic Nuclear Proteins

Concerning the basic nuclear proteins, the species-specificity and tissue-specificity of histones, particularly the tumor-specificity of the RP2-L subfraction of histone have been discussed by several reviews[22-25] recently, but the antigenicity, or antigenic specificity, has been considered negligible or negative, because of relatively low molecular weight and to say nothing of the cancer specific antigenicity of histones. To make sure, we have eluted basic nuclear proteins from isolated nuclei which were carefully prepared by Chauveau's method, 2.2 M sucrose—even by this method we could not separate some cytoplasmic contamination, particularly in the case of adhesive cancerous materials—and then subfractionated the so-called arginine-rich histone I and lysine-rich histone II which were extracted from normal rat liver, DAB-hepatoma and AH tumors. In these determinations, the method of Ui[15] was applied exactly according to his description which requires the use of sulphuric acid. Antisera against nuclear basic proteins, so-called Ui's histones, from DAB-hepatoma, abbreviated as anti/DAB·histone, reacted with the relatively purified test antigens of so-called Ui's histone I and II and resulted in producing some precipitating lines. Two out of these lines remained in the histone II fraction of DAB-hepatoma by the absorption test in which the anti/DAB·histone absorbed with NRL·histone was used so as to remove cross-reacting components, and as a result they can be considered highly DAB-specific components of the basic nuclear protein.[16] So far as

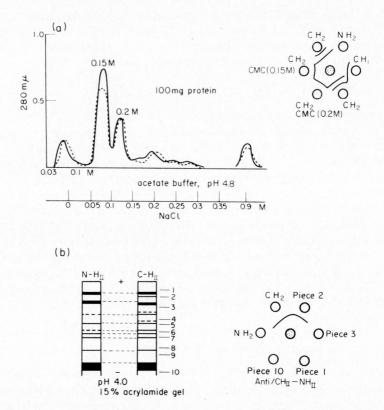

FIG. 2A. Highly tumor-specific factor of histone II from DAB-induced hepatoma, detected by immune diffusion technique. Anti/DAB·H–NH$_{II}$ (rabbit antiserum against DAB·H; that is, against Ui's histone of DAB-induced hepatoma, absorbed with N·H$_{II}$, i.e. histone II of normal rat liver) reacted with DAB.H$_{II}$, i.e. histone II of DAB-induced hepatoma, resulting in producing 2 precipitating lines, which could be hardly detected in the case of using N·H$_{II}$. These highly tumor-specific lines could be ascertained in DAB·H$_{II}$.P$_3$ (chromatographic peak 3 of histone II of DAB-hepatoma). Chromatographic fractionation was carried out in the columns of DEAE cellulose. 100 mg of protein histone II was applied to column 1.5 × 20 cm. Stepwise elution at 4°C and pH 4.8 with 0.01 M phosphate buffer and 0.1 M NaCl. Fractions of 5 ml were collected. The above chromatogram was recorded at 280 mμ.

As for histones of AH66F and AH127 tumor cell and human gastric cancer materials, individual tumor specific factors could be eluted by similar methods respectively and each tumor has its own specific factor. Fig. 2B shows disc electrophoretic patterns of N·H$_{II}$ (histone II of non-cancerous material in this case) and C·H$_{II}$ (histone II of gastric cancer material). Electrophoresis was carried out in the column of polyacrylamide gel, prepared according to Beisfeld et al[20] and the bands are indicated by numbers for C·H$_{II}$. The above mentioned highly tumor-specific factor could be detected in the second band, but in neither of the other bands of C·H$_{II}$ nor in every fraction of N·H$_{II}$.

the purity of histones is concerned, Busch has reported at this symposium on the absolutely pure f_3 subfraction of histone.[17]

This highly tumor-specific component of cancerous histone II could be eluted in the third fraction of CM cellulose column chromatography, by the stepwise elution with acetate buffer pH 4.8 as Fig. 2A indicates and we have termed this factor DAB-specific histone II-3 factor. The subfraction which contained this specific factor showed a single peak of Svedberg, value 1.2 ultracentrifugically. This DAB- specific factor could be eluted by disc electrophoresis similarly as in the human gastric cancer case of Fig. 2B. The DAB-specific histone II-3 could be ascertained in the first band from the anodic side. This factor, being not detected in normal rat liver, began to increase at the precancerous stage; that is, at the end of third month and then reached a maximum at the fifth cancerous month after feeding the rats on the dye. This specific factor, therefore, can perhaps be regarded as the chemical indicator of cancerogenesis.

AH127 and AH66F tumor cells similarly contained their own highly tumor-specific components of histone II which were different from that of DAB-hepatoma immunologically and which could be also isolated by means of CM cellulose, respectively.

In regard to the human gastric cancer, we have also detected by the immune-diffusion method its own particular antigenic components from histone II, as compared with the one of the non-cancerous gastric materials. This particular specific factor may be cancer-specific and moreover will be regarded to some degree as gastric cancer-specific, because it could not be detected in the non-cancerous materials and scarcely in the cancerous materials of other visceral organs, for example, the rectum, mammary gland and lung.

The disc electrophoretic patterns of histone II from the cancerous and the non-cancerous gastric materials are shown in Fig. 2B. Comparison of the two yielded no remarkable difference between them, but the immunologically cancer-specific factor could be ascertained in the second band of cancerous material by the immune-diffusion method.

There are two procedures for preparing histones, namely, the one by sulphuric acid extraction according to Ui and the other by the hydrochloric acid extraction of Butler[18,19] and the latter is most usually used. The cancer-specific antigenic factor from lysine-rich histone II which was extracted by Ui's method was also detected in the lysine-rich f_1 and f_{2b} subfraction of histone extracted by Butler's procedure.

In the case of DAB-hepatoma and AH tumor cells, this individual tumor specific factor of histone II was detectable in the basic proteins prepared from the nuclei, but scarcely detectable in the cytoplasm and in the basic protein from the cytoplasmic ribosome which were obtained by the same method as with nuclear histone. Accordingly, it is assumed that this tumor-specific basic protein factor may be what mainly originated from the nucleus.

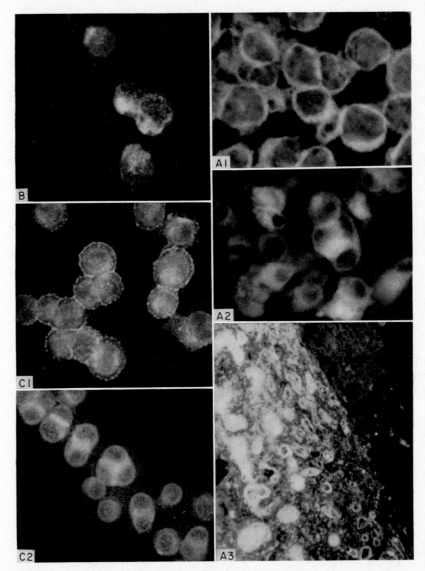

F<small>IG</small>. 3

F<small>IG</small>. 3. Immune-fluorescent pictures showing the intracellular distributions of highly tumor-specific factors, prepared from subcellular structures of cancer cells; that is, specific factors from acidic nuclear globulin, basic nuclear proteins (histone II by sulphuric acid extraction method), microsomes, mitochondria and cytoplasmic membranes of tumor cells. Our other reports have explained that these factors have been isolated chromatographically and examined immunologically. The pictures are stained by the blocking method; that is, cells were treated at first with antibody against individual fractions from normal tissue material and then

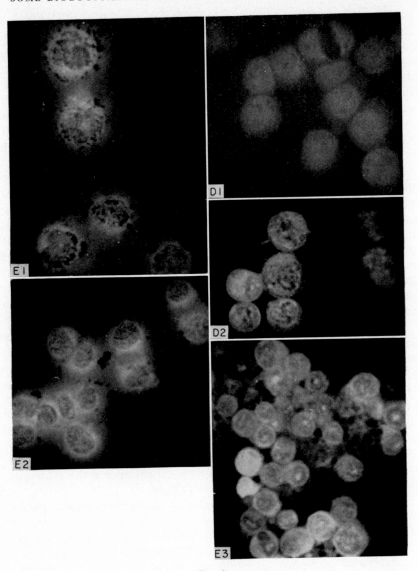

FIG. 3

were stained with fluorescein-labeled antibody against the one from cancer
material. A: mitochondria (AH66F), A2: mitochondria (AH127), A3: mitochondria (DAB-hepatoma, bright cancerous, dark non-cancerous portions), B: micro-
some (AH66F), C1: cytoplasmic membrane (Ehrlich ascites tumor cells), C2:
cytoplasmic membrane (AH127), D1: basic nuclear protein on the chromosome-
net (AH66F), D2: basic nuclear protein (AH127), E1: acidic nuclear protein in
the nuclear sap (human gastric cancer cells, exfoliative), E2: acidic nuclear
protein (AH66F), E3: acidic nuclear protein (AH127).

But in the case of human gastric cancer, the tumor-specific protein factor from the nucleus could be discovered by the immune-diffusion method in the cytoplasm and also in the basic protein of the cytoplasmic ribosome of cancer cells, but it was not found in those of non-cancerous cells. We cannot decide whether it was due to some cytoplasmic contamination, due to the similarity of the chemical constituents of cytoplasmic basic protein and of nuclear histones, occasioned by the migration of nuclear histones into the cytoplasm, or due to the biosynthesis of the nuclear histone in cytoplasmic ribosome.[26–38] Concerning these possibilities, a new experimental approach is needed.

Fluorescein-labeled antibody is quite a useful tool to investigate the intracellular distribution of particular antigenic components. When AH127 tumor cells were stained by the fluorescent antibody against AH127- specific histone II component, especially in the case of using the blocking technique, the fluorescence was positive on the chromosome-net, corresponding to the originated sites of the nuclear histones, sometimes strongly positive in the nucleolus, which has so far been considered the synthesizing site of nuclear histone, almost negative in the nuclear sap and weakly positive in the cytoplasm of tumor cells, and, on the contrary, almost negative everywhere in the control cells of normal rat liver, as shown in Fig. 3D.

Next, in the case of human gastric cancer cells stained by the fluorescent antibody against the tumor-specific histone II factor, the fluorescence is positive on the chromosome-net and also diffusely positive in the cytoplasm. According to these discoveries, it is almost certain that some parts of the basic proteins which we have extracted from the isolated nuclei, originated from the chromosome, because of the immune-fluorescence on the chromosome-net; that is, these parts are of nuclear or chromosomal origin and some other parts of them were contaminated by amounts of cytoplasm during our preparative procedure according to Chauveau's technique and hence the results of the immune-fluorescence in the cytoplasm; that is, these parts may be of cytoplasmic origin.

For reference, here is shown in Fig. 3 the immune-fluorescent pictures explaining the distribution of particular antigenic components of other subcellular structures of cancer cells; that is, of cancerous mitochondria, microsomes and cytoplasmic membranes. We have stained the tumor cells with the fluorescent antibodies against the cancer-specific factor of an individual subcellular structure which were prepared by us.[39–43] The fluorescence was positive in fine granules in the cytoplasma as to mitochondria, diffusely positive as to microsomes, and spottedly positive on the cell surface as to cytoplasmic membranes, corresponding to the site from which each component originated we have reported elsewhere on the immunological properties of these specific components.

In the experiments mentioned, we have analysed the particular antigenicities of the nuclear substances such as acidic nuclear globulin, DNA and basic nuclear proteins (histones) of the cancer cells by the immune diffusion method, and have concluded that every constituent of the nucleus contains its own highly tumor-specific factor which is supposed to participate in building up to the peculiar properties of nuclei of the cancer cells.

Our second subject is related to the biological significances of the nuclear substances, particularly of the basic nuclear proteins from the cancer cells. As histones interact with DNA and, described more in detail, as the specific kinds of histones are required for the interaction with each of functional units of DNA, we are almost sure that histones perhaps regulate the functions of DNA.

The Regulatory Effect of Cancer DNA on the Polyoma Virus Multiplication

This was co-experimented with Dr. Hatano.[44,45] As mentioned above, there is no detectable difference in immunological properties of DNA between the cancerous and non-cancerous materials and though no difference was found immunologically, we have examined for some particular biological activities of the cancer DNA. Polyoma virus, when inoculated into the

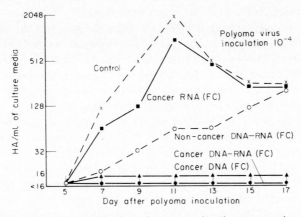

FIG. 4. DNA effect on multiplication of polyoma virus in mouse embryonic (ME) cell. Primary culture of trypsinized mouse (Swiss or ICR) embryonic cells was carried out in Hanks solution containing 0.5% lactoalbumin hydrolysate and 10% calf serum. Two days later, the nutrient fluid was replaced by the same medium and if needed, the cells were subcultured. Three or four day after the culture was initiated, or the subculture was started, the cell sheets were completed and ready to be used for all experiments. Polyoma virus, the virus strain 4B5-6, was grown in ME cells, cultured in a nutrient medium of M-199 and 2% horse serum. The culture fluid was removed from the ME cells, and to which each nucleic acid sample was added, and the above ME cells were cultured with M-199 containing 10% calf serum for 6 days. This medium was replaced by two-tenth of polyoma virus diluted in M-199 with the addition of 2% horse serum. Five days after the addition of the polyoma virus, the hemagglutination (HA) test, using portions of culture fluid, was continued every two days up to the 17th or 19th day.

cultured embryonic cells of mice, replicated exponentially during the first 12 hr as represented in Fig. 4. This was measured by the hemagglutination test. The embryonic cells were then pretreated with various kinds of nucleic acids from different sources which were purified by the 12-time fluorocarbon treatment. In the case of the pretreatment by the DNA–RNA mixture from human gastric cancer, there was observed a remarkable inhibition of viral multiplication, compared with the one from the non-cancerous gastric materials; the cancerous DNA which was prepared from the DNA–RNA mixture by RNase exhibited an inhibition as remarkable as the DNA–RNA mixture's. On the contrary, the cancerous RNA is isolated from the DNA–RNA mixture by DNase was not so effective. These observations suggest that the DNA from human gastric cancer may promise to produce some growth inhibitory substances against the viral multiplication. Yet cancerous DNA and RNA and the non-cancerous RNA–DNA mixture did not exhibit such growth inhibitory effects. This suggested functional differences of DNA between the cancerous and the non-cancerous materials.

Histone Effect on the Mitochondrial Functions

This experiment is concerned with the effects of histones on the oxidative phosphorylation reactions in the intact mitochondria of rat liver.[46–51] On the second day of this symposium, Dr. Seno pointed out the uncoupling effect of the arginine-rich histone subfraction from calf thymus and Dr. Busch explained Dr. Schwartz's experiments[52,53] in his laboratory which have suggested the possibility of the migration of histones from the nucleus to the acidic sites of the mitochondrial membrane and the possibility of these serving some function in mitochondrial control as well.

Our experiment using histones I and II of Ui demonstrated two types of responses after the addition of histones to the mitochondrial suspension of normal rat liver in the basic respiratory medium which contains inorganic phosphate, ADP and succinate as substrate. The one is that, at low concentration it uncoupled oxidative phosphorylation accompanied with stimulation of oxidation, decrease of respiratory index control and of ADP/O, ard with increase of ATPase activity (Mg^{++} type which is activated by Mg^{++}), and, behaving somewhat in a different way when compared with the uncoupler 2,4-dinitrophenol, with increase of latent ATPase. The other is that, at higher concentration it inhibited the activities concerning oxygen consumption, respiratory index control and ADP/O ratio, except for ATPase activity which was remarkably activated. This function controlling of the acceleration and the repression may be attributed to the regulatory function of histones on mitochondria.

Figure 5A, B represents an oxygen electrode tracing of the effects of histones from AH66F at various concentration levels on the respiration of rat liver mitochondria and, taking one example concerning the effect of the respiratory release in Table 1, the dose to cause the respiratory release is about 50 mg/g mitochondria of the histone of normal rat liver and about 250 mg/g mitochondria of the histone of the human gastric materials. The above results compared, it is obvious that the histone of rat liver is more

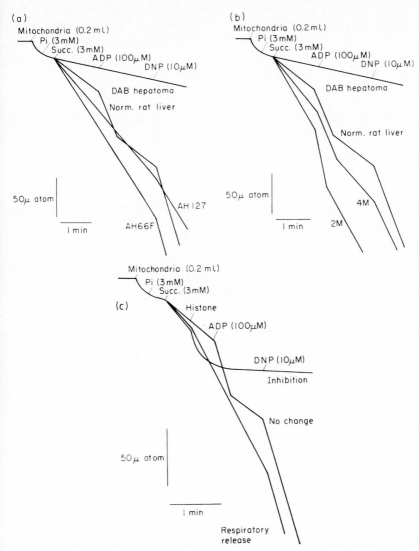

FIG. 5A, B. Effects of various concentrations of histone II on oxidative phosphory-lation of mitochondria. Mitochondria isolated from rat liver were incubated in 2 ml of basic reaction mixture at 25 °C. Platinum electrode trace. *Arrows* show the addition of substances. FIG. 5C. Oxygen consumption rate and ADP/0 ratio of mitochondria from various AH tumor cells and DAB-induced hepatoma during 6 months of carcinogenesis. They are divided into 2 main types of respiratory release and of respiratory inhibition.

sensitive to the mitochondria of the isologous rat liver, about 6–7 times as sensitive as the heterologous histone from the human gastric materials. These facts seem to suggest the species-specificity or organ-specificity of histones.

In our consideration of the effectiveness of each fraction of histones; that is, histones I and II from the cancerous and non-cancerous materials, there were also observed some differences between each fraction; generally speak-

TABLE 1. DOSE $\dfrac{\mu g \text{ OF HISTONE}}{mg \text{ PROT OF MITOCHONDRIA}}$ TO CAUSE RESPIRATORY RELEASE AND RESPIRATOY INHIBITION

		Respiratory release	Respiratory inhibition
Normal rat liver	H_1	250 μg	to excess
	H_2	100	250 μg
	f_{1+2b}	to excess	to excess
	f_{2a}	150	750
	f_3	100	250
DAB-	H_1	500	1000
hepatoma	H_2	250	500
AH66F	H_1	100	750
	H_2	50	100
AH127	H_1	750	to excess
	H_2	250	550
Gastric	H_1	to excess	to excess
ulcer	H_2	705	to excess
Gastric	H_1	480	to excess
cancer	H_2	600	to excess

ing, histone II of Ui is more effective than histone I and above all, histone II of AH66F is most distinguishable.

(N.B. Dr. Seno[54] has reported arginine-rich histone f_3 to be a true uncoupler. The histone II of Ui, in our experiments, was as effective as the f_3 of Bulter.)

Figure 5C indicates oxygen consumption rates and P/O ratios of mitochondria from AH66F and AH127 tumor cells respectively, and in comparison with those of normal rat liver, it is worth noting that mitochondria of tumor cells, particularly those of AH66F were somewhat in the uncoupled state. These differences might be caused by some inhibitory mechanism or inhibitory factor, probably by some basic protein like histones in the mitochondrial membranes of the cancer cells, and will be cited in more detail in the following paper.

(N.B. There is another type of basic protein: the toxohormone of Dr. Nakahara,[55,56] a characteristic catalase inhibitor in cancer cells. But on oxidative phosphorylation of mitochondria, this toxohormone prepared according to Dr. Fujii[57] did not produce any discernible effect.)

Histone Effect on the Biophysical Properties of the Mitochondrial Membrane

As histone attacks the acidic sites of mitochondrial membranes, this histone-mitochondrial interaction may cause alterations in the biophysical properties of membranes, accompanying the disorders of ion-transport systems and moreover resulting in some effect on the mitochondrial activities, for example, on oxidative phosphorylation already indicated.

The biophysical properties of the membrane, whether it is intact or damaged, may be represented electrophysiologically[58] by our equivalent circuit formula.[59] Consequently, the recordings of equivalent circuits denote the biophysical state of membranes quantitatively. Apart from the further details reported elsewhere, a brief explanation is as follows: when certain rectangular waves were applied to the membranes of mitochondria, i.e. to the mitochondrial suspension in the basic reaction medium, so-called transient phenomena were observed and judging from these transient curves, we can calculate the value of each component of the circuit by our formula. A representative example is shown in Fig. 6 which is related to the effects of various histones on intact mitochondria, and histone of tumor cells being added, there were observed, in principle, characteristic changes of each component of the equivalent circuit, particularly an increase of R_1 and decrease of C_1 value. Some differences could be noted according to the kinds of histones and the effect of histone II from AH66F was above all most remark-

TABLE 2. EQUIVALENT-CIRCUIT-GRAM OF MITOCHONDRIA

	$R_1 \times 10^3$ Ω	$R_2 \times 10^3$ Ω	$r \times 10^3$ Ω	$C_1 \times 10^{-6}$ μF	$C_2 \times 10^{-6}$ μF
Basic sucrose medium only	100	9.85	39.4	0.0630	2.96
Normal rat liver suspended in basic medium	328	13.6	45.5	0.0165	5.12
DAB, 1 month on dye	328	13.2	24.7	0.0165	6.12
DAB, 2 months on dye	328	11.9	27.4	0.0203	7.73
DAB (liver cirrhosis), 3 months on dye	281	16.6	36.0	0.0248	3.90
DAB-hepatoma 5 months on dye	273	20.15	50.7	0.0411	4.85
AH127	328	12.7	34.7	0.0165	4.29
AH66F	328	11.9	35.8	0.0265	4.90

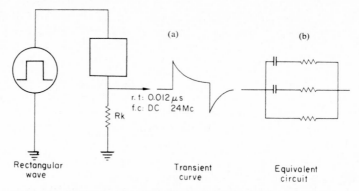

Fig. 6. A rectangular voltage (rise-time 0.1 μsec, duration-time 1.0 msec, 2.0 volt, 20 cps) was applied to mitochondria suspended in a basic reaction medium consisting of sucrose 0.05 M, KCl 0.02 M, Tris-HCl buffer 0.01 M and EDTA 0.1 mM, pH 7.4 in 2 ml, and then deformations of the curve, that is, transient phenomena were observed and recorded by the synchroscope (rise-time 0.012 μsec, frequency characteristic DC ~ 24Mc). Analysing this transient curve by our formula

$$I = A \frac{(\gamma + \alpha)\,\varepsilon^{\alpha t} - (\gamma + \beta)\,\varepsilon^{\beta t}}{\alpha - \beta}$$

we could calculate the value of each component of our equivalent circuit. According to our concept, this biological membrane consists of C- and R-components, which are arranged in parallel coupling of several C, R series, as Fig. 6b indicates. This equivalent circuit represents the electrophysical state of the biological membrane. Details have been reported elsewhere.

able (Table 2). Though the definite significance of C- and R-components have been partially explained in another of our reports, the pattern of the equivalent circuits may indicate the histone-mitochondrial interaction quantitatively.

In summary we have detected highly tumor- specific antigenic components in each subcellular structure, particularly in the nuclear substances, such as acidic nuclear globulin and basic proteins like histone associated with DNA and we have also discussed their biological significances, particularly the effects of histones on the mitochondrial activities from our standpoint.

REFERENCES

1. J. CHAUVEAU, Y. MOULE and C. H. ROUILLER, *Exptl. Cell Res.*, **11**, 313, 1956.
2. J. R. DAVIS and H. BUSCH, *Cancer Res.*, **19**, 1157, 1959.
3. G. PATEL and T. Y. WANG, *Exptl. Cell Res.*, **34**, 120, 1964.
4. S. DE CARVALHO, *J. Lab. Clin. Med.*, **56**, 333, 1960.
5. J. S. COLTER and K. A. O. ELLEM, *Nature*, **190**, 550, 1961.
6. V. P. BUTLER, S. M. BEISER, B. F. ERLANGER, S. W. TANENBAUM, S. COHN, and A. BEN-DICH, *Proc. Nat. Acad. Sci., U.S.*, **48**, 1597, 1962.
7. S. DE CARVALHO, H. J. RAND and J. R. UHRICK, *Experimental and Molecular Pathology*, **1**, 96, 1962.

8. D. STOLLAR and L. LEVINE, *Arch. Biochem. Biophys.*, **101**, 335, 1963.
9. D. STOLLAR and L. LEVINE, *Arch. Biochem. Biophys.*, **101**, 417, 1963.
10. S. W. TANENBAUM and S. M. BEISER, *Proc. Nat. Acad. Sci., U.S.*, **49**, 662, 1963.
11. V. D. TIMAKOV, A. G. SKAVRONSKAYA, N. B. BORISOVA and L. A. ZAMCHUK, *Federation Proc.*, **22**, 6, T 1029, 1963.
12. B. F. ERLANGER and S. M. BEISER, *Proc. Nat. Acad. Sci., U.S.*, **52**, 68, 1964.
13. S. PEREZ-CUADRADO, S. HABERMAN and G. J. RACE, *Federation Proc.*, **23**, 451, 1964.
14. O. J. PLESCIA, W. BAUN and N. C. PALCZUK, *Proc. Nat. Acad. Sci., U.S.*, **52**, 662, 1964.
15. N. UI, *Biochim. Biophys. Acta*, **25**, 493, 1957.
16. Y. SAEKI, *J. Juzen Med. Society*, **74**, 51, 1966.
17. L. S. HNILICA and H. BUSCH, *J. Biol. Chem.*, **238**, 918, 1963.
18. E. JOHNS and J. A. V. BUTLER, *Biochem. J.*, **82**, 15, 1962.
19. D. J. R. LAURENCE, P. SIMSON and J. A. V. BUTLER, *Biochem. J.*, **87**, 200, 1963.
20. R. A. REISFELD, U. J. LEWIS and D. E. WILLIAMS, *Nature*, No. 4838, 283, 1962.
21. J. BONNER and P. O. P. T'SO (Ed.), *The Nucleohistone*. Holden-Day Inc. San Fransisco, London, Amsterdam, 1964.
22. H. BUSCH and W. G. STEELE, Nuclear Histone of Neoplastic Cells. *Adv. in Cancer Res.* **8**, 42, 1964. Academic Press, New York, London.
23. H. BUSCH, *Histones and Other Nuclear Proteins*. Academic Press, New York, London, 1965.
24. H. BUSCH, P. BYVOET and J. R. DAVIS, Nuclear Proteins of Neoplastic Tissues. In *Molecular Basis of Neoplasma*, Univ. of Texas Press, Austin, 1965.
25. K. MURRAY, The Basic Proteins of Cell Nuclei. In *Annual Review of Biochem.*, **34**, 209, 1965. Annual Reviews, Inc. California.
26. P. O. P. T'SO, J. BONNER and H. DINTZIS, *Arch. Biochem. Biophys.*, **76**, 225, 1959.
27. J. A. W. BUTLER, P. COHN and P. SIMPSON, *Biochim. Biophys. Acta*, **38**, 386, 1960.
28. R. R. HUANG and J. BONNER, *Proc. Nat. Acad. Sci., U.S.*, **48**, 1216, 1962.
29. Y. G. ALLFREY, V. C. LITTAU and A. E. MIRSKY, *Proc. Nat. Acad. Sci., U.S.*, **49**, 414, 1963.
30. M. L. BIRNSTIEL and W. G. FLAMM, *Science*, **145**, 1435, 1964.
31. W. G. FLAMM and M. L. BIRNSTIEL, *Exptl. Cell Res.*, **33**, 616, 1964.
32. V. C. LITTAU, V. G. ALLFREY, J. H. FRENSTER and A. E. MIRSKY, *Proc. Nat. Acad. Sci., U.S.*, **52**, 93, 1964.
33. B. B. REID and R. D. COLE, *Proc. Nat. Acad. Sci., U.S.*, **51**, 1044, 1964.
34. H. G. SCHWEIGER, R. W. P. MASTER and S. G. A. ALIVISATOS, *Federation Proc.*, **23**, 382, 1964.
35. R. DAVENPORT and J. C. DAVENPORT, *Exptl. Cell Res.*, **39**, 74, 1965.
36. W. G. NIEHAUS and C. P. BARNUN, *Exptl. Cell Res.*, **39**, 435, 1965.
37. J. F. WHITEFIELD and T. YOUDALE, *Exptl. Cell Res.*, **40**, 421, 1965.
38. D. T. LINDSAY, *Arch. Biochem. Biophys.*, **113**, 687, 1966.
39. T. ISHIKAWA, *Japan J. Exp. Morph.*, **17**, 8, 1963.
40. T. ISHIKAWA and K. INOUE, *Intercellular Membraneous Structures*, p. 435. Ed. by S. Seno, 1965.
41. H. MORITA, *J. Juzen Med. Soc.*, **71**, 537, 1965.
42. Y. HONDA, *Ibid.*, **73**, 146, 1966.
43. T. HOKO, *Ibid.*, **74**, 30, 1966.
44a. M. HATANO, Reported at the 24th Annual Meeting of the Proceedings of the Japanese Cancer Association, 1965 (in press).
44b. M. HATANO, Reported at the 17th Symposia. Jap. Cell Biol. 1965 (in press).
45. Z. ROTEM, R. A. COT and A. ISAACS, *Nature*, **197**, 564, 1963.
46. J. R. BRONK and W. W. KIALLY, *Biochim. Biophys. Acta*, **24**, 440, 1957.
47. C. COOPER and A. L. LEHNINGER, *J. Biol. Chem.*, **224**, 547, 1957a.
48. J. B. HANSON and H. R. SWANSON, *Biochem. Biophys. Res. Commun.*, **9**, 442, 1962.
49. W. L. RIVENBARK and J. B. HANSON, *Biochem. Biophys. Res. Commun.*, **7**, 318, 1962.

50. V.G.ALLFREY, *Exptl. Cell Res.* Suppl. **9**, 183, 1963.
51. V.G.ALLFREY, *Ibid.*, **9**, 418, 1963.
52. A.SCHWARTZ, *J. Biol. Chem.*, **240**, 939, 1965.
53. A.SCHWARTZ, *Ibid.*, **240**, 944, 1965.
54. K.UTSUMI and G.YAMAMOTO, *Biochim. Biophys. Acta*, **100**, 606, 1965.
55. W.NAKAHARA and F.FUKUOKA, *Japan Med. J.*, 1271, 1948.
56. W.NAKAHARA and F.FUKUOKA, *Gann*, **41**, 47, 1950.
57. S.FUJII, T.KAWACHI, H.OKUDA, B.HAGA and Y.YAMAMURA, *Gann*, **51**, 223, 1960.
58. K.S.COLE, *J. Gen. Physiol.*, **15**, 641, 1932.
59. T.ISHIKAWA and S.ODASHIMA, *Symposia Cell. Chem.*, **16**, 219, 1966.

DISCUSSION

S. SENO: This is a very exciting presentation, but I would like to make some comments. Concerning the antibody to histone, for one year Dr. Kimoto tried extensively in my laboratory to get antibody to calf thymus histones by injecting them into rabbits, but up to the present he has had no success. Probably, the method he employed for histone extraction is inadequate, and I would like to test again with the histones obtained by the method you used.

Apart from the problem, I am interested in the point that the lysine-rich histone has so strong an uncoupling effect on the oxidative phosphorylation of isolated rat liver mitochondria, for in our observations the L-rich histone from calf thymus was rather weaker in that effect than arginine-rich histone. It might be specific to the tumor histone. According to Busch, A-L histone has the strongest uncoupling effect on oxidative phosphorylation. We have never observed the effect of this fraction but it seems to be well worth trying with this fraction.

S. H. HORI: Several data have shown that there might be important changes in the amount and/or the nature of phospholipids in mitochondria during carcinogenesis. Your today's representation showed unequivocally that histones have an inhibitory effect on mitochondrial oxidation processes. Since phospholipids are known to be indispensable constituents for mitochondrial oxidative enzymes, I wonder if you have some information on the possible effect of your histone fractions on mitochondrial phospholipids.

T. ISHIKAWA: Some reports indicate that the histone-mitochondrial interaction may be attributed to the acidic groups of phospholipid. It is assumed that by this interaction histone uncouples oxidative phosphorylation and alters the bioelectrical state of the mitochondrial membrane. As for oxidative enzymes, histone suppresses the activity of cytochrome C oxidase *in vitro*, as reported in some papers. This enzyme contains phospholipid as the main constituent, as you mention.

H. HANAFUSA: In your experiment about the effect of cancer DNA on the replication of polyoma virus in ME cells, what kind of cancer DNA was used? To what extent was this DNA preparation purified? Do you have evidence on the entry of cancer DNA into the cells? Do you explain your results by assuming that your cancer DNA acts as a histone inducer in the cells?

T. ISHIKAWA: We used DNA material from human gastric cancer, prepared by the fluorocarbon 12-time treatment. By this method, we obtained almost pure nucleic acid samples, which, normal rat liver being concerned, contained RNA 150–200 µg/ml, DNA 40–50 µg/ml and negative biuret reaction, and in the case of DAB-hepatoma and some AH tumor cells, contained RNA 120–180 µg/ml, DNA 55–60 µg/ml and negative biuret reaction.

In our experiments, there is no direct evidence that DNA should enter cells.

CELL AUTORADIOGRAPHY
AND BIOCHEMICAL STUDIES
OF CANCERIGENIC VIRAL INFECTIONS

YSIDRO VALLADARES*

Laboratory of Biology and Biochemistry of Cancer, National Cancer Institute, Ciudad
Universitaria, Madrid 3, Spain

THE infectious process of cancerigenic viruses at an intracellular level offers
a very interesting field of study that was realized by employing a virus
isolated in our laboratory (Cancervirus). The virus was first developed from
DNA of Ehrlich ascites cancer cells maintained by routine passages in SWR
mice. DNA was kept in contact with mouse embryo primary tissue cultures
and HeLa cell established cultures. Later on, viruses having the same properties were obtained with DNA from leukemic organs of AKR mice, sarcomas induced with 20-methylcholanthrene, and some human tumors.[6−9]
Cancervirus used in these experiments originated from DNA of a human
lung tumor after several passages in mouse embryo primary tissue cultures.

MATERIALS AND METHODS

Tissue cultures. Strain L-M cells (mouse fibroblasts) were grown in suspension cultures; subcultures were made by seeding 2×10^5 cells per ml of
fresh medium. Depending on the experiment, 50 ml capacity (25×150 mm)
or 20 ml capacity (16×150 mm) screw-capped tubes (Kimax brand) were
employed. Routine subcultures were prepared every 48 hr.

Growth medium. McCoy's medium[4] with 3% glucose, 0.5% lactalbumin
hydrolysate, and 20% calf serum was used; 25 ml of medium were added
to the 50 ml capacity tubes, and 10 ml to the 20 ml tubes.

Synchronized cultures. 24-hr-old cultures were centrifuged to remove the
old medium, and the cells resuspended in fresh medium warmed to 37°C
containing 450 μg amethopterin (4-amino-N^{10}-methyl folic acid, a folic acid
antimetabolite) and 13.5 mg adenosine per ml. The cultures were incubated
at 37°C and after 17 hr 10 μg of thymidine per ml were added. This resulted

* Scientific Collaborator of the "Highest Council of Scientifical Investigations"
(Consejo Superior de Investigaciones Científicas), Madrid.

in the synchronization of the cell cycle. The S phase (DNA synthesis) began immediately after thymidine addition, and continued for 6 hr.

DNA precursor, thymidine antimetabolite, puromycin, and mitomycin C. When indicated, 1μ Ci tritiated thymidine (TDr-H^3) per ml, 5-bromodeoxyuridine (BrUDr) at a concentration of 25 μg/ml, 10 μg/ml puromycin as protein synthesis inhibitor, or 5 μg/ml mitomycin C were added to the cultures.

Autoradiography. 100 μg colchicine per ml of cell suspension were added 16 hours before collecting the cells. These were centrifuged, the growth medium decanted, then the cells resuspended in 10 ml of Hank's balanced salt solution diluted to 1/4 with distilled water for hypotonic shock during 10 min in order to spread the chromosomes. After 7 min the pellet was fixed with acetic alcohol (methanol-acetic acid, 3:1) for 30 min. The cells were dispersed with a capillary pipette, a drop of cell suspension deposited on a microscope slide, and then air-dried to flatten the cells and further spread the chromosomes.

Nuclei were stained with acetic orcein. Two drops were placed on the cells and covered with a microscope slide. Acetic orcein was prepared with 2 g orcein and 45 ml boiling acetic acid; when cooled at room temperature, 55 ml distilled water were added. The staining solution was filtered before use.

To observe interphase nuclei, colchicine and hypotonic treatments were omitted, the washed cells being directly fixed and stained.

The coverslips were detached and the cells exposed to Kodak AR-10 autoradiographic stripping film, and developed and fixed after 7 to 14 days. The slides were air-dried and a coverslip mounted using diluted Permount.

Photomicrography. Representative cells were photographed using a model M20 Wild microscope with phase-contrast optics and Agfa Isopan IF (17° DIN, 40 ASA) film.

Virus. Cancervirus originally derived from DNA of a human lung cancer and maintained in primary mouse embryo cultures, was used in the experiments. Passages were made by inoculating new cultures with 0.4 ml of growth medium from 18-day-old cultures, 14 days after infection.

Virus was titrated by plaquing in 60 mm diameter Petri dishes; 1×10^6 primary mouse embryo cells were grown to a confluent layer, and then received agar overlay medium. Virus inoculation was realized using a multiplicity of 100 PFU per cell into 24-hr-old suspension cultures.

For the maintenance of the remaining virus, the following systems were used: chicken embryo fibroblast cultures for Rous sarcoma virus (RSV), HeLa cell cultures for Adenovirus 12, and LLC-MK$_2$ cell cultures for simian vacuolating agent (SV40)* and for Reovirus 3. The employed growth medium was prepared with McCoy's medium supplemented with 10% human blood serum, 0.5% lactalbumin hydrolysate, 2 ml of 1% phenol red per 1000 ml

---

* An SV40 strain adapted by us to LLC-MK$_2$ der. was used.

of medium, 100 I.U. penicillin and 100 μg streptomycin per ml (or 100 μg neomycin/ml), and 0.07% NaHCO$_3$ to bring the pH to 7.4.

RSV was maintained in cultures of chicken embryo fibroblasts prepared from 10-day-old embryonated eggs. Decapitated embryos were washed with Hank's balanced salt solution, minced, and dispersed with a 0.2% trypsin solution in a magnetic stirrer for two hours. The gauze-filtered cell suspension was centrifuged, trypsin solution decanted, and the cell pellet resuspended with 20 volumes of McCoy's growth medium suplemented with 0.5% lactalbumin hydrolysate and 10% calf serum; 0.5 ml of the resulting cell suspension was seeded into 120 ml capacity Owen's ovals, 9.5 ml of growth medium added, and the cultures incubated at 37°C. Cultures were inoculated with RSV when they were three days old, using 0.4 ml of infected culture medium containing about 10^6 PFU/ml.

RESULTS AND CONCLUSIONS

Growth curve. A typical growth curve of L-M cells in suspension may be seen in Fig. 1. They have a generation time of 16 hr (M phase, 1 hr; G1 phase, 4 hr; S phase, 6 hr; and G2 phase 5 hr).

TDr-H^3 uptake. TDr-H^3 was added at a concentration of 1 μCi/ml. Figure 2 indicates the time of mitosis labeling. The number of labeled mitoses increased when TDr-H^3 was present in the medium between 1 and 5 hr before, indicating that TDr-H^3 uptake and, consequently, DNA synthesis took place

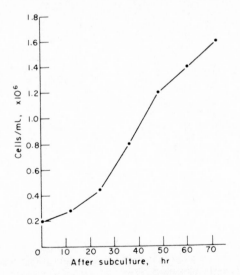

FIG. 1. Growth curve of L-M strain cells growing in suspension cultures. For subculture, 200,000 cells were seeded per ml of McCoy's suspension growth medium. The generation time was 16 hours. Routine subcultures were made every 48 hr. At this time the cells numbered 1,200,000 or six times more than initially.

between 5 and 11 hr before the onset of mitosis. *This corresponded to the S phase (6 hr), while 5 hr elapsed without labeling between the end of the labeling period and the onset of mitosis represented the G2 period.*

Effect of the antimetabolite. Amethopterin (450 μg/ml) prevented thymidine incorporation into the nuclei and cell growth (Fig. 3). As soon as the antimetabolite was removed and thymidine added, TDr-H³ was incorporated into the nuclei in a synchronized way. Amethopterin-treated cells were incubated at 37 °C for 17 hr and then 10 μg of thymidine/ml added to promote a synchronized S period, which lasted 6 hr. When TDr-H³ (1 μCi/ml) was added at several intervals during the S phase it was observed that a differential labeling occurred among the several chromosomes with rich and poor-labeling and early and late-labeling zones. The specific rate of chromosomal DNA synthesis took place during the first 3 hr, while in the remaining 3 hr the labeling was roughly proportional to the chromosomal length of the unlabeled portion at that time.

Tritium labeling. Tritium is a low energy radioisotope allowing the localization of the disintegrating atom within a zone of approximately 900 mμ. Taking this into consideration, each chromosome was divided in microns along its length for chromosomal topology.

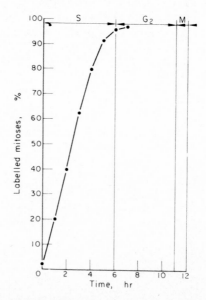

FIG. 2. Time of mitosis labeling (thymidine-H³ incorporation). When TDr-H³ uptake was studied in relation to mitosis, it was observed that labeled mitoses increased when TDr-H³ had been present in the growth medium 1 to 5 hours before. That is to say, the last TDr-H³ was incorporated into the chromosomes 5 hours before the onset of mitosis; this time was considered as the G2 phase preceding mitosis. The six-hour period during which the cells incorporated the TDr-H³ that appeared in the next mitosis, was taken as the S phase.

Labeling during the first 10 min of the S period was followed by TDr-H³ incorporation from all the chromosomes, but each chromosome incorporated the radioactive material at a different zone and with a different intensity.

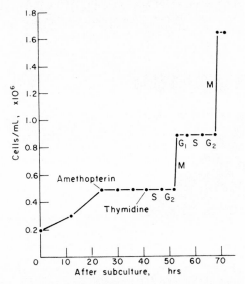

FIG. 3. When amethopterin is added to the cultures, thymidine incorporation into the nuclei and cell growth are strongly inhibited. The removal of the antimetabolite and the addition of thymidine is immediately followed by a synchronized cell cycle starting with the S phase. Under adequate conditions cell synchronization may last several cell generations.

After labeling for the first 60 min of the S period, all the chromosomes showed TDr-H³ incorporation. This labeling was not uniform among the different chromosomes and in several portions of each chromosome, indicating that the *incorporation proceeded at different rates in each chromosome.* The label was the same in the two halves of every individual chromosome.

Labeling during the first 3 hr of the S period indicated the existence of TDr-H³ uptake along the whole length of all the chromosomes, but rich and poor-labeled portions could be distinguished.

When the labeling began 4 hr after the beginning of the S period, it was possible to see that the zones labeled during the first 2 hr took little or no TDr-H³, while the unlabeled or poor-labeled regions during the first 2 hr were richly labeled in the last hours of the S phase.

Labeling of the cells during the whole S period showed that the final label of isotope in each chromosome was proportional to the length of the individual chromosomes, but in each one rich and poor-labeled portions were observed. *The differential labeling, then, lasted 3 hours, while a proportional labeling was verified during the remaining 3 hours.*

Removal of thymidine during the S phase. When thymidine was eliminated from the cultures at any time during the 6 hr of the S period, mitosis was arrested. This happened when thymidine was removed 10 min, 1 hr, 2 hr, 4 hr, and $5\frac{1}{2}$ hr after the onset of the S period. This indicated that *mitosis took place only after the complete endowment of DNA molecules had been biosynthesized* (Fig. 4).

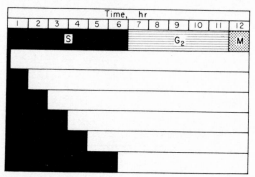

Fig. 4. Effect of thymidine removal on cell DNA synthesis and cell reproduction.— Removal of thymidine at any moment during the S phase results in the stopping of the cell cycle. This figure shows the effect of thymidine removal 10 minutes, 1 hr, 3 hr, 4 hrs, and $5\frac{1}{2}$ hrs after the onset of the S period. DNA has to be completely synthesized in order for the cell cycle to continue.

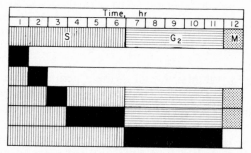

Fig. 5. Effect of puromycin on cell DNA synthesis and cell reproduction.—Cell DNA synthesis is a puromycin-sensitive process during the first two hours, but not during the last 4 hours. G2 phase is also a puromycin-sensitive process, and mitosis does not follow when the drug has been present during it. This indicates the need for enzyme and protein synthesis during the first two hours of the S phase, and the complete G2 phase.

Effect of puromycin. Puromycin (10 μg/ml) inhibits protein synthesis, but permits RNA synthesis. Puromycin acting during the first 60 min of the S period, and between the first and the second hour of the S period, strongly inhibited the TDr-H³ uptake (Fig. 5).

Puromycin did not inhibit the TDr-H³ uptake between the second hour and the end of the S period. *Protein synthesis, probably of enzymatic nature, is then necessary for DNA synthesis and for cell division during the first 2 hr of the S phase, but not in the remaining 4 hr* (Fig. 5).

Puromycin inhibited cell mitosis if added in the G2 phase after the S period was completed. Again, *protein synthesis is necessary for cell division during the G2 period*, after the completion of the S period (Fig. 5).

Effect of 5-bromodeoxyuridine (BrUDr). In the presence of BrUDr (25 μg/ml) at any moment during the S phase, no radioactivity was shown in the DNA thymine, because BrUDr blocked the conversion of dUMP to dTMP, and suppressed DNA synthesis (Fig. 6).

On the other hand, BrUDr allowed the synthesis of RNA and protein to continue.

Virus infection in non-synchronized cell cultures. 100 PFU of virus were added per cell to 24-hr-old L-M cell suspension cultures. One hour after infection, TDr-H³ was added for 2 hr. Autoradiography showed that 75% of the nuclei were labeled, against 42% in the control cells (Fig. 7).

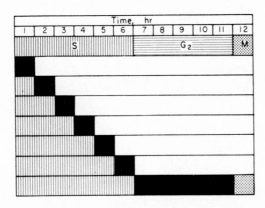

FIG. 6. Effect of 5-bromodeoxyuridine on cell DNA synthesis and cell reproduction. —BrUDr blocks the conversion of dUMP to dTMP, and suppresses DNA synthesis. The presence of BrUDr at any time during the S phase, stops the continuation of the cell cycle.

When TDr-H³ was added between 2 and 12 hr after infection, and allowed to act for 2 hr, the cultures showed practically 100% of labeled nuclei, against 30 to 50% in the corresponding control cells (Fig. 7).

Two hours after infection, TDr-H³ was accumulated *within the nucleus and towards the nuclear membrane*, in one or two different portions, and *reaching the vicinity of the nucleolus* without entering in it. The form of the label was ring-, falciform- or butterfly-shaped in the infected cells, while the control cells showed a diffuse nuclear labeling.

Six hours after infection, the radioactivity tended to be disposed *more centered in a ring-shaped way*. This disposition coincided with the disappearance of the nucleolus.

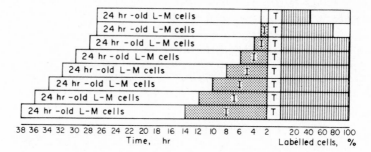

24 hr -old L-M cells		
24 hr -old L-M cells		
24 hr -old L-M cells		
24 hr -old L-M cells		
24 hr -old L-M cells		
24 hr -old L-M cells		
24 hr -old L-M cells		
24 hr -old L-M cells		

38 36 34 32 30 28 26 24 22 20 18 16 14 12 10 8 6 4 2   20 40 60 80 100
Time,   hr   Labelled cells,   %

FIG. 7. Virus infection in non-synchronized cells.—Addition of TDr-H³ for two hours, after 1, 2, 4, 6, 8, 10 and 12 hours of infection with 100 PFU of Cancervirus showed that the precursor was incorporated by 75% of the cells that had been infected for one hour, and by all the cells that had been infected from 2 to 12 hours. The corresponding control cells showed TDr-H³ incorporation in 30 to 50 per 100 of the cells.

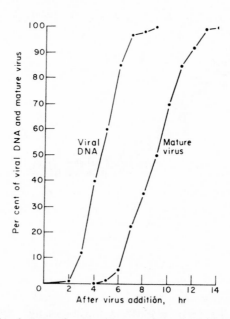

FIG. 8. Relation between the time of appearance of viral DNA and mature vegetative virus.—98.5% of the viral DNA was formed during 6 hours, beginning 2 hours after infection. Mature virus began to be formed 5 hrs. after infection (lag phase), when 60% of the viral DNA had been produced. Nine hours after infection 50% of the mature virus had been formed, in coincidence with the completion of the viral DNA synthesis. Mature virus was produced between the 5th and the 13th hour after infection (8 hours).

Twelve hours after infection, radioactivity was *evenly distributed in the nucleus*, without invading the dense chromatin zones.

When cultures were grown in the presence of TDr-H^3 for 15 hr, the old medium removed, the cells washed, new medium containing virus added for one hour, and then again virus-containing medium changed for fresh medium, the cells showed the TDr-H^3 in the chromatin, especially in the peripheral portions.

Virus infection in cells treated with BrUDr. When virus was added, the cells washed, BrUDr (25 µg/ml) added to one group of cells, the cultures incubated at 37°C for 18 hr, and the virus titrated, the following observations were made: (a) virus grew well in the infected cells; (b) no virus grew in the infected cells treated with BrUDr; (c) virus grew normally in infected cells simultaneously treated with BrUDr and thymidine.

It is evident that BrUDr stopped viral synthesis. Therefore, when BrUDr was added at different intervals, the DNA of any virus formed had to be synthesized before the BrUDr addition. The relative amount of DNA was calculated from the number of grains in autoradiography. The mature virus was calculated by plaque titration.

The results obtained appear in Fig. 8.

Viral DNA began to be formed 2 hours after infection and was rapidly synthesized during the next 6 hours (98.5%), with only 1.5% of the viral DNA being produced in the next hour. *Infectious virus began to be formed 5 hours after infection*, corresponding to the lag phase, when 60% of the viral DNA had been formed. *Nine hours after infection 50% of the mature virus had been formed, in coincidence with the completion of the viral DNA synthesis.* Mature vegetative virus was produced between the 5th and the 13th hour after infection (8 hr). *The synthesis of viral DNA took place 3 hr before the building of new infective virus.* The slope of the infective virus curve and that of the virus DNA curve are slightly different.

Virus infection in synchronized cells (Fig. 9). Amethopterin (450 µg/ml) was added to 24-hr-old L-M cell suspension cultures, and allowed to act for 17 hr, then the cultures were infected with virus and 5 min later thymidine (10 µCi/ml) was added.

When TDr-H^3 (1 µCi/ml) was added during the first 20 min of infection, neither cell nor virus showed TDr-H^3 incorporation (Fig. 9b).

When TDr-H^3 was added for 10 min, 20 min after infection, viral DNA could be seen in the periphery of the cell nucleus (Fig. 9c).

When TDr-H^3 was added for 10 min, 30 min after infection, a big increase of viral DNA was observed, tending to occupy the periphery of the nucleus (Fig. 9d).

Control cells showed an increased diffuse labeling starting when amethopterin was substituted by thymidine (Fig. 9a, and Figs. 10 and 11). The diffuse labeling was not observed in the infected cultures during the first 45 min of infection (Fig. 9e). *Virus, then, delayed cell DNA synthesis, the onset of*

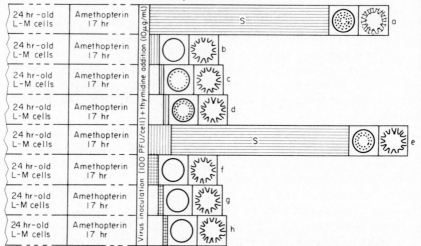

24 hr-old L-M cells	Amethopterin 17 hr			a
24 hr-old L-M cells	Amethopterin 17 hr			b
24 hr-old L-M cells	Amethopterin 17 hr			c
24 hr-old L-M cells	Amethopterin 17 hr			d
24 hr-old L-M cells	Amethopterin 17 hr			e
24 hr-old L-M cells	Amethopterin 17 hr			f
24 hr-old L-M cells	Amethopterin 17 hr			g
24 hr-old L-M cells	Amethopterin 17 hr			h

FIG. 9. Virus infection in synchronized cells.—Amethopterin followed by thymidine were used to synchronize the cells. a) Diffuse nuclear labeling and complete chromosomal labeling of non-infected cells; b) Absence of nuclear and chromosomal labeling during the first 20 minutes of infection, indicating absence of viral DNA synthesis and delay of cell DNA synthesis; c) Peripheral nuclear labeling due to viral DNA synthesis, when TDr-H^3 was added for 10 minutes, 20 after infection; the cell DNA synthesis is not yet observed; d) Increase in nuclear labeling due to viral DNA synthesis; cell DNA synthesis is not observed; some viral DNA appears, incorporated into certain chromosomes; e) Viral DNA is synthesized before cell DNA and cell DNA synthesis shows a delay of 45 minutes; some viral DNA appears, incorporated into certain chromosomes. Viral DNA is a puromycin-sensitive process whether puromycin acts during the first 20 minutes of infection (f), for 10 minutes, 20 minutes after infection (g), or for 10 minutes, 30 minutes after infection (h).

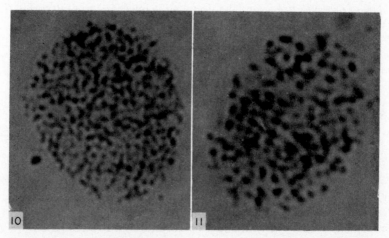

FIGS. 10 and 11. Evenly diffused labeling of the cell nucleus. Grains correspond to the presence of radioactivity after TDr-H^3 incorporation into cell DNA.

S period being 45 minutes later than in control cultures (Fig. 9e). *Viral DNA was synthesized before cell DNA*, the synthesis taking place in the nucleus and around the nucleolus, but not in contact with dense chromatin and most of it out of the chromosomes (Figs. 12 to 15). However, *a small amount of label*

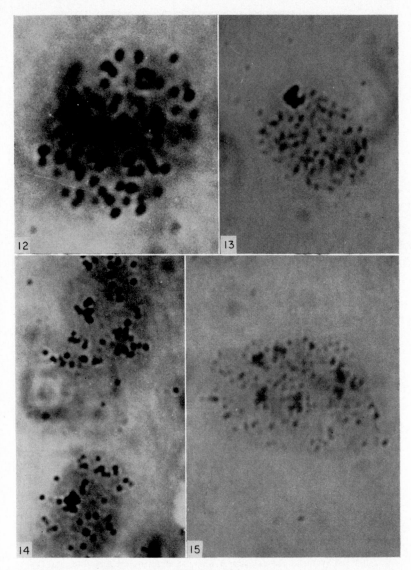

FIGS. 12 to 15. Labeling pattern of cell nuclei belonging to cells infected with Cancervirus. Grains correspond to TDr-H³ uptake during viral DNA synthesis. Observe the scattered labeling and the presence of one or several foci of grain accumulation, usually in the vicinity of the nucleolus.

appeared in the mitotic chromosomes (Figs. 16 to 18), *at specific sites in one or two chromosomes, and it could correspond to an episomal or proviral (lysogenic) incorporation.*

Karyotype studies should be necessary to see precisely the pattern of the

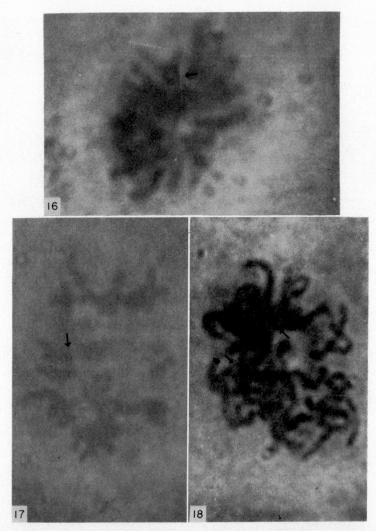

FIGS. 16 to 18. Aspect of three mitoses from cells infected with Cancervirus. TDr-H³ was present in the medium for 45 minutes, during which viral DNA was synthesized but cell DNA synthesis was delayed. At some time during the period of viral DNA synthesis or the S phase of the cell, viral DNA was incorporated into the chromosomes, as it appeared when studying forthcoming mitosis. The incorporation of viral DNA seems to take place at specific locations, and it was generally found at a single place (Figs. 16 and 17), and a few times in two chromosomes (Fig. 18).

differential chromosomal labeling, and to know the exact sites of viral DNA incorporation. The L-M cells studied had a modal number of 60 chromosomes, and about 86% of the cells contained marker chromosomes, but a careful karyologic study was not realized. The DNA content was calculated in an average of 16.8 p.g. per cell.

Puromycin (10 µg/ml) interfered with the synthesis of viral DNA whether present during the first 20 min of infection (Fig. 9f), for 10 min, 20 min after infection (Fig. 9g), or for 10 min, 30 min after infection (Fig. 9h). *The synthesis of viral DNA is then a puromycin-sensitive process indicating that the simultaneous synthesis of proteins, probably enzymes, was a requisite for it.*

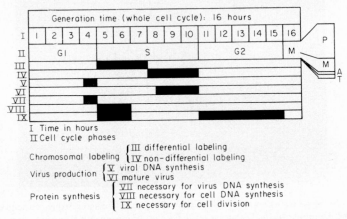

FIG. 19. Correlation between the several observed phenomena, as indicated in the text.

Figure 19 shows the correlations between some of the observed phenomena.

A group of 24-hr-old L-M cells were synchronized with amethopterin (450 µg/ml); after 17 hours the cells were infected using 100 PFU of Cancervirus per cell. Thymidine (10 µg/ml) was added immediately after infection to start a synchronous DNA synthesis. In control cells, this event marked the onset of the S phase (cell DNA synthesis). In the presence of virus, only viral DNA was synthesized during the first 45 min, and then started cell DNA synthesis. TDr-H³ (1 µCi/ml) was added for the first 45 min, then the cells washed, and the cell cycle allowed to continue through the G2 phase, M phase, and next G1 phase by replacing medium containing only unlabeled thymidine.

Mitomycin C (5 µg/ml) and then TDr-H³ (1 µCi/ml) were added to the cultures for the period of 45 min prior to the S phase. Cells were washed to remove TDr-H³.

Control cultures, untreated with mitomycin C, showed viral DNA production in the nucleus as usual while a discrete amount of label was seen in specific chromosomes (Figs. 16 to 18).

Mitomycin C-treated cultures showed the normal virus production and, in addition, some cells presented a peculiar accumulation of grains at a discrete zone at the beginning of the S phase (Fig. 20).

This fact was interpreted as a probable virus multiplication starting on viral DNA incorporated into the genome, after the oncolysogenic cell had been induced by mitomycin C.

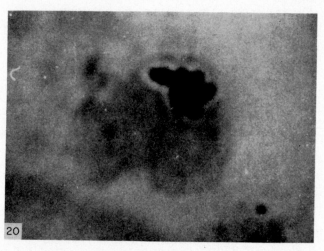

FIG. 20. Appearance of an accumulation of grains at a discrete zone of the nucleus at the beginning of the S phase after the action of mitomycin C on cells containing viral DNA incorporated into the chromosomes.

Effect of Viral Nucleic Acids on Cell DNA

Rous sarcoma virus (RSV). First of all I wish to recall some experiments reported in 1964, to emphasize the importance of cell DNA in the cancerous character. Rous sarcoma virus (RSV), a single-stranded RNA virus acting at a cytoplasmic level, was maintained in cultures of chicken embryo fibroblasts prepared from 10-day-old embryonated eggs. Seven days after infection, cell DNA was extracted following Kirby's procedure. The samples were treated with trypsin (10 μg/ml) plus chymotrypsin (10 μg/ml) for 30 min, and then with RNase (50 μg/ml) for another 30 min. Orcinol reaction showed the complete absence of RNA. Inoculation of DNA from RSV-infected cells promoted the neoplastic transformation of chicken embryo fibroblasts, and *de novo*-formed RSV appeared in the cultures after several passages by mixing cells of the previous culture with new primary cultures of chicken embryo fibroblasts.

This relation between RNA and DNA is in accordance with previous findings, about the cancerigenic action of RNA from cancer cells, reported by us in 1959 for Ehrlich ascites cancer cells of the mouse.

Taking into consideration the effect of RSV, and the apparent inductive action of mitomycin C, some experiments were realized employing Adenovirus 12, a 70 mμ DNA virus reported as oncogenic by J.J.Trentin, Y. Yabe, and G.Taylor (1962)[5] and by R.J.Huebner, W.P.Rowe, and W.T.Lane (1962)[3], simian vacuolating virus (SV40), a 45 mμ DNA virus also reported as oncogenic by B.Eddy, G.S.Borman, G.B.Grubbs, and R.B.Young (1962)[1] and Reovirus 3, a non-cancerigenic 75 mμ two-stranded RNA virus according to P.J.Gomatos and I.Tamm (1963).[2]

Adenovirus 12. Adenovirus 12 was inoculated into HeLa cell cultures. After 12 days the infected cells showed large foci of round degenerated cells with the typical aspect of bunches of grapes. Hematoxylin-eosin staining demonstrated the presence of intranuclear inclusions.

Cell DNA was extracted from cultures 12 days after infection. In order to inhibit any free virus, the extract was treated with rabbit antiserum against growth medium from infected cultures, which contains an Adenovirus group soluble antigen.

DNA (50 μg) was used to inoculate new HeLa cell cultures. After 3 days the medium was changed, the fresh medium containing mitomycin C (5 μg/ml). Twelve days later, the cultures offered the CPE which are typical of Adenovirus 12 infection, and reproduced mature virus.

*SV40.** SV40 virus was inoculated into LLC-MK$_2$ cell cultures where it produced vacuolation of the entire cell population within 5 days.

DNA was isolated from cultures 8 days after infection. Free virus was inhibited by treatment with rabbit anti-SV40 serum, and DNA (50 μg) was added to new LLC-MK$_2$ cultures. Medium containing mitomycin C (5 μg/ml) was added after 3 days. Vacuolation of the cell population was observed 7 days later. SV40 virus was recovered from the cultures.

Reovirus 3. This virus was inoculated into 4-day-old LLC-MK$_2$ cell cultures. Ten days after infection, typical CPE consisting of intracytoplasmic inclusions were observed after staining with hematoxylin-eosin.

DNA was extracted from LLC-MK$_2$ cells ten days after infection. The extracts were treated with trypsin (10 μg/ml) plus chymotrypsin (10 μg/ml) for 30 min, and then with RNase (50 μg/ml) for another 30 min to eliminate virus and any possible RNA contamination. Absence of RNA was tested by the orcinol reaction.

Cell DNA (50 μg) was added to new LLC-MK$_2$ cell cultures. Mitomycin C (5 μg/ml) was added three days later. CPE appeared 12 days after DNA addition; intracytoplasmic inclusions were observed in preparations stained with hematoxylin-eosin.

Some concluding remarks. Cancervirus, Adenovirus 12 and simian vacuolating virus 40 seem to be incorporated in a lysogenic way. In the case of RSV and Reovirus 3, we cannot suppose a lysogenic incorporation, but a

---

* SV40 virus strain adapted to *Macaca mulatta* kidney cells.

transduction of the RNA genetic code of the viruses into the DNA of the genome seems to exist. Considering this, an alternative form of viral DNA action upon the cell genome could be a transcription rather than a true lysogeny. Studies are in progress in our laboratory to decide between these two alternatives, or if both of them may exist.

If the results indicated here for RSV and Reovirus 3 as examples of RNA viruses (a single-stranded and a two-stranded virus, respectively), and for Cancervirus, Adenovirus 12 and SV40 as representatives of DNA viruses can be generalized, and we think this may be the case, the following conclusions would be drawn.

Every virus, whether of RNA or DNA composition seems to imprint its complete genetic code into the cell genome. The genetic code of the virus acts not only as a template for its autocatalytic replication, and as a director of cell metabolism to organize the synthesis of new enzymes needed for the synthesis of viral proteins and nuclei acid, but, in addition to this, the genetic message is fully translated by the cell DNA. This may be done in several ways; viral DNA would be incorporated into the cell genome as proviruses to form lysogenic cells and/or the genetic message would be transcribed into the cell DNA. As for RNA viruses, the RNA genetic code would be read and transduced into the cell DNA in a process of reversal translation. Each combination of ribonucleotides would be translated into the language of a DNA codon, with the corresponding combination of deoxyribonucleotides. In all cases, new genes or wild genes with new genetic information would be formed.

Lysogenic cells may produce mature virus by induction, that is to say by luxating the provirus from the genome. DNA resulting from a transcription could also luxate to form new vegetative viruses. DNA resulting from a reversal translation would give rise to RNA viruses by acting as a template for complementary RNA, and organizing mRNA carrying the information for the assembling of the amino acids transferred by sRNA into specific polypeptide chains for the viral capsomers at the level of the ribosomes (rRNA).

Viral DNA could also exist as an episome, and viral RNA as a cytoplasmic hereditary factor.

SUMMARY

Employing a cancerigenic virus isolated in our laboratory (Cancervirus), that was formed starting on cancer cell DNA, we studied the infective process at an intracellular level in suspended cultures of L-M cells. Simultaneous protein synthesis was necessary during the first 2 hr of the S phase (6 hr) in order for the DNA synthesis and cell reproduction to be verified, and during the G2 phase (5 hr) in order for the cell multiplication to follow. Mitosis (1 hr) began only after the whole DNA endowment had been synthesized.

Virus was synthesized within the nucleus. The onset of viral DNA synthesis was 2 hr after infection; 98.5% of it was formed during 6 hr, and only 1.5% in the next hour. Vegetative virus underwent a lag phase of 6 hr, during which 60% of viral DNA was produced; 9 hr after infection, 50% of mature virus had been formed, and the synthesis of viral DNA was completed. Viral DNA synthesis began 4 hr before the onset of vegetative virus formation.

In synchronized cells the viral infection delayed by 45 min, cell DNA synthesis. Viral DNA was synthesized before cell DNA. The synthesis took place within the nucleus and most of it out of the chromosomes, but a small fraction of TDr-H^3-labeled viral DNA appeared within the mitotic chromosomes, probably incorporated into specific locations as an episome or a provirus (lysogenic cell). Viral DNA synthesis needed the simultaneous synthesis of proteins.

The importance of cell DNA in the cancerous character is emphasized by the fact that DNA from cultures of chicken embryo fibroblasts infected with Rous sarcoma virus possesses cancerigenic properties and is able to regenerate the RNA virus.

The results derived from the studies made with Cancervirus, Adenovirus 12, simian vacuolating virus (SV40), Rous sarcoma virus, and Reovirus 3 indicate that every virus, whether of RNA or DNA composition imprints its complete genetic code into the cell genome.

REFERENCES

1. B. EDDY, G. S. BORMAN, G. B. GRUBBS and R. B. YOUNG, Identification of the oncogenic substance in Rhesus monkey kidney cell cultures as simian virus 40. *Virology*, **17**, 65–75, 1962.
2. P. J. GOMATOS and I. TAMM, The secondary structure of Reovirus RNA. *Proc. Nat. Acad. Sci., U.S.*, **49**, 707, 1963.
3. R. J. HUEBNER, W. P. ROWE and W. T. LANE, Oncogenic effects in hamsters of human Adenovirus types 12 and 18. *Proc. Nat. Acad. Sci., U.S.*, **48**, 2051, 1962.
4. T. A. McCOY, M. MARWELL and P. F. KRUSE, Amino acid requirements of the Novikoff hepatoma *in vitro*. *Proc. Soc. Exptl. Biol. Med.*, **100**, 115–118, 1959.
5. J. J. TRENTIN, Y. YABE and G. TAYLOR, The quest for human cancer viruses. *Science*, **137**, 835–841, 1962.
6. Y. VALLADARES, Experimental production of leukemia and polycythemia vera and immunotherapia assays upon cancer and leukemia. *Proc. Ninth Cong. of the Intern. Soc. of Hemat.* (Mexico City), Vol. II, pp. 571–587, 1962.
7. Y. VALLADARES, Estudios metabólicos y genéticos empleando DNA de un virus cancerígeno (Cancervirus) sintetizado enzimáticamente. *Rev. Esp. Oncol.*, **11**, 265–302, 1964.
8. Y. VALLADARES, Leucemia, policitemia vera y cánceres experimentales. *Sangre*, **10**, 14–56, 1965.
9. Y. VALLADARES, Cancerigenic viruses neoformed in *in vitro* cultured cells starting on DNA from human cancer cells. *Med. Pharmacol. Exp.*, **16**, 311–324, 1967.

DISCUSSION

I. YAMANE: (1) About what percentage of your virus labeled with thymidine-H^3 did attach to the chromosome?

(2) Does such a virus particle attach to any special chromosome of all the cellular chromosomes?

Y. VALLADARES: (1) For your first question, we made our studies on a cellular basis rather than on a viral basis. We did not count the number or percentage of virus that attached to the chromosomes because we inoculated it at a multiplicity of 100 PFU per cell. The cells we used had a modal number of 60 chromosomes; therefore, they belong to a triploid line, considering they are mouse cells. The measured amount of DNA was 16.8 p.g. $(16.8 \times 10^{-12}$ g) per cell. On the cellular basis, virus labeled with TDr-H^3 was observed in about 10% of the cells, presumably those that will suffer the malignant transformation after this supposedly lysogenic incorporation.

(2) As for your second question, a single viral DNA molecule seems to become incorporated into a special chromosome, and at most into two different chromosomes. Unfortunately careful karyological studies that we have not yet made, are necessary to detect precisely the specific chromosome receiving the viral DNA.

F. H. KASTEN: (1) When we first began chemical synchronization studies, we used the amethopterin technique on epithelial tumor cell lines and were somewhat disappointed in the degree of synchrony achieved. For example, Dr. J. Adams in our laboratory never obtained a mitotic burst greater than about 15%, as I recall. We finally abandoned this technique in favor of the excess thymidine method which yields mitotic bursts as high as 70% and a tritiated thymidine incorporation rate of 98% during most of the S phase with a drop off to 10% during the G2 phase (Kasten and Strasser, *Nature*, 1966). You have probably had far better success with the amethopterin technique than we ever achieved. What is the degree of synchrony achieved during the cell cycle in your experiments and what are your criteria for assessing the degree of synchrony?

(2) In your autoradiographs, how do you distinguish between host and viral DNA?

(3) Do your experiments answer the question as to whether host DNA may break down to form viral DNA?

(4) With regard to the question raised by the previous discussant, Miss Susan House and I found with time-lapse photography that cells in mitosis which contain viral inclusions (human Adenovirus type 1) may complete mitosis and transmit these inclusions to daughter cells.

Y. VALLADARES: (1) To answer the first question of Dr. Kasten, I can say that the degree of synchronization you may obtain in tissue cultures depends on several factors, according to the particular system that is used.

First, you speak of epithelial tumor cell lines in monolayer; our experiments were realized with L-M cells (mouse fibroblasts) growing in suspension as a monodispersed culture.

Second, it is necessary to take into consideration the concentration of amethopterin which is added to the medium, and the time of its action in relation to the cell cycle of any particular cell line; we used amethopterin at a concentration of 450 μg/ml during 17 hr in a cell strain having a cell cycle in suspension of 16 hr.

Third, we have found that the degree of cell synchrony that is achieved depends on the simultaneous presence of some other substances or metabolites in addition to amethopterin. Originally, synchronization was attained with amethopterin (a folic acid antagonist) plus FUdR (a thymidylate synthetase inhibitor) (Rueckert and Mueller, 1960). We found a better synchronization using amethopterin at the indicated concentration together with adenosine (13.5 mg/ml), and this is actually the method we used for the experiments reported here. Under our conditions, cell synchrony lasted several cell generations. We were able to get a synchronization for 90% of our rapidly growing population, and 95% of the cells showed TDr-H³ incorporation into the nuclei during the 6 hr of the S phase that followed immediately the removal of the antimetabolite and the addition of 10 μg of thymidine per ml. During the G2 phase only 5% of the cells (probably nonsynchronized) showed TDr-H³ incorporation.

(2) As for the second question of Dr. Kasten, when 24-hr-old cultures were treated with amethopterin (450 μg/ml) during 17 hr, then the antimetabolite removed and thymidine (10 μg/ml) added, an increasing labeling of the nuclei was observed during the following 6 hr of the synchronous S phase. The autoradiographic grains were abundant and evenly distributed. On the other hand, in the infected cultures viral DNA is synthesized before cell DNA, as we have demonstrated; when 10 μg thymidine and 1 μCi TDr-H³ per ml were added immediately after infecting with 100 PFU of Cancervirus per cell, and just prior to the following synchronous S period, a peculiar distribution of the grains was observed instead of the evenly distributed labeling, and the chromosomes appeared unlabeled when TDr-H³ was not allowed to act for more than 45 min. Under these conditions, only the single grain in a particular chromosome was observed, as shown in the projected slides. Viral infection then delayed cell DNA synthesis in 45 min. When TDr-H³ was present after this interval, the usual evenly distributed grains and their chromosomal location were observed.

(3) In his third question, Dr. Kasten wishes to know whether the host DNA may break down to form viral DNA. We have no definite direct proof that this does not happen. However, viral DNA is a specific molecular species that multiplies only after the necessary enzymes have been synthesized. My belief is that there is a *de novo* viral DNA formation according to the current knowledge of how the viral multiplication takes place.

(4) As for your fourth and the last question, according to your own find-

ings, as reported here yesterday, mitomycin C stops the cell cycle at the meta-phase stage, but cell DNA synthesis takes place before that stage, and viral DNA still earlier. In the conditions of our experiments, mitomycin C seems to promote an induction phenomenon in an oncolysogenic cell, employing the word "induction" with the meaning it has in microbiology. This is per-haps the first experimental demonstration of the fact that induction may exist in a mammalian cell. I do not think than that the viral reproduction I have indicated derives from any inclusions; these are actually produced in the infected cells where virus multiplies, but not in the oncolysogenic cells.

M. E. KAIGHN: What is the nature of your evidence that a DNA "imprint" or "provirus" of the viral RNA genome is made in RSV infected cells?

Y. VALLADARES: As I have indicated, the tested DNA viruses may become incorporated into the cell genome in a lysogenic way, or their genetic mes-sage may be transcribed or copied by the DNA of the host. But Rous sar-coma virus and Reovirus 3 are RNA viruses and we cannot think of either in a lysogenic incorporation nor in a transcription, but only in a reversal trans-lation from the viral RNA to the host DNA, a fact never reported until now. We can deduce this because we started from cell DNA isolated after a cer-tain time of infection, and we treated the product with proteolytic enzymes (trypsin and chymotrypsin) or with ribonuclease in order to eliminate any possible RNA contamination. So what we inoculated was a rather pure cell DNA, and it was responsible both for the appearance of the cytopathic effects that were typical for the original virus, and also for the *de novo* forma-tion of the original RNA virus that could be recovered from the DNA-treated cells after a certain time.

ONCOGENIC PROPERTIES
OF DNA DERIVED FROM SV-40 VIRUS

Tetsuo Kimoto and James T. Grace

Department of Pathology, Okayama University Medical School
and Roswell Park Memorial Institute, Buffalo, New York, U.S.A.

Concerning nucleic acids of SV-40, Borison,[1] with the use of SV-40 grown on baban cells, extracted nucleic acid by Weil's[2] modification of the Kirby[3] method. More recently, Gerber[4] described that infectious DNA derived from SV-40 could be extracted by the cold phenol method of Gierer and Schramm[5] and nucleic acids produced a cytopathic effect identical with the one caused by an intact SV-40 propagated in *cercopithecus* kidney cell cultures. Weil and Eddy[6] also reported that infectious virus could be recovered by incubating the nucleic acid with *cercopithecus* kidney cell cultures. Our studies are concerned with the oncogenic capacity of DNA partially purified from SV-40 virus, to elicit malignant cell transformation *in vivo* and *in vitro*. The findings of carcinogenesis of nucleic acids are of considerable interest for the attempt to resolve the mechanism of cell transformation caused by the incorporation of DNA into a host cell as subviral carcinogenesis.

MATERIALS AND METHODS

Extraction of Nucleic Acids from BS-C-1 Cell Infected by SV-40 Virus

Various methods of extraction were tried in attempts to extract infectious nucleic acids from BS-C-1 cells inoculated with SV-40; cold phenol extraction[7] was carried out by the modified method of Gierer, Schramm[5] and Ito.[8]

At least 200 test tubes containing the monolayer cells of BS-C-1 inoculated with SV-40 were routinely used for extraction of nucleic acids each time. On the other hand, nucleic acids from BS-C-1 cell uninfected with SV-40 were extracted as the control material. The PBS (phosphate buffered saline, pH) or 0.88 M NaCl suspension of BS-C-1 cells infected with SV-40 was homogenized slightly once using the Potter homogenizer in ice water. Then two volumes of 88% cold phenol and 1/10 M ethylene–diamine sodium tetraacetate (5.6×10^{-4} M solution in PBS) were added quickly and the mixer was run for another 2 min. The homogenate was then transferred to a flask placed in a magnetic

421

mixer and ground at 4 °C for 1 hr. The homogenate was then transferred to a flask placed in 56 °C water bath for 5 min and, after immediate cooling, extraction was continued at 4 °C for 30 min. After centrifugation at low speed, the milky suspension was separated from the tissue debris and was extracted with an equal volume of 80 % phenol at 4 °C for 30 min. The same procedures were repeated on the withdrawn aqueous layer and interphase material. The final aqueous phase was extracted with ether and then the residual ether was evaporated by bubbling nitrogen through the suspension. The spectrum of the ultraviolet adsorption was characteristic of the solution containing nucleic acids, with the maximum adsorption around 270 mμ. In addition, calorimetric quantitative determination for DNA was made by the Burtan method.[9]

Assay of Infectivity of Nucleic Acid Preparations

Monolayer cells grown in the T type flask 30 (Pyrex Co.) and test tubes were available as the susceptible cells and they were rinsed twice with PBS prior to use. Monolayer cells grown in the T type flask were inoculated with 0.5 ml of extracted nucleic acid mixed in 4.5 ml of 199 medium (Difco) containing 2 % calf serum, and in the case of the monolayer cells grown in test tubes the cells were inoculated with 0.2 ml of nucleic acid mixed in 0.8 ml of 199 medium containing 10 % calf serum, the mixture being allowed to adsorb for 24 hr at 37 °C. The inoculum was removed and replaced with 199 medium containing 10 % fetal calf serum. For identification of the infectivity of nucleic acids, DNAse (20–100 μg/ml) diluted in PBS was also treated with infectious nucleic acids. The infectivity of nucleic acids was completely destroyed by exposure for 30 min at room temperature to DNAse in the presence of 0.005 M MgSO$_4$. The treatment with RNAse (100 μg/ml) and specific anti-SV-40 rabbit serum had no effect on the infectivity of nucleic acids. On the other hand, the infectious DNA proven to possess the specific infectivity on the susceptible cells was inoculated into newborn hamsters within 24 hr after birth in the following manner.

Fluorescent Antibody Staining Technique

The indirect fluorescent antibody technique was used for the experiments. Anti-SV-40 serum was obtained from rabbits given intravenous injection of SV-40 PBS suspension containing $10^{8.5}$ TCID 50/0.1 ml. Fluorescein-conjugated anti-rabbit goat globulin solution was adsorbed with powder prepared from the liver of a mouse. The monolayer cell cultures of BS-C-1 cells were set up in Leighton tubes with 2 ml of 199 medium containing 10 % fetal calf serum, and when monolayers had been formed, the cells were washed with PBS and then incubated with 1.0 ml PBS containing 0.2 ml nucleic acid preparations, whose contents were to be determined, 0.1 ml cell-free filtrate inoculated with DNA and 0.1 ml intact SV-40 virus ($10^{8.5}$ TCID 50/0.1 ml) as the control group. The coverslips were taken from the Leighton tubes at intervals of 4, 24 and 48 hr, and 3, 5, 7, 10 and 14 days after the inoculation of these agents. They were rinsed in PBS at pH 7.4, dried in air and fixed for 10 min in acetone. One or two drops of 3-fold dilution of the anti-SV-40 rabbit serum were placed on the coverslips and incubated in a humidity chamber at 37°C for 30 or 60 min. They were then washed for 15 min in PBS on the magnetic stirrer, and 1 or 2 drops of the 3-fold dilution of fluorescein-conjugated globulin solution were added. After incubation for 40 min at 37°C they were washed in PBS for 15 min on the magnetic stirrer and mounted in phosphate buffered glycerine on slides.

Hemagglutination and Hemagglutination Inhibition Test

As the agents used for the tests, SV-40 virus, nucleic acids and cell-free filtrates of BS-C-1 cells inoculated with DNA were examined for hemagglutination using washed guinea pig erythrocytes. Tumor bearing hamsters induced by infectious DNA, SV-40 virus, and non-inoculated normal hamsters were bled and their sera were tested for hemagglutination inhibition.

The Permanent Human Cell Line (Intestine 407)[10]

Intestine 407 (Henle and Deinhardt[10]) isolated from the jejunum and ileum of a two-month human embryo was used and clone forming epithelium-like cells were observed in the cell lines morphologically (Fig. 9). 30 T type flask (1×10^6/ml cell number) cultures and 12 bottle cultures (16 oz bottle, each bottle containing 3×10^6 cells of initial culture) were used fully developed. In the T type flasks each culture received 5 ml of diluted virus (3.0 ml of 199 medium supplemented with 2% calf serum: 2.0 ml of undiluted virus suspension: $10^{8.5}$ TCID 50/0.1 ml) and 20 ml of diluted virus (12 ml of 199 medium containing 2% calf serum supplemented with 8.0 ml undiluted virus suspension: $10^{8.5}$ TCID 50/0.1 ml) were added to each bottle (16 oz bottle) culture. 24 hours later the cells were replaced by $1 \times$ Eagle medium supplemented with 10% fetal calf serum and 100 units penicillin. Medium fluids were changed every day or every other day.

OBSERVATIONS AND RESULTS

Depending upon the concentration of nucleic acids in the inocula, the magnitude and rapidity of such cytopathic effects (CPE) varied and CPE was usually detectable within 14 days. CPE could be detected in the BS-C-1 inoculated with nucleic acids in doses of 393, 208, 161, 152, and 106 μg per ml (Figs. 1, 2). Generally, the development of CPE was prolonged compared with the intact viral effect. The CPE was identical with the one caused by an intact virus. The infectivity of these nucleic acids was completely destroyed by exposure to DNAse as mentioned above, and from the findings it has been revealed that the infectivity of the nucleic acids was related to DNA.

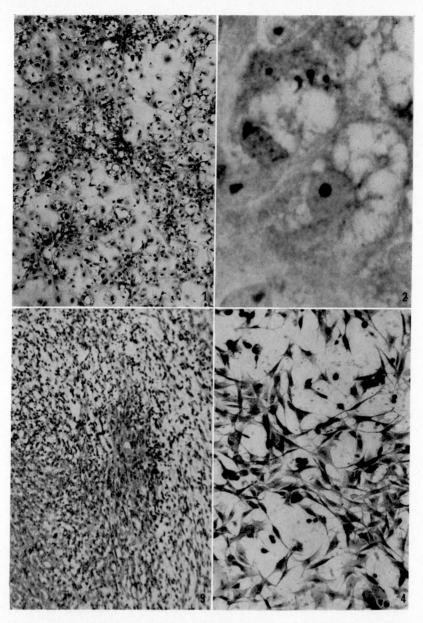

FIG. 1. Cytopathic effect on BS-C-1 inoculated with SV-40 DNA. Note a typical vacuolation of BS-C-1 cell line which is identified to be the one by intact SV-40 virus. Wright-Giemsa staining.

FIG. 2. SV-40 virus. 15 days post-inoculation of DNA. Wright-Giemsa staining.

FIG. 3. Histologic appearance of subcutaneous tumor induced by SV-40 DNA. Note fibrosarcoma arises from fibroblasts. 185 days post-inoculation. H-E. staining.

FIG. 4. 50 passage cultured cells from tumor induced by DNA. Histologic appearance shows fibroblastic cells. Wright-Giemsa staining.

Induction of Tumors in Newborn Hamsters Inoculated with DNA

235 hamsters received a single subcutaneous injection of 0.1 ml or 0.2 ml DNA preparations within 24 hr of birth. 123 newborn hamsters were used for the experiment since cannibalism was a serious problem. On 30 days after the inoculation, very remarkable changes occurred in the liver, lung, and the injected subcutaneous tissues. Particularly, cellular responses of vascular cells were predominant, and at the later stage of 50 and 120 days, small hemorrhagic lesions and formation of microhemangioma appeared in the livers of a few cases. However, degeneration of the liver parenchymal cells was a relatively inconspicuous change. On 100 days following the inoculation of DNA, proliferation of fibroblastic cells appeared in the portal spaces of the liver and these cellular responses were found predominantly in the newborn hamsters inoculated with cell-free filtrates of BS-C-1 infected with DNA. At 30 and 131 days post-inoculation, typical viral pneumonia developed in the alveolar septum of the lung.

3 hamsters of these animals inoculated with DNA subcutaneously developed tumors (2.4%). Two of them appeared in the subcutaneous area, 130 days post-inoculation. Characteristic of the cellular composition of tumors were differentiated fibrosarcomas and these were presumably related to proliferation of the vascular cells (Figs. 3, 5 and 6). It seems to suggest that the perivascular cell is one of the mother cells developing to malignant transformation. Morphologically, in the cellular patterns of tumors there are two types, one of which shows a fibroblastic cell and the other a reticulum cell pattern accompanied by multinucleated giant cells. These transformed cells can also be cultured *in vitro* (Fig. 4) and they are transplantable in newborn hamsters. In the other case, tumors were induced in the liver, lung, intestinal ducts and abdominal surface at 126 days post-inoculation subcutaneously. The cytological appearance showed multiple hemangiosarcoma combined with proliferation of vascular cells (Figs. 5 and 6).

Induction of Tumors in Newborn Hamsters Inoculated with Cell-free Filtrates of BS-C-1 Infected with DNA

39 hamsters (30%) developed tumors within about 200 days post-inoculation. Sarcomas were common and they were confined to subcutaneous tissues in 35 hamsters and to the peritoneum in others by 2 subcutaneous inoculation of the filtrates. However, the intestinal adenocarcinomas were induced at early days post-inoculation in two hamsters of the litter which developed subcutaneous sarcomas in three newborn hamsters by inoculation of the same agent (Fig. 8). Although essential histologic findings were similar to the response to DNA, cellular responses to DNA were more highly stimulated by inoculation of the filtrate than by DNA alone.

The most characteristic change was vascular cell response combined with active proliferation of fibroblasts. It also was interesting to note that liver

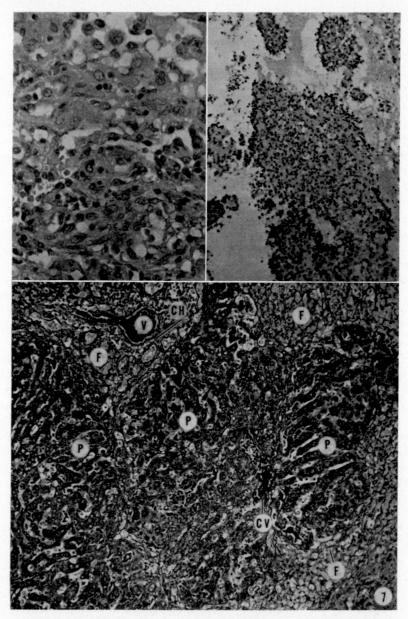

FIG. 5. Hemangioma of the liver induced by SV-40 DNA. 126 days post-inoculation. Note that transformed cells are associated with vascular cells. H-E. staining.

FIG. 6. Hemangiosarcoma of the liver induced by SV-40 DNA. 126 days post-inoculation. PAS staining.

FIG. 7. Primary liver cirrhosis without damage of liver cells. Note proliferation of blood capillaries, bile ducts (CH) and connective fibers (F) located in the portal spaces of the liver. 51 days post-inoculation. Azan staining. CV: Central vein. CH: Cholangiole. V: Portal vein. P: Liver parenchymal cell.

cirrhosis accompanied by active new formation of capillary vessels and proliferation of bile ducts was observed in infected hamsters (Fig. 7). Including tumor bearing hamsters, viral pneumonia was found in all hamsters inoculated with the filtrates. Most of the tumors found in hamsters were fibrosarcomas composed predominantly of well-differentiated spindle-shaped or stellate cells and some of the lesions had an abundant fibrous matrix in which a myxomatoid structure predominated.

It was very significant that carcinoma in primates (Fig. 8) could be induced as one of viral carcinogenesis originated from DNA. Cancer cells are well differentiated and the lamina muscularis is destroyed by the invasion of cancer cells into the submucosa and subserosa.

Cell Transformation of Established Human Cell Line (Intestine 407) Induced by Viral Carcinogenicity Developed from DNA

From *in vivo* experiments an assay for cell responses of human cells to SV-40 DNA was investigated to obtain more information on the relationship between carcinogenesis in mammalian and human cells. In this investigation, it has been noted that clone forming epithelial-like cells of the intestine 407 were more susceptible to viral agents developed from DNA than diploid fibroblasts in primary cell cultures and human cell lines. In infected primary cultured human cells, about 1% of the cells exhibited a very faint specific nuclear fluorescence when stained 10 days after viral inoculation. On the other hand, the human intestine 407 showed high susceptibility to virus and about 20% of cells exhibited specific nuclear fluorescence after 24 hr of viral inoculation and about 80% of cells were infected at 7 days after the inoculation.[11] When these cells were exposed to SV-40 virus this did not cause immediate morphological changes but it enhanced growth potentials of the cell lines for 1 or 2 weeks of SV-40 inoculation (Phase 1 in Fig. 13) provided the culture was maintained without attempting to transfer the cells to a new culture. However, following the first cell transfer, cells with vacuolated cytoplasm appeared either singly or in groups 10 days after the inoculation. As shown in Fig. 13, a culture of human virus shows a marked cytocidal effect within 5 weeks of the infection, and it was characterized by vacuolating cells, clumping of chromatin and formation of inclusion bodies.[11] With the fluorescent staining, it was possible to observe various changes in the cells about 2 or 5 weeks after the infection of a clone. The infected cells of a clone exhibited specific nuclear fluorescence and antigen was released from the nucleus into the cytoplasm through the nuclear membrane, cytoplasmic inclusions were observed in some clone cells within 4 weeks after SV-40 inoculation. Although many of the cells die (lysis), many survive (proliferation), and the establishment of such a combination state of lysis and proliferation revealed that not all the cells in a culture respond equally to infection with virus.

In 5 weeks, cell transformation from epithelial-like cells to fibroblastic

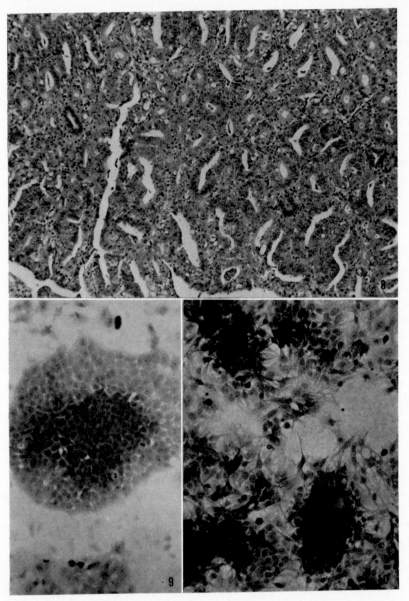

FIG. 8. Well differentiated adenocarcinoma of the intestine induced by cell-free filtrate of BS-C-1 infected with SV-40 DNA. 59 days post-inoculation. H-E. staining.

FIG. 9. Uninfected cultured cell of human cell line (Intestine 407). Note clone forming epithelial-like cells. × 100. H-E. staining.

FIG. 10. Fibroblastic cell transformation. Note cells heaped up in the center of clones and loss of contact-inhibition. Phase I (5 weeks post-infection). × 100. H-E. staining.

cells occurred but it was different from the cellular patterns of human diploid cells in that the central portion of such clones of fibroblastic cells consisted of a thick area of many cells and the cells had lost cell contact-inhibition (Fig. 10). They appear to be readily damaged by extended trypsin treatment, and by adverse conditions of cultivation, e.g., alkaline or acid medium, infrequent medium changes.

The observations of the dynamics of this cellular response indicate that SV-40 is capable of converting epithelial-like cells to fibroblastic cells, morphologically. Although some epithelial-like cells degenerated due to the sequence of metabolic disturbances, the cells infected by virus showed the fibroblastic cell patterns in some portion of the clones, but had no proliferative capacity in further cultures. In the phase of morphological fibroblastic

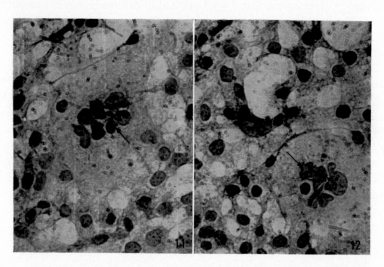

FIGS. 11 and 12. Giant cells accompanied by chromosomal abnormalities infected with SV-40 virus (arrow) at Phase III (13 weeks post-infection). The cells show fibro-reticulum cell formation and resemblance to syncytial cells. H-E. staining.

cells transformed from epithelial-like cells, the rate of the cell death gradually declined after several weeks of growth. At the end of this stage of crisis the virus becomes unnoticeable in many clone cells and only a small proportion of the clones are still infected by the virus. These findings show that virus production is maintained by a population phenomenon called carrier state by Dulbecco[12,13] concerning cell transformation by polyoma virus. The stage of the crisis is also connected with resistance of the cells to the virus and the transformation of the clonizing cells by the virus may also be connected with resistance. At the stage deviating from crisis to recovery the extensive reticulum cellular patterns were predominant in the fibroblastic altered cell (Phase III: fibro-reticulum cell formation). These reticulum cellular patterns reconstructed from fibroblastic cells were accompanied by giant cells which have chromosomal abnormalities after 10 or 15 weeks of growth (Figs. 11 and 12). The reticulum-formed cells associated with fibroblastic cells show a marked proliferation after the stage of crisis (Phase IV) and the phenomenon of contact-inhibition was lost in clonizing cells.

DISCUSSION

Histologically it has been confirmed that interstitial cell responses were highly susceptile to SV-40 DNA and virus. Liver cirrhosis in primates and extensive proliferation of fibroblasts and connective tissues were specific and striking changes *in vivo*. Microhemangioma and fibrosis of the liver at early stages were also related to its potential carcinogenecity. Considering cellular responses to these agents, various fibrosarcomas of the liver were closely associated with fibroblastic proliferation arising from the vascular cells.

According to the results, the initation of an infection depends on whether at least one DNA can penetrate into a target cell, particularly a fibroblast *in vivo*, where it is safe from the usual host defense mechanisms. Once a cell is penetrated by DNA and a complete virus is replicated from DNA in a host cell, the spread of the infection in tissue is determined by the same forces plus the added, but little understood, feature of cell-to-cell transmission of the nucleic acids, possibly without its exposure to extracellulai insults, for example DNAse as a defense mechanism. On the other hand, it has been recognized that SV-40 DNA also has oncogenic properties in epithelial cells as well as in fibroplastic cells *in vivo*.

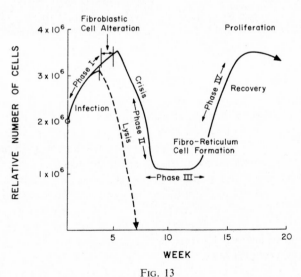

FIG. 13

In addition, the difference in the susceptibility to SV-40 virus as observed in intestine 407, may depend on alteration of the cell membrane by acquisition of inheritable properties different from those of the original cells. Although the proliferation phase has been continuing for about 12 months after infection, the ability of transformed cells to multiply seems to be limited in the established cell lines under various conditions. The altered growth potential and morphological transformation of the established cell lines may be attributable to alterations of the nuclear apparatus (Figs. 11 and 12) and the mechanism for biosynthesis resulting from infection of subviral DNA.

ACKNOWLEDGEMENTS

This work was done at the Roswell Park Memorial Institute, Buffalo, New York, U.S.A. We wish to express our thanks to Drs. Edith Sproul, Edwin Mirand, Thomas Stim of Roswell Park Memorial Institute, Buffalo, New York, U.S.A., and to Prof. Satimaru Seno, Dept. of Pathology, Okayama University Medical School.

REFERENCES

1. M. Borison, C. Paolettic, M. Thomas, J. P. Rabiere and J. Bernhard, C.R. Acad. Sci., 254, 2097, 1962.
2. P. Weil, Virology, 1, 46, 1961.
3. K. S. Kirby, Biochem. J., 66, 495, 1957.
4. P. Gerber, Virology, 16, 96, 1962.
5. A. Gierer and G. Schramm, Z. Naturforsch., 11 b, 138, 1956.
6. P. Weil and B. E. Eddy, Progr. Exptl. Tumor Res., 4, 1, 1964.
7. T. Kimoto and J. T. Grace, Acta Med. Okayama, 20, 1, 1966.
8. Y. Ito and C. A. Evans, J. Exptl. Res., 114, 485, 1961.
9. K. Burton, Biochem. J., 62, 315, 1956.
10. G. Henle and F. Deinhardt, J. Immun., 79, 54, 1957.
11. T. Kimoto and J. T. Grace, Acta Med. Okayama, 20, 215, 1966.
12. R. Dulbecco, J. Am. Med. Assoc., 190, 721, 1964.
13. R. Dulbecco, Science, 142, 932, 1963.

DISCUSSION

H. Hanafusa: Have you examined the recovery of SV-40 from transformed cells?

T. Kimoto: Although I did make attempts to examine the recovery of SV-40 by irradiation and transplantation on the BS-C-1, I could not succeed in detecting an SV-40 from transformed cells induced by SV-40 DNA.

VIRUS–HOST CELL INTERACTION IN CELLULAR PROLIFERATION INDUCED BY POXVIRUS

Shiro Kato, Hiroyuki Miyamoto, Kohei Ono, Keiichiro Tsuru, Masanobu Mantani and Takehiko Tanigaki

Research Institute for Microbial Diseases, Osaka University

At present, poxvirus has attracted rather little attention from the standpoint of oncogenesis. However poxgroup viruses are known to show marked proliferative effects upon cells. In fact most poxviruses have the ability to produce pocks as a result of cell proliferation and some cell infiltration. According to Rivers,[1] the first phenomenon observed in the infection of rabbit corneal epithelial cells by vaccinia virus is cell proliferation. Several viruses of this group can produce tumors such as the Shope fibroma and the Yaba monkey tumor. And it has been known that the Shope fibroma virus is capable of inducing even fibrosarcomatous tumors under certain circumstances.

During our studies on the inclusion bodies formed by poxviruses, it has become apparent that the inclusions of poxvirus should be classified into two types: "A" and "B", and that all poxviruses produce a common type of cytoplasmic inclusion, the "B" type, once viral multiplication occurs.[2-4] The morphology, development, and the staining characteristics of the "B" type inclusions of all the poxviruses are alike. The inclusions are reddish purple with Giemsa stain, hematoxylinophilic and slightly eosinophilic with hematoxylin-eosin stain, and histochemically Feulgen positive. However, the appearance and the characteristics of "A" type inclusions depend upon the kind of poxvirus.

Subsequently, by the methods of autoradiography and fluorescent antibody technique, we found that the sites of viral DNA synthesis and the sites of viral antigen of all poxviruses correspond exclusively to the "B" type inclusions.[2-9] Thus "B" type inclusions can be used as a suitable indicator of virus multiplication at cellular level. In this way the poxviruses show a great advantage as experimental agents over other animal viruses in that the viral DNA synthesis is easily distinguishable from the nuclear DNA synthesis of cells.

The present paper gives a summary of our experiments which have been carried out to clarify the relationship between viral multiplication and cellular proliferation in the several poxvirus and host cell systems. Since the length of the paper is limited, further details may be found in the cited papers.

MATERIALS AND METHODS

Our studies began with cowpox virus-FL cell system. Under normal conditions, the cultured cells show active cellular proliferation. Therefore the *in vitro* system is suitable for observation of the suppressive effect of poxvirus upon nuclear DNA synthesis. On the other hand, most of the cells in *in vivo* systems are, as a rule, under a certain biological regulation and do not show any abnormal hyperplasia or growth. Consequently, *in vivo* systems were used to observe the cytokinetic effects of poxvirus upon regulated cells *in vivo*.

In Vitro Systems

1. Cowpox virus-FL cells.
2. Yaba monkey tumor virus (Yaba virus)-LLC-MK2 cell.
3. Primary culture of Yaba virus-induced tumor.
4. Shope fibroma virus (SFV)-FL cell.
5. SFV-primary culture of rabbit kidney cell.
6. Primary culture of SFV-induced sarcomatous tumor.
7. Primary organ culture of molluscum contagiosum.

These cells were cultured on coverslips in Leighton tubes. Various concentrations of ^{3}H-thymidine solution were added to these systems after virus infection. In the primary cultures of virus-induced tumor, ^{3}H-thymidine solution was added at various intervals after cultivation. One hour later, the samples on the coverslips were removed, and autoradiography carried out on them using Kodak NTB2 nuclear track emulsion.

In Vivo Systems

1. Rabbit corneal epithelial cells, infected with vaccinia virus. These cells do not show active nuclear DNA synthesis under normal conditions.
2. Ectodermal cells of the chorioallantoic membrane (CAM) of the 10-day-old embryonated egg, infected with vaccinia virus. These cells are rather active in nuclear DNA synthesis under normal conditions.
3. Yaba virus-induced tumor in the monkey.
4. SFV-induced tumor in the rabbit.
5. SFV-induced sarcomatous tumor in the rabbit.

These subcutaneous tumors (Nos. 3, 4, 5) could be regarded as a result of proliferation of either subcutaneous histiocytes or fibroblasts which are in a rather dormant state under normal conditions.

In the rabbit corneal system, ^{3}H-thymidine solution was inoculated into the anterior chamber of the eyeball at various intervals after virus infection. One hour later the cornea was excised and autoradiography was carried out on both smear and section preparations. For the ectodermal cells of the CAM of the embryonated egg, ^{3}H-thymidine solutions were placed on the surface of the CAM at various intervals after infection. In the poxvirus-induced tumors, ^{3}H-thymidine solutions were inoculated intravenously into tumor-bearing animals. One hour later tumors were excised. Then autoradiography was carried out both on the smear and section preparations of the tumors.

The fluorescent antibody procedure was also used in order to confirm the specificity of the inclusions.

RESULTS

Cowpox Virus-FL Cell System[10]

Cowpox virus multiplies in FL cells, forming "B" type inclusions. The infected cells were kept in the medium containing ^{3}H-thymidine for 1 hr. Then autoradiography was carried out on the samples. The "B" type inclusions were well labeled, while nuclear DNA synthesis was definitely suppressed.

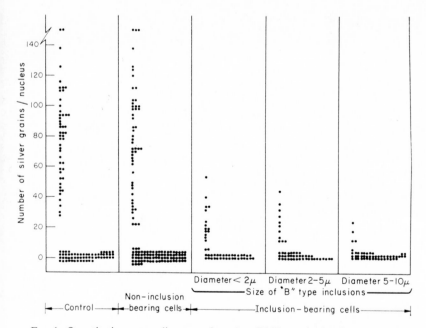

FIG. 1. Quantitative autoradiogram of nuclear DNA synthesis of cowpox virus-FL cell system. Virus-infected cells were kept in the medium containing ^{3}H-thymidine (1 μc/ml) for 1 hour. Then autoradiography was carried out. The exposure time was 48 hours. The number of silver grains per nucleus was counted and plotted.[10]

Figure 1 shows the quantitative analysis of the autoradiogram of ³H-thy-midine carried out in this system. Cells were divided into two groups: non-inclusion-bearing cells and inclusion-bearing cells. Inclusion-bearing cells were further divided into four groups according to the size of inclusions, which indicates the stage of infection. Then 50 cells of each group were randomly chosen, and the number of silver grains in each nucleus counted and plotted. A definite difference was found between the number of silver grains in the nuclei of non-inclusion-bearing cells and in the nuclei of inclusion-bearing cells. Regardless of the size of the "B" type inclusions, DNA synthesis in the nuclei of inclusion-bearing, virus-producing cells, was definitely suppressed.

The Other Systems

Work with the other systems is based on the definite phenomenon found with the cowpox virus-FL cell system *in vitro* as noted above.

Figure 2 shows the representative pattern of quantitative autoradiogram produced by the culture system of FL cell and Shope fibroma virus.[9]

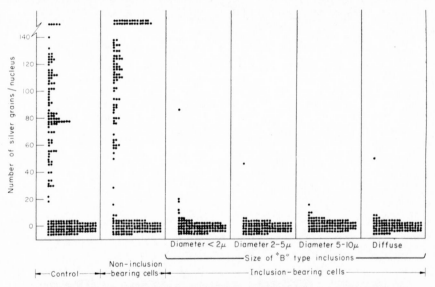

FIG. 2. Quantitative autoradiogram of nuclear DNA synthesis of Shope fibroma visus-FL cell carrier culture system. The system was kept in the medium containing ³H-thymidine (1 µc/ml) for 1 hour. Then autoradiography was carried out. The exposure time was 72 hours. The number of silver grains per nucleus was counted and plotted.[9]

Inclusion-bearing cells were divided into three groups according to the size of inclusions. Cells bearing inclusions with diameters more than 10 µ were rarely encountered. Then 100 cells of each group were randomly chosen, and the number of silver grains on each nucleus was counted and plotted. The number of silver grains were compared with those of non-inclusion-bearing cells in the carrier culture system and in normal FL cells.

A definite difference was found again between the number of silver grains in the nuclei of non-inclusion-bearing cells and inclusion-bearing cells. Thus, regardless of the tumorigenicity of poxvirus, DNA synthesis in the nuclei of inclusion bearing cells was definitely suppressed.

The same is true with the other poxvirus-cell systems examined (Table 1).

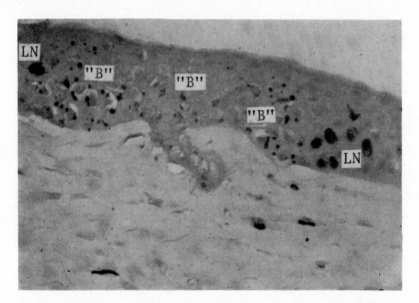

Fig. 3. Autoradiogram of rabbit corneal epithelial cells infected with vaccinia virus (24 hr after infection), stained with hematoxylin-eosin. The central infected area consists of many cells bearing "B" type inclusions which are well labeled radiographically. The "B"-bearing cells do not show any nuclear DNA synthesis. Nuclear DNA synthesis is visible in several cells surrounding the infected focus.

Figure 3 shows the autoradiogram of rabbit corneal epithelium infected with vaccinia virus.[11,12] The "B" type inclusion-bearing cells in the infected focus do not show any active nuclear DNA synthesis. To our interest, active nuclear DNA synthesis of several non-inclusion-bearing cells adjacent to the focus was induced. Recently we confirmed the similar morphology of the cytoplasmic inclusions in the early stage of the molluscum contagiosum infection to the "B" type inclusions of poxvirus. These inclusions were well labeled with ^{3}H-thymidine, while the nuclei of these inclusion-bearing cells did not show any active DNA synthesis, as shown in Fig. 4.[13] Non-inclusion-bearing cells surrounding the infected foci were induced with active nuclear DNA synthesis again.

TABLE 1. RESULTS OF THE EXPERIMENTS WITH AUTORADIOGRAPHY OF ³H-THYMIDINE ON POXVIRUS-HOST CELL SYSTEMS

Virus	Host cell	Condition of experiment	Suppression of NucDNA synthesis of VP cells	Induction or accentuation of NucDNA synthesis of NVP cells*	Phenomenon produced by virus	References
Cowpox	FL cell	in vitro	+		Total degeneration	10
Vaccinia	Rabbit corneal epithelial cell	in vivo	+	+	Hyperplasia	11, 12
	Ectodermal cell of CAM	in vivo (embryonated egg)	+	+	Pock	12
Yaba monkey	Subcutaneous histiocyte or fibroblast of monkey	in vivo	+	+ +	Histiocytoma	8
		in vitro (primary culture)	+		Carrier culture	19
	MK2 cell	in vitro	+		Carrier culture	8, 19
	FL cell	in vitro	+		Carrier culture	9, 12
	Rabbit kidney cell	in vitro (primary culture)	+		Total degeneration	20
Shope fibroma	Rabbit subcutaneous fibroblast	in vivo	+	+ +	Fibroma	9, 20
	Sarcomatous cell of rabbit	in vivo	+	+ +	Sarcomatous tumor	14
		in vitro (primary culture)	+		Carrier culture	12
Molluscum contagiosum	Epidermal cell of human	in vitro	+	+	Nodule	13

NucDNA: nuclear DNA. VP cell: virus producing cell ("B" inclusion-bearing cell). NVP cell: non-virus producing cell (non-inclusion-bearing cell). CAM: chorioallantoic membrane of embryonated egg. *: NVP cells in the area adjacent to the VP cells.

Thus, the hyperplastic or tumorous conditions in poxvirus infection must be due to the reactive proliferation of non-inclusion-bearing cells which are adjacent to the infected cells.

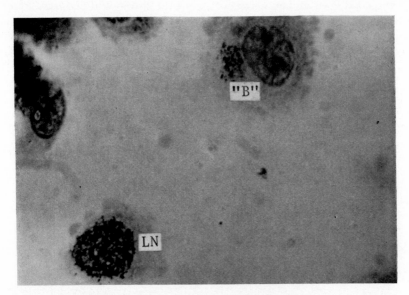

FIG. 4. Autoradiogram of two primary cultured epidermal cells of molluscum contagiosum, stained with Giemsa solution. The "B" type inclusion of the right upper cell is well labeled, while the nucleus is not labeled. The nucleus of the left lower cell which is free from any inclusion, is well labeled.

Virus-host Cell Interaction in Fibrosarcomatous Tumor Produced by Shope Fibroma Virus[14,11,12]

If viral multiplication is necessary for cellular proliferation, a certain immunosuppressive condition might cause the inhibition of the production of antiviral antibody, leading to progressive growth of the tumor. Modification of the immunological response of the rabbits in various ways often caused progressive sarcomatous growth of tumor after SFV inoculation.[15–18]

As an immunosuppressant, various doses of Co irradiation were used. Some irradiated rabbits were subsequently treated with 100 mg of prednisolone. In the non-irradiated controls of the present experiments, fibromata developed normally and reached a maximum size in about 9 days. By the 30th day the fibromata had largely disappeared. No secondary tumors were noted. Regardless of post-treatment with prednisolone, tumors appeared in all irradiated rabbits at the site of virus inoculation about 3 days after virus inoculation, and most of these tumors grew progressively over a 9- to 30-day period. In a few animals regression of the macroscopic tumors began about the 9th day, in others much later.

Ten out of 17 irradiated rabbits had several pea-sized tumors in uninoculated areas of the skin in the 3rd or the 4th week. These secondary tumors grew progressively to some extent. Histologically, most of these secondary tumors were diagnosed as fibrosarcomas. The evolution and histological features of these secondary tumors have a great resemblance to those in the rabbits receiving 6-mercaptopurine described by Hurst.[17] Smear preparations of these secondary tumor tissues, stained with Giemsa solution, showed that more than

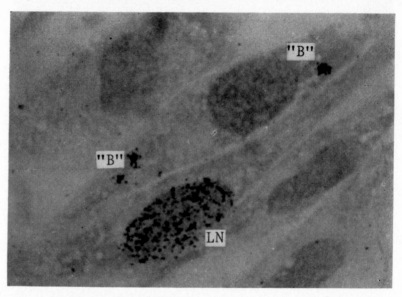

FIG. 5. Autoradiogram of primary cultured fibroblasts of Shope fibroma virus-induced sarcomatous secondary tumor of the rabbit, stained with Giemsa solution. The central fibroblast bears bipolar "B" type inclusions which are well labeled. The nucleus of the fibroblasts is not labeled. The neighboring, lower fibroblasts which is free from any inclusion, is well labeled in the nucleus. "B": "B" type inclusion. LN: labeled nucleus.

sixty percent of the sarcomatous cells bore typical "B" type inclusions of Shope fibroma virus. Virus from the various tissues of secondary tumor-bearing rabbits was titrated by intradermal inoculation of serial dilutions and calculation of the rabbit tumor forming unit $_{50}$(RTFU$_{50}$). All secondary tumors examined contained more than 4×10^4 RTFU$_{50}$ per gram of tumors. A secondary tumor-bearing rabbit was inoculated with ^{3}H-thymidine solution intravenously, then autoradiography was carried out on the smear preparation of the secondary tumor. Most of the inclusions were well labeled, while the nuclear DNA synthesis of these inclusion-bearing cells was suppressed. Many cells with no inclusion showed marked nuclear DNA synthesis. The same was true with primary culture of secondary tumor tissue (Fig. 5).[12] Results are summarized in Table 1.

Thus, regardless of the malignant appearance of SV-induced tumors, these tumors contain both virus-producing cells, which never multiply, and non-inclusion-bearing cells, some of which show active nuclear DNA synthesis. The true malignancy of these sarcomatous secondary tumor cells remains uncertain.

DISCUSSION

Throughout our experiments, it was evident that cellular proliferation in any kind of poxvirus infection was always accompanied by active virus multiplication, and that nuclear DNA synthesis in cells showing viral multiplication of any kind of poxvirus was definitely suppressed. Inclusion-bearing cells (virus-producing cells) showed mitotic figures only rarely, and many cells with diffuse well-developed inclusions appeared to be degenerating. This suggested that virus-producing cells did not multiply, but degenerated. If virus-producing cells were eliminated, as by increase of neutralizing antibody against the virus, the cellular proliferation ceases. There must be two biological cycles in the inflammatory areas, one of virus propagation, and one of cellular proliferation, as shown for poxvirus in Fig. 6. Some unknown factor must induce dormant cells to proliferate. It is plausible that this growth promoting factor might be released from degenerating virus-producing cells.

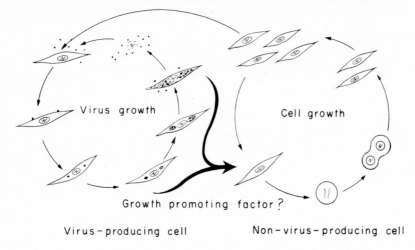

FIG. 6. Induction of cellular proliferation by poxvirus: possible mechanism.

The process of cellular proliferation in tumor formation by poxvirus has been shown to be considerably different from that produced by other groups of tumor viruses. We believe that poxvirus provides one of the unique tools to clarify the relationships between inflammation, benign tumor formation and malignant tumor formation by viruses.

SUMMARY

1. Poxviruses, whether oncogenic or non-oncogenic, multiply in the cytoplasm of the infected cells, forming "B" type inclusions which are the sites of viral DNA synthesis and of viral antigen.

2. Cellular proliferation, whether hyperplasia, benign growth, or sarcomatous tumor, was always accompanied by active virus multiplication.

3. Nuclear DNA synthesis in cells showing any type of poxvirus multiplication was definitely suppressed.

4. The relationship between virus multiplication and cellular proliferation in poxvirus infection was discussed.

REFERENCES

1. T. M. RIVERS, *Amer. J. Path.*, **4**, 91, 1928.
2. S. KATO, M. TAKAHASHI, S. KAMEYAMA and J. KAMAHORA, *Biken's J.*, **2**, 353, 1959.
3. S. KATO and J. KAMAHORA, *Symposia Cell. Chem.*, **12**, 47, 1962.
4. S. KATO, *Acta Pathol. Jap.*, **14**, 189, 1964.
5. S. KATO, S. KAMEYAMA and J. KAMAHORA, *Biken's J.*, **3**, 135, 1960.
6. S. KATO, S. KAMEYAMA and J. KAMAHORA, *Biken's J.*, **3**, 183, 1960.
7. S. KATO, M. TAKAHASHI, H. MIYAMOTO and J. KAMAHORA, *Biken's J.*, **6**, 127, 1963.
8. S. KATO, K. TSURU and H. MIYAMOTO, *Biken's J.*, **8**, 45, 1962.
9. S. KATO, H. MIYAMOTO and K. TSURU, *Symposia Cell Chem.*, **15**, 75, 1965.
10. S. KATO, M. OGAWA and H. MIYAMOTO, *Biken's J.*, **7**, 45, 1964.
11. S. KATO, *Igaku no ayumi*, **58**, 122, 1966.
12. S. KATO, *Progress in Virology* (in Japan), 98, 1966.
13. T. TANIGAKI and S. KATO, *Biken's J.*, **10**, 41, 1967.
14. S. KATO, K. ONO, H. MIYAMOTO and M. MANTANI, *Biken's J.*, **9**, 51, 1966.
15. C. H. ANDREWES and C. G. AHLSTROM, *J. Path. Bact.*, **47**, 87, 1938.
16. J. HAREL and TH. CONSTANTIN, *Bull. Ass. Franc. Cancer*, **41**, 482, 1954.
17. E. W. HURST, *J. Path. Bact.*, **87**, 29, 1964.
18. J. CLEMMESEN, *Am. J. Cancer*, **35**, 378, 1939.
19. K. TSURU, H. MIYAMATO and S. KATO, unpublished.
20. H. MIYAMOTO and S. KATO, unpublished.

DISCUSSION

F. H. KASTEN: Do you have any evidence to indicate whether or not the cytoplasmic DNA virus material may be derived from nuclear DNA?

S. KATO: Cells were previously labeled with ^{3}H-thymidine. Then the cells were challenged with cowpox virus in radioisotope-free medium. The autoradiography was carried out. Although the infected cells produced many DNA-containing inclusions in the cytoplasm, silver grains were exclusively localized in the nucleus. Therefore, I think that nuclear DNA was not transferred into viral DNA.

ABSENCE OF INFECTIOUS VIRUS
IN CHICKEN AND MAMMALIAN TUMORS
INDUCED BY ROUS SARCOMA VIRUS*

H. HANAFUSA and T. HANAFUSA

The Public Health Research Institute of the City of New York, Inc., New York, N.Y.

VIRUS-HOST cell interaction in Rous sarcoma has been studied extensively as a model of viral carcinogenesis, mainly because a direct and rapid response of cells to viral infection can be obtained both *in vivo* and *in vitro*. When cultured chick embryo cells are infected with Rous sarcoma virus (RSV), transformed cells generally appear two to three days after infection. These transformed cells assume the typical rounded appearance of cultivated Rous sarcoma cells, and they have all the attributes of malignant tumor cells.

One of the interesting features of this system is that RSV contains RNA as the carrier of its genetic information, and therefore the mechanism of carcinogenesis for RSV could differ from that of DNA tumor viruses. Infection of cells with a DNA tumor virus results in two different kinds of interaction, first, the lytic interaction accompanied by viral replication, and second, cellular transformation occurring without viral replication. A remarkable characteristic of DNA tumor virus infections is that once cells undergo malignant transformation, infectious virus particles usually cannot be detected. By analogy with the lysogenic state of temperate bacteriophages, it has been suggested that at least a piece of the viral genome integrates into host cell chromosome. This suggestion was borne out by recent studies with transformed cells induced by three different DNA tumor viruses (Benjamin, 1966; Fujinaga and Green, 1966; Reich *et al.*, 1966).

A similar non-virus producing state of virus-induced tumor cells has been found with RSV infected cells, but avian cells differ from mammalian cells in their basic requirements for establishing a non-virus producing state following transformation by one of the avian tumor viruses. This report concerns the nature of the underlying mechanisms responsible for the differences observed in these two systems.

* This investigation was supported by U.S. Public Health Service Research Grant No. CA-08747 from the National Cancer Institute.

443

Non Virus-producing Chicken Tumor Cells Induced by RSV

When chick embryo cells are infected with low doses of the Bryan strain of RSV, cellular transformation takes place, but the transformed cells do not release any detectable RSV progeny. RSV production from such cells was achieved only by infecting them with any one of the avian leukosis viruses as a helper virus. Non-virus producing transformed cells, hereafter called NP cells, retain their specific rounded morphology during subsequent growth, and every descendant cell can produce RSV upon superinfection with helper virus. It is clear that the Bryan strain of RSV is a defective virus since it is unable to reproduce without the aid of helper virus (Hanafusa et al., 1963). Further studies have shown that the defectiveness of RSV stems from its inability to produce its own viral envelope in infected cells (Hanafusa et al., 1964b). Thus, no virus-specific surface antigen is made in the NP cells, and the role played by the helper virus is to provide its viral envelope to RSV genomes. As a consequence, RSV acquires an envelope identical to that of helper virus, and those properties of RSV which depend on its surface structure are determined by avian leukosis virus acting as a helper virus (Hanafusa et al., 1964a, b; Hanafusa, T., et al. 1964; Hanafusa, 1965).

Obviously, such NP cells can be formed only with defective RSV strains. Some of the other strains of RSV appear to be non-defective (Hanafusa, 1964; Dougherty and Rasmussen, 1964; Hanafusa and Hanafusa, 1966b) and do not need the aid of helper virus. Both cell transformation and viral reproduction occur in cells infected with a non-defective RSV strain.

Non Virus-producing Mammalian Tumor Cells Induced by RSV

Absence of infectious virus characterizes *mammalian* cells that have been transformed by RSV, irrespective of whether or not the virus is defective (Svoboda, 1962; Ahlström and Forsby, 1962; Svoboda et al., 1963). Thus, the Schmidt-Ruppin strain of RSV (SR-RSV) which is non-defective in chicken cells is one of the most active producers of mammalian tumors. Even with the most sensitive assay technique, infectious RSV could not be detected in mammalian tumor cells cultivated *in vitro* (Vigier and Svoboda, 1965).

In order to determine the possibility of participation of helper virus in RSV production, avian and murine leukemia viruses were added to the cultivated mammalian tumor cells that had been induced by RSV. No RSV production was observed (Svoboda, 1964; Šimkovič et al., 1965). RSV production was found, however, when these cells were implanted into chickens or were cultured with normal chick embryo cells (Svoboda et al., 1963; Svoboda, 1964; Šimkovič, 1964). Further studies have shown that direct physical contact between mammalian tumor cells and chicken cells is essential to obtain RSV, and this virus production may have resulted from the transfer of the viral genome from mammalian tumor cells to chicken cells through contact (Svoboda, 1964). Most, if not all, of the mammalian NP cells cultivated *in vitro* had the capacity to produce RSV when they were brought into direct contact with living chicken cells, indicating that the RSV genome is transmitted into daughter cells through many generations (Svoboda, 1964).

The difference between mammalian and chicken NP cells may be summarized as follows: (1) Mammalian NP cells may be induced by *any* strain of RSV, whereas chicken NP cells can be obtained only with *defective* strains of RSV. (2) Failure of RSV to replicate in mammalian cells might be due either to a dependence of the virus on certain host functions which are missing in mammalian cells, or because mammalian cells inhibit certain RSV functions. Once the RSV genome is transferred to chicken cells, the deficiency or inhibition no longer operates, and RSV completes its growth cycle. On the other hand, as already discussed, failure of defective RSV to replicate in chicken cells is due to the defect in the viral genome, and the missing function of RSV can be complemented by superinfecting helper virus.

Infectivity of RSV in mammals. In 1957 it was shown that tumor production by RSV was not restricted to domestic fowl, and the oncogenic effect could be obtained in mammals (Zilber and Kryukova, 1957; Svet-Moldavsky and Skoreekova, 1957). Once the assumption of strict host specificity was broken, findings of tumor production by RSV were reported with a variety of mammals (Svoboda, 1960; Ahlström and Forsby, 1962; Munroe and Windle, 1963). From these and other studies (Ahlström and Jonsson, 1962; Klement and Svoboda, 1963; Harris and Chesterman, 1963; Munroe and Southam, 1964) it has become apparent that certain strains of RSV are active in tumor production in mammals and others are inactive. Among them, non-defective SR-RSV was active, and the defective Bryan strain of RSV, B-RSV, was inactive. As the difference is so marked, it seemed to be an inherent characteristic of RSV strains. However, as will be shown below, our recent studies showed that the difference is dependent on the nature of the viral envelope. Defective B-RSV is infectious for mammalian cells if it has been given a coat of proper host specificity (Hanafusa and Hanafusa, 1966a).

Because a number of avian leukosis viruses can act as helper virus for defective RSV, and because defective RSV depends on the helper virus for its envelope, it is now possible to prepare a number of RSV's which have the same envelope characteristics as helper virus, by simply infecting chicken NP cells with various helper viruses. For example, B-RSV obtained from NP cells superinfected with RAV-1† has several properties identical to RAV-1, and was designated as B-RSV(RAV-1). B-RSV(RAV-2) can be prepared from the same line of NP cells by adding RAV-2. It has been found that there is genetic heterogeneity among chickens for susceptibility to these two kinds of RSV (Hanafusa, T., *et al.*, 1964; Vogt, 1965; Hanafusa, 1965).

† Abbreviation for Rous associated virus-1, an avian leukosis virus isolated from a stock of Bryan strain of RSV.

Domestic fowl can now be genetically classified into four groups:† susceptible to only RSV(RAV-2), susceptible to only RSV(RAV-1), susceptible to both, and resistant to both, as shown in Table 1. Since the two kinds of RSV can be obtained from sister NP cultures, and the host range of a given RSV is identical to that of helper virus for that RSV, it is most likely that host range specificity of RSV is determined by the viral envelopes supplied from respective helper viruses.

TABLE 1. PLATING EFFICIENCY OF AVIAN TUMOR VIRUSES ON CHICK EMBRYO CELLS OF DIFFERENT PHENOTYPE*

Viruses	Genetic type of cells			
	K or C/O	C/A†	K/2 or C/B	C/AB†
RAV-1	1	$< 10^{-4}$	1	$< 10^{-4}$
B-RSV(RAV-1)	1	$< 10^{-4}$	1	$< 10^{-4}$
RAV-2	1	1	$< 10^{-4}$	$< 10^{-4}$
B-RSV(RAV-2)	1	1	$< 10^{-4}$	$< 10^{-4}$

* Plating efficiency of viruses on various genetic types of cells was compared to that on K or C/O type.

† Data on C/A and C/AB cells were obtained from Vogt and Ishizaki (1965).

If helper-control of RSV-host range can be applied to mammalian infection by RSV, an RSV strain which can infect mammals should have a specific envelope. An attempt was made to exchange the viral envelope between B-RSV and SR-RSV and to test the tumor producing activity of the artificially modified viruses in baby hamsters.

Since RAV-1 is the most typical helper virus associated with Bryan RSV stocks, SR-RSV coated with RAV-1 envelope, designated as SR-RSV-(RAV-1), was made by mixed infection of chick embryo cells with SR-RSV and RAV-1. Parental SR-RSV, which was present in minute quantity in the SR-RSV(RAV-1) preparation, was removed by specific neutralization by antiserum for SR-RSV. Various tests confirmed that the SR-RSV(RAV-1) preparation was composed of SR-RSV genomes coated temporarily by RAV-1 envelopes like defective RSV or phenotypic mixing of other animal viruses (Hirst and Gotlieb, 1953; Granoff, 1959; Holland and Cords, 1964; Wecker and Lederhilger, 1964; Ikegami et al., 1964). In order to define the nature of its viral envelope, the parental SR-RSV will be designated hereafter as SR-RSV(SR). Tumor production in hamsters by SR-RSV(SR), SR-RSV-(RAV-1) and B-RSV(RAV-1) is shown in Table 2. It can be seen that both SR-RSV(RAV-1) and B-RSV(RAV-1) had no oncogenic action even though

† From genetic studies by Crittenden et al. (1964) and Rubin (1965) it was shown that susceptibility of cells to one virus is controlled by a single gene with two alleles, of which susceptibility to RSV is dominant, and resistance is recessive.

they were inoculated in a greater amount than SR-RSV(SR). The inactivity of SR-RSV(RAV-1) was not due to RAV-1 which existed in the stock of SR-RSV(RAV-1), because simultaneous inoculation of RAV-1 with SR-RSV(SR) did not diminish the activity of the latter.

The opposite direction of exchange, i.e. preparing B-RSV coated with an envelope of SR-RSV was found to be more difficult. Such a virus, B-RSV(SR), could be produced by adding SR-RSV(SR) to Bryan RSV-induced NP cells, but the preparation always contains some SR-RSV(SR), and these two viruses could not be separated as they had the same envelope. However, this difficulty was fortunately overcome by isolation of a new helper virus, designated as RAV-50, from a crude stock of SR-RSV. RAV-50 had characteristics of the avian leukosis virus: it multiplied in chick embryo cells without causing significant morphological alteration in the cells, but it induced a high degree of resistance to challenge infection with RSV. B-RSV(RAV-50),

TABLE 2. TUMOR PRODUCTION BY RSV IN NEW-BORN HAMSTERS

Viruses	Dose inoculated	Hamsters with tumors at 30 days
SR-RSV(SR)	10^3FFU*	17/17
SR-RSV(RAV-1)	10^5	0/20
SR-RSV(SR) + RAV-1	$10^3 + 10^6$	10/10
B-RSV(RAV-1)	10^5	0/31
B-RSV(RAV-2)	10^4	0/16
B-RSV(RAV-50)	10^3	24/24
RAV-50	10^4	0/22

* FFU = Focus forming unit.

prepared by infecting B-RSV-induced NP cells with RAV-50, was found to be very similar to SR-RSV(SR) in its envelope dependent properties (Tables 3 and 4). As shown in Table 2, B-RSV(RAV-50) was very active in inducing tumors in hamsters, whereas B-RSV(RAV-1) or B-RSV(RAV-2) prepared from the same NP line as well as RAV-50 itself were completely inactive. Both the acquisition of tumor producing activity in mammals by B-RSV following activating by RAV-50 and the loss of activity of SR-RSV by coating in RAV-1 envelope support the hypothesis that the main determining factor for infectivity of RSV in mammals is the viral envelope. It seems that the ability or inability of RSV to infect mammalian tissues is determined by some early stage of virus-cell interaction where the viral envelope plays the most important role. Once the viral genome enters the cells, RSV of either strain can induce specific malignant changes. The primary importance of viral surface structure in tissue tropism and species specificity of virus infection has also been shown in poliovirus infection (De Somer et al., 1959; Mountain and Alexander, 1959; Holland, 1962; Cords and Holland, 1964).

TABLE 3. SENSITIVITY OF RSV TO INTERFERENCE BY HELPER VIRUSES*

Challenge virus	Efficiency of RSV infection on cells previously infected with:			
	None	RAV-1	RAV-2	RAV-50
B-RSV(RAV-1)	1.0	0.00012	1.0	1.0
B-RSV(RAV-2)	1.0	5–30	0.00017	0.02
B-RSV(RAV-50)	1.0	5–30	0.058	0.0024
SR-RSV(SR)	1.0	5–30	0.055	0.0020

* K type chick embryo cells were infected with about 1×10^5 infectious units of helper viruses. After two transfers for RAV-1 and RAV-2 and after four transfers for RAV-50, they were challenged with various RSV's. The number of Rous sarcoma foci in the helper infected cultures was compared with that in control cultures uninfected with helper viruses.

TABLE 4. ANTIGENIC SPECIFICITY OF RSV*

Virus	Surviving fraction of RSV after incubation with			
	Anti-RAV-1	Anti-RAV-2	Anti-RAV-50	Anti-SR-RSV(SR)
B-RSV(RAV-1)	0.0002	1.1	1.1	1.1
B-RSV(RAV-2)	1.2	<0.0005	0.54	0.94
B-RSV(RAV-50)	1.2	0.29	0.044	<0.002
SR-RSV(SR)	1.2	0.28	0.033	<0.002

* One milliliter of about 1×10^5 FFU of various RSV was mixed with 0.1 ml of ten-fold dilution of each antiserum and incubated at 37° for 40 min. The surviving RSV fraction was determined by assay for FFU.

Cell to cell transfer of RSV genome. As mentioned before, the mammalian tumor cells induced by SR-RSV produce no detectable infectious RSV unless they are mixed with living chicken cells. These observations were confirmed in our studies (Hanafusa and Hanafusa, 1966a) on SR-RSV induced tumors (Table 5). In Bryan RSV(RAV-50)-induced hamster tumors, no infectious RSV was found, and mixed cultivation of these hamster tumor cells with normal chick cells did not result in formation of transformed chicken cells or production of infectious RSV. However, when RAV-1 was added to the mixed culture, B-RSV(RAV-1) appeared in the culture medium (Table 5).

Several possible mechanisms are conceivable for RSV production following interaction between mammalian NP cells and normal chicken cells. (1) *Infectious* RSV might be produced in mammalian cells in undetectable quantities, and immediate contact with the more susceptible chicken cells could facilitate RSV production. (2) RSV genome, *non-infectious* by itself, may be transferred from mammalian cells to chicken cells through the medium, or (3) at a point of direct contact, or (4) by fusion of the two types of cells. (5) Infectious RSV may be produced in the mammalian cell itself upon contact with chicken cells which supply factor(s) essential for completion of RSV.

The first and second possibilities were excluded by careful examination of RSV-induced rat tumor cells (Svoboda, 1964; Vigier and Svoboda, 1965). The requirement of helper virus for B-RSV production from mixed cultures indicated that production of infectious RSV takes place in chicken cells, eliminating the last possibility. The two alternatives, (3) and (4), cannot be excluded by our studies. However, absence of transformed chicken cells after continued culture of the B-RSV-induced hamster tumor cells with normal chicken cells in the absence of helper virus suggests that chick cell genome does not survive after fusion.

If transfer of RSV genome from mammalian to chicken cells is mediated by a fusion mechanism, the same type of transfer could also be expected between chicken cells themselves. This possibility was examined in the following experiment utilizing host range specificity of RSV. Chicken NP cells of K/2 phenotype were mixed with K type normal chick embryo cells. The mixed cultures were incubated in the presence of RAV-2. Since RAV-2 infects

TABLE 5. RECOVERY OF RSV FROM HAMSTER TUMOR CELLS (HTC)*

	RSV titer (FFU/plate) in the culture fluid at:			
	1st	2nd	3rd	4th transfer
A. *From SR-RSV induced tumors*				
HTC + hamster cells	0	0	0	—
HTC + chick cells	20	6×10^2	1×10^4	—
B. *From B-RSV(RAV-50) induced tumors*				
HTC + hamster cells	0	0	0	0
HTC + hamster cells + RAV-1	0	0	0	0
HTC + chick cells	0	0	0	0
HTC + chick cells + RAV-1	0	0	5×10^2	1.5×10^4

* Cells were trypsinized from hamster tumors induced either by SR-RSV or by B-RSV-(RAV-50). About 10^4 hamster tumor cells were plated on 10^6 normal hamster or chick embryo cells. They were subcultured in the presence or absence of RAV-1, and RSV production was determined by assay of the culture fluid.

only K type cells, RSV production occurs only when the RSV genome in K/2-NP cells is transferred to K type normal cells or these two type cells fused together. As shown in Table 6, RSV production did occur in such a mixed culture after 8 or 12 days cultivation in the presence of RAV-2.

Although these experiments do not give an exact estimate of the frequency of the RSV-productive interaction of two types of cells, it may be concluded that a *productive* transfer of the RSV genome in this system was a rare event.

TABLE 6. TRANSFER OF RSV GENOME AMONG CHICK EMBRYO CELLS*

Mixture of cells and virus			RSV in culture fluid (FFU/ml):						
NP cells (10⁴)	Normal cells (10⁶)	Virus (10⁶)	Experiment 1				Experiment 2		
			0	1	2	3	0	1	2
				(Transfer)				(Transfer)	
K/2 + K/2		None	0	0	0	0	0	0	0
K/2 + K/2		+ RAV-1	2.4 × 10⁴	–	–	–	1.2 × 10⁵	–	–
K/2 + K/2		+ RAV-2	0	0	0	0	0	0	0
K/2 + K		None	0	0	0	0	0	0	0
K/2 + K		+ RAV-1	4.0 × 10⁴	–	–		5.1 × 10⁴	–	–
K/2 + K		+ RAV-2	0	20	1.5 × 10³	1.8 × 10⁴	0	0	210
K + K		+ RAV-1	2.0 × 10⁴				1.5 × 10⁴		
K + K		+ RAV-2	2.5 × 10³				5.5 × 10³		

* About 10^4 NP cells of K/2 type were mixed with 10^6 normal chick embryo cells of either K or K/2 type. They were grown in the presence or absence of RAV-1 or RAV-2, and titer of RSV in the culture fluids was determined.

One RSV productive interaction occurred when more than 10^4 NP cells were kept in contact with normal cells during 8 to 12 generation times. However, if cell fusion is involved in RSV production, the frequency would largely depend on the conditions of cultivation requisite for fused cells. This may explain the variation in the length of period of cultivation of mixed cultures required for RSV production in different experiments.

DISCUSSION

There are significant similarities between chicken and mammalian NP cells as well as the fundamental differences discussed above. Both cells acquire malignant properties, but fail to produce complete infectious RSV. RSV genomes replicate in both kinds of cells and are transmitted to every daughter cell. The presence of the RSV genome is demonstrable by production of infectious RSV either following helper virus infection or following direct contact with living chicken cells. The RSV strain differences in ability to produce mammalian tumors reside in the capacity of the viral envelope to interact with mammalian cells. Once the cell has been entered, any strain of RSV may cause malignant changes in either mammalian or chicken cells.

These facts strongly suggest that both the mechanism of cellular transformation induced by RSV infection and the state of RSV genome in transformed cells are essentially the same in chicken and mammalian NP cells. \\

In addition to these similarities, neither NP cell when injected into living homologous animals gives rise to virus neutralizing antibodies for the RSV which has been used in producing the respective NP cells (Hanafusa et al., 1964b; Hanafusa and Hanafusa, unpublished). This indicates that in both kinds of NP cells RSV fails to induce the synthesis of viral specific coat antigens which are an integral part of viral envelope. Since SR-RSV can replicate in chicken cells, failure of SR-RSV to induce the synthesis of viral coat only in mammalian cells could be attributed to either suppression of genetic expression of SR-RSV for this function in the latter cells, or dependence of this function somehow on a certain cellular function which is not properly provided in the mammalian cells. In any event, it seems very instructive that the same process in viral replication, i.e. the synthesis of viral components of the surface envelope, is blocked, and this results in no RSV production in either chicken or mammalian NP cells.

It should be added here that even though SR-RSV is not defective in chicken cells, RSV production is somewhat restricted (Hanafusa and Hanafusa, 1966a, b). Addition of helper virus to SR-RSV infected cells always results in increased production of SR-RSV. Evidence has accumulated now to indicate that this restriction is due to the limited production of viral envelope by SR-RSV (Hanafusa and Hanafusa, unpublished), and it may be concluded that this is a general phenomenon in RSV-infected chicken and mammalian cells.

Almost nothing is known about the state of the viral genome in the NP cells. Temin (1964a, b) proposed a theory that the genetic information of RSV is maintained in transformed cells as a form of provirus, a DNA molecule complementary to viral RNA. However, there seems to be no good evidence for or against this hypothesis.

In studies on clonal lines of hamster cells (BHK 21) transformed by SR-RSV, Macpherson (1965) has found the appearance of morphologically normal cells from clones of morphologically transformed cells. While transformed cells produced tumors by transplantation into chickens, the morphologically normal, revertant cells failed to produce tumors in chickens, suggesting that the latter cells had lost the RSV genome. These findings have important implications. First, they suggest that *a few copies* of the viral genome may exist in transformed hamster cells, and since they are *not integrated* with host chromosomes very intimately, when the rate of cell growth exceeds the rate of replication of viral genome, cells lacking the latter might appear. Second, the reversion of altered morphology to normal morphology (and loss of ability to grow in agar medium†) was accompanied by loss of the RSV

† Only transformed cells acquire the ability to grow into a colony in agar medium. This was used to distinguish transformed cells from normal cells.

genome as detected by tumor production upon transplantation in chickens. This would mean that even if the cells had acquired malignant properties, once the viral genome is lost from these cells, the characteristics of malignant cells may also be lost. In other words, transformed morphology or possibly other malignant properties of transformed cells may be expressed only under the influence of viral genome existing in the cells.

No such revertants have been found among clones of transformed chicken cells (Trager and Rubin, 1966). Further clonal analyses are required with both chicken and mammalian NP cells for thoroughly testing the reversion hypothesis suggested by Macpherson's results.

A new type of transfer of non-infectious viral genome from cell to cell has been revealed in analysis of interaction between RSV-induced mammalian tumor cells and chicken cells (Svoboda et al., 1963; Svoboda, 1964). As shown in this paper, the same type of transfer occurs even with Bryan RSV which is defective in chicken cells. It seems clear now that the viral genome has to be transferred first into chicken cells, either intact or fused with mammalian cells, before RSV production begins. It is of great interest that a very similar interaction was found between green monkey cells and hamster tumor cells induced by SV-40, a DNA-containing tumor virus (Gerber and Kirchstein, 1962; Gerber, 1966). It is hoped that studies on the nature of the interaction may provide useful information about the state of viral genome in transformed cells.

ACKNOWLEDGEMENTS

The authors wish to thank Drs. George K. Hirst and Robert W. Simpson for their critical reading of the manuscript.

REFERENCES

C. G. AHLSTRÖM and N. FORSBY (1962), Sarcomas in hamsters after infection with Rous chicken tumor material. J. Exptl. Med., 115, 839–852.
C. G. AHLSTRÖM and N. JONSSON (1962), Induction of sarcoma in rats by a variant of Rous virus. Acta Path. Microbiol. Scand., 54, 145–172.
T. L. BENJAMIN (1966), Virus-specific RNA in cells productively infected or transformed by polyoma virus. J. Mol. Biol., 16, 359–373.
C. E. CORDS and J. J. HOLLAND (1964), Alteration of the species and tissue specificity of poliovirus by enclosure of its RNA within the protein capsid of Coxsackie B 1 virus. Virology, 24, 492–494.
L. B. CRITTENDEN, W. OKAZAKI and R. H. REAMER (1964), Genetic control of responses to Rous sarcoma and strain RPL 12 viruses in the cells, embryos, and chickens of two inbred lines. Natl. Cancer Inst. Monograph, 17, 161–177.
P. DE SOMER, A. PRINZIE and E. SCHONNE (1959), Infectivity of poliovirus ribunucleic acid for embryonated eggs and unsusceptible cell lines. Nature, 184, 652–653.
R. M. DOUGHERTY and R. RASMUSSEN (1964), Properties of a strain of Rous sarcoma virus that infects mammals. Natl. Cancer Inst. Monograph, 17, 337–350.
K. FUJINAGA and M. GREEN (1966), The mechanism of viral carcinogenesis by DNA mammalian viruses: Viral-specific RNA in polyribosomes of adenovirus tumor and transformed cells. Proc. Nat. Acad. Sci., U.S., 55, 1567–1574.

P. GERBER (1966), Studies on the transfer of subviral infectivity from SV 40-induced hamster tumor cells to indicator cells. *Virology*, **28**, 501–509.

P. GERBER and R. L. KIRCHSTEIN (1962), SV 40-induced ependymonas in new-born hamsters. 1. Virus-tumor relationships. *Virology*, **18**, 582–588.

A. GRANOFF (1959), Studies on mixed infection with Newcastle disease virus. II. The occurrence of Newcastle disease virus heterozygotes and study of phenotypic mixing involving serotype and thermal stability. *Virology*, **9**, 649–670.

H. HANAFUSA (1964), Nature of the defectiveness of Rous sarcoma virus. *Natl. Cancer Inst. Monograph*, **17**, 543–556.

H. HANAFUSA (1965), Analysis of the defectiveness of Rous sarcoma virus. III. Determining influence of a new helper virus on the host range and susceptibility to interference of RSV. *Virology*, **25**, 248–255.

H. HANAFUSA, T. HANAFUSA and H. RUBIN (1963), The defectiveness of Rous sarcoma virus. *Proc. Nat. Acad. Sci., U.S.*, **49**, 572–580.

H. HANAFUSA, T. HANAFUSA and H. RUBIN (1964a), Analysis of the defectiveness of Rous sarcoma virus. I. Characterization of the helper virus. *Virology*, **22**, 591–601.

H. HANAFUSA, T. HANAFUSA and H. RUBIN (1964b), Analysis of the defectiveness of Rous sarcoma virus. II. Specification of RSV antigenicity by helper virus. *Proc. Nat. Acad. Sci., U.S.*, **51**, 41–48.

H. HANAFUSA and T. HANAFUSA (1966a), Determining factor in the capacity of Rous sarcoma virus to induce tumors in mammals. *Proc. Nat. Acad. Sci., U.S.*, **55**, 532–538.

H. HANAFUSA and T. HANAFUSA (1966b), The role of helper virus in virus-induced tumors. In "Subviral Carcinogenesis" (Y. Ito ed.) pp. 283–295. The Editorial Committee for the 1st International Symposium on Tumor Viruses, Nagoya.

T. HANAFUSA, H. HANAFUSA and H. RUBIN (1964), Differential responsiveness of Rous sarcoma virus stocks to specific cellular resistance induced by avian leukosis viruses. *Virology*, **22**, 643–645.

R. J. C. HARRIS and F. C. CHESTERMAN (1963), An analysis of the action of Rous sarcoma virus *in vivo*. *Proc. Roy. Soc. Med.*, **56**, 307–308.

G. K. HIRST and T. GOTLIEB (1953), The experimental production of combination forms of virus. I. Occurrence of combination forms after simultaneous inoculation of the allantoic sac with two distinct strains of influenza virus. *J. Exptl. Med.*, **98**, 41–51.

J. J. HOLLAND and B. H. HOYER (1962), Early stages of enterovirus infection. *Cold Spring Harbor Symp. Quant. Biol.*, **27**, 101–111.

J. J. HOLLAND and C. E. CORDS (1964), Maturation of poliovirus RNA with capsid protein coded by heterologous enteroviruses. *Proc. Nat. Acad. Sci., U.S.*, **51**, 1082–1085.

N. IKEGAMI, H. J. EGGERS and I. TAMM (1964), Rescue of drug-requiring and drug-inhibited enteroviruses. *Proc. Nat. Acad. Sci., U.S.*, **52**, 1419–1426.

V. KLEMENT and J. SVOBODA (1963), Induction of tumours in Syrian hamsters by two variants of Rous sarcoma virus. *Folia Biol. (Prague)*, **9**, 181–188.

I. MACPHERSON (1965), Reversion in hamster cells transformed by Rous sarcoma virus. *Science*, **148**, 1731–1733.

I. M. MOUNTAIN and H. E. ALEXANDER (1959) Study of infectivity of ribonucleic acid (RNA) from type 1 poliovirus in the chick embryo. *Fed. Proc.*, **18**, 587.

J. S. MUNROE and W. F. WINDLE (1963) Tumors induced in primates by chicken sarcoma virus. *Science*, **140**, 1415–1416.

J. S. MUNROE and C. M. SOUTHAM (1964) Oncogenicity of two strains of chicken sarcoma virus for rats. *J. Natl. Cancer Inst.*, **32**, 591–623.

P. R. REICH, P. H. BLACK and S. M. WEISSMAN (1966), Nucleic acid homology studies of SV-40 virus-transformed and normal hamster cells. *Proc. Nat. Acad. Sci., U.S.*, **56**, 78–85.

H. RUBIN (1965), Genetic control of cellular susceptibility to pseudotypes of Rous sarcoma virus. *Virology*, **26**, 270–276.

D. ŠIMKOVIČ (1964), Interaction between mammalian tumor cells induced by Rous virus and chicken cells. *Natl. Cancer Inst. Monograph*, **17**, 351–364.

D. Šimkovič, J. Svoboda and N. Velentova (1965), Induction of formation and release of infectious Rous virus by cells of rat tumor XC *in vitro*. *Folia Biol.* (*Prague*), **11**, 350–358.

G. J. Svet-Moldavsky and A. Skoreekova (1957), The development of multiple cysts in rats after inoculating them with Rous sarcoma virus. *Vop. Onkol.*, **6**, 673–677.

J. Svoboda (1960), Presence of chicken tumour virus in the sarcoma of the adult rat inoculated after birth with Rous sarcoma virus. *Nature*, **186**, 980–981.

J. Svoboda (1962), Further findings on the induction of tumors by Rous sarcoma in rats and on the Rous virus-producing capacity of one of the induced tumours (XC) in chicks. *Folia Biol.* (*Prague*), **8**, 215–220.

J. Svoboda (1964), Malignant interaction of Rous virus with mammalian cells *in vivo* and *in vitro*. *Natl. Cancer Inst. Monograph*, **17**, 277–298.

J. Svoboda, P. Chyle, D. Simkovic and I. Hilgert (1963), Demonstration of the absence of infectious Rous virus in rat tumour XC, whose structurally intact cells produce Rous sarcoma when transferred to chicks. *Folia Biol.* (*Prague*), **9**, 77–81.

H. M. Temin (1964a), Nature of the provirus of Rous sarcoma. *Natl. Cancer Inst. Monograph*, **17**, 557–570.

H. M. Temin (1964b), The participation of DNA in Rous sarcoma virus production. *Virology*, **23**, 486–494.

G. W. Trager and H. Rubin (1966), Rous sarcoma virus production from clones of nontransformed chick embryo fibroblasts. *Virology*, **30**, 266–274.

P. Vigier and J. Svoboda (1965), Étude en culture, de la production du virus de Rous par contact entre les cellules du sarcome XC du Rat et les cellules d'embryon de Poule. *C.R. Acad. Sci., Paris*, **261**, 4278–4281.

P. K. Vogt (1965), A heterogeneity of Rous sarcoma virus revealed by selectively resistant chick embryo cells. *Virology*, **25**, 237–247.

P. K. Vogt and R. Ishizaki (1965), Reciprocal patterns of genetic resistance to avian tumor viruses in two lines of chickens. *Virology*, **26**, 664–672.

E. Wecker and G. Lederhilger (1964), Genomic lacking produced by double infection of HeLa cells with heterotypic polioviruses. *Proc. Nat. Acad. Sci., U.S.*, **52**, 705–708.

L. A. Zilber and I. N. Kryukova (1957), Haemorrhagic disease in rats caused by Rous sarcoma virus. *Vop. Virus*, **4**, 239–243.

DISCUSSION

M. Kawakami: (1) Can you see any spontaneous elimination of these defective viruses during the propagation of NP cells?

(2) I am wondering whether your system corresponds to true lysogeny or carrier (or pseudolysogeny) of bacteriophage infection. Do you have any idea?

H. Hanafusa: In chicken NP cells induced by RSV, the removal of viral genome from transformed cells has not been observed. One may recall the studies made by Dr. Macpherson on hamster tumor cells induced by SR-RSV. He found that from clonal lines of transformed hamster cells, normal appearing cells developed. These normal looking revertant cells lost the tumor inducing ability in chickens, indicating the loss of viral genome accompanying the reversion of transformed morphology.

As to the state of viral genome in NP cells, we have little information. We only know that since every cell retains the capacity to produce RSV even after 5 months cultivation of NP cells, viral genome or genetic information for RSV replicates in these cells continuously. The state of the genome and the number of its copies will be the most important task for future work.

S. MITSUHASHI: We are studying a bacterial episome which is able to confer drug resistance and is transmissible from cell to cell by conjugation. This episome is termed R factor.

We can get quite easily defective R factor which is not transferable and is attached to the host chromosome. When a host bacterium carrying defective R factor is infected with other episomes such as a wild type of R factor, F factor, defective acquired transmissibility, results from the formation of recombinant episomes such as F-dR (di defective) or R.dR factor.

Have you got the recombinant form of defective virus with helper virus?

H. HANAFUSA: We have not got any recombinant between defective RSV and helper virus. One of the difficulties is the lack of proper genetic markers. It seems to be easier to study non-defective RSV and helper viruses for this purpose.

CYTOCHEMISTRY OF VIRUS-INDUCED CELL TRANSFORMATION IN VITRO*

B. THORELL

Karolinska Institutet, Department of Pathology, Stockholm 60

THE paper describes observations on the process of cell transformation, induced by some RNA-viruses of the chicken. Intracellular changes in content and distribution of nucleic acids have been recorded by ultraviolet microspectrophotometry. As an introduction, Fig. 1 shows a bone marrow sinusoid 48 hr after infection with erythroleukemia virus *in vivo*. The target cells, in this case the erythroid stem-cells at the periphery of the sinusoid, show an

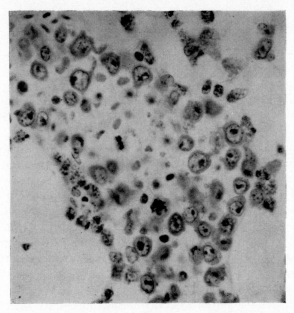

FIG. 1. Chicken bone marrow sinusoid 48 hours after intravenous injection of erythroleukemia virus. Early changes in peripherally located erythroid stem-cells: Swelling of nucleus and increased nucleolar volume and density. Htx-E. ×250.[1,2]

* Supported by the Swedish Cancer Society (Proj. No. 64:176, 65:89), the Mrs. Tora Wåhlins Fund and the Wetterkulla Medical Center.

increase in size and density of their nucleoli. This is followed by an accumulation of cytoplasmic RNA concomitant with the morphological development into the typical, hemoglobin-containing erythroleukemia cell.[1,2] The general conclusion from these *in vivo* observations is that the viral epigenetic action imposed on the developmental potentialities of the target stem-cell results in the characteristic tumor cell.

To study details of early stages in the virus-induced cell transformation, we have chosen the *in vitro* system of chicken fibroblasts infected by Rous sarcoma virus. The experimental work has been conducted by Dr. Sundelin at our laboratory, and for details of material and technique, I refer to his original papers.[3-4]

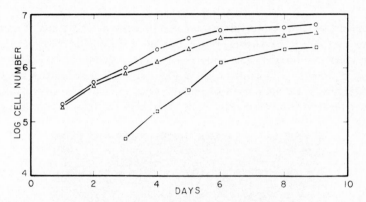

FIG. 2. Growth curves of uninfected (open circles) and infected (open triangles) chick embryo fibroblasts *in vitro*. Curve with open squares represents number of morphologically transformed cells in the cultures infected by Rous sarcoma virus.[3]

Figure 2 summarizes the growth characteristics of the fibroblast cultures. On the third day after infection there are the following three classes of cells in the cultures: (1) morphologically normal fibroblasts, negative with respect to new viral antigens when tested with the fluorescent antibody technique, (2) morphologically normal fibroblasts in which the fluorescent antibody technique demonstrates the presence of newly formed viral antigen, and (3) morphologically transformed fibroblasts, all exhibiting viral antigens. Microcinematography of the development of transformed cell foci reveals the transition of type 2 into type 3, and the different classes of cells thus represent a sequential event.

Compared with the morphologically normal fibroblasts, the transformed cells show an increase of total cytoplasmic RNA, on an average more than double. This is in accordance with the *in vivo* observations referred to in the first paragraph.

Analyses of a large number of fibroblasts of the different classes can be displayed in the form of histograms which characterize the above mentioned cell types as individual populations. Class 2, before the cells are morphologically transformed and have increased their cytoplasmic RNA (Fig. 3), shows significantly higher amounts of RNA in the nucleus (Fig. 4). The DNA complement as well as the amount of cytoplasm are still within the normal

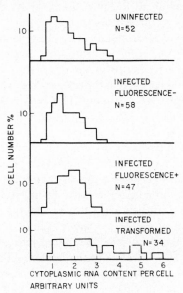

FIG. 3. Cytoplasmic RNA contents in single fibroblasts of the different (sequential) populations described in text. The morphologically transformed cells exhibit a significant increase.[3]

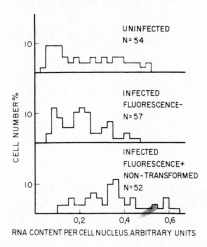

FIG. 4. Nuclear RNA contents in the different populations of morphologically normal fibroblasts. The cells with new viral antigens exhibit a significant increase.[3]

ranges. This phenomenon, which precedes the increase of cytoplasmic nucleic acids concomitant with the morphological transformation, has in turn been preceded by the synthesis of new viral antigens (according to the definition of this particular cell population). Thus, with regard to developmental sequence, the events leading to morphological (and functional) cell transformation are secondary to the synthesis of new virus. This might be taken as an indication of early intracellular virus multiplication by a (specific) replicase, presumably localized in the cytoplasm. Eventually, as is the case of chicken fibroblasts infected by Rous sarcoma virus, the virus-replication triggers off normally repressed parts of the cell genome, which can be observed as an increase of the nuclear RNA content. This in turn leads to formation of new cytoplasmic RNA, and in a functional respect the acquisition of neoplastic properties.

If we again turn back to the *in vivo* system, it is of particular interest to observe normal differentiation processes in spite of the epigenetic viral action on the undifferentiated target stem-cell. In the case of erythroleukemia, synthesis of hemoglobin can be recorded parallel with the neoplastic cell transformation.[5,6] This hemoglobin consists of the two normal types of the chicken, i.e. the hemoglobin synthesis of the developing leukemia cell is governed by the ordinary structural genes in a normal fashion. Eventually, as seems to be the case in Rauscher erythroleukemia of the mouse,[7] the normal erythroid cell differentiation might completely take over and supersede the otherwise obligatory and dominating influence of the virus. How this happens, whether for example the viral genome is eliminated from a daughter cell by the preceding cell division or if it somehow becomes repressed, can today only be speculated upon. However, the phenomenon presents an interesting aspect on the relationship between viral epigenetic action and cell differentiation.

REFERENCES

1. J. PONTÉN and B. THORELL, The histogenesis of virus-induced chicken leukemia. *J. Nat. Can. Inst.*, **18**, 443, 1957.
2. B. THORELL, Interaction between virus and target cell during development of leukemia. In *Cellular Control Mechanismus and Cancer*. Ed. by P. Emmelot and O. Mühlbock. Amsterdam, 1964.
3. P. SUNDELIN, Microspectrophotometric determination of DNA and RNA in single chick embryo fibroblasts during morphological transformation induced by Rous sarcoma virus. *Exp. Cell Res.*, **46**, 581–592, 1967.
4. P. SUNDELIN, DNA and RNA content at successive postmitotic intervals in chick embryo fibroblasts infected with Rous sarcoma virus. *Exp. Cell Res.*, **50**, 233–238, 1968.
5. J. COWLES, J. SAIKKONEN and B. THORELL, On the presence of hemoglobin in erythroleukemia cells. *Blood*, **13**, 1176–1184, 1958.
6. E. AMBS and B. THORELL, On the type of hemoglobin in the neoplastic cell of virus-induced fowl erythroleukemia. *J. Nat. Can. Inst.*, **25**, 685, 1960.
7. K. YOKORO and B. THORELL, Cytology and pathogenesis of Rauscher virus disease in splenectomized mice. *Cancer Res.*, **26**, Part 1, 536, 1966.

MOUSE ASCITES SARCOMAS INDUCED BY ROUS SARCOMA VIRUS

Tadashi Yamamoto

The Institute for Infectious Diseases, The University of Tokyo,
(Shiba-Shirokane, Tokyo)

Mammalian tumor cells or *in vitro* transformed cells due to infection with some variants of Rous sarcoma virus (abbreviated as RSV) usually contain no directly demonstrable infectious viruses, but when the cells are inoculated back into young chickens or mixed-cultured with chicken fibroblasts, the infectious viruses are generally detectable from the chicken cells, indicating the thus-induced mammalian cells do contain the virus in masked form (cf. a review of Vogt[1]).

Recently, three lines of transplantable mouse ascites sarcomas induced by the Schmidt-Ruppin strain of RSV(SR-RSV) have been available in our laboratory, each being established in different strains of mice. Needless to say, such transplantable ascites sarcomas provide a convenient source of large amounts of cells, which might enable us to do more easily, mass-treatment studies such as the mode of localization of the masked virus as the possible relation of the masked virus to malignant properties of the cell, etc. On the other hand, SR-RSV itself can multiply into the matured form in chicken cells and in this sense it is purely non-defective, while the defective nature of the Bryan strain of RSV (B-RSV) was at first and clearly demonstrated by Hanafusa *et al.*, introducing a new concept of "defectiveness" of RSV[2-4] (cf. also Thorell's article). Thus, SR-RSV-induced mammalian tumor cells, although resembling the B-RSV-induced chicken NP-cells, are conceptually quite different: rather the "defectiveness" exists in the mammalian host side, where some host-controlled steps of virus multiplication might be lacking. Possible differences of such host-controlled factors might also be comparable amongst various sarcomas thus-induced in different strains of mice with different genetic backgrounds. Anyhow, it is a very fascinating idea that only a few earlier steps of virus multiplication are enough to make the cell neoplastic.

The present paper describes some characteristics of our three lines of mouse ascites sarcomas induced by SR-RSV in different strains, C3H/He, DDD and C57BL/6 mice as well as the possible correlation of the masked virus or allied antigen(s) to the tumor cell nucleus.

15a NAM

Outline of Three Lines of Mouse Ascites Sarcomas

Table 1 shows the outline of the characteristics of these three lines of tumors. The ascites sarcoma induced by SR-RSV in C3H/He mouse is abbreviated as SR-C3H/He, and SR-DDD and SR-C57BL/6 are the same sort of abbreviation.

TABLE 1. OUTLINE OF 3 LINES OF SR-RSV-INDUCED MOUSE ASCITES SARCOMAS

	Source	Primary tumor	Ascitic conversion	At present	Wing web test
SR-C3H/He	Ch-T-cells	091464 ₹ 100664 (20)	5th gen. (90)	60 gen. (>500)	+
SR-DDD	Ch-T-cells	101265 ₹ 110765 (26)	9–11th (191)	24th (>320)	+
SR-C57BL/6	CAM-sup	022665 ₹ 062165 (115)	8–11th (126)	37th (>318)	−

Numbers in parenthesis mean duration in day.

In the case of SR-C3H/He, the conversion into an ascitic form occurred rather easily at its 5th generation. The primary tumor was obtained from a mouse, neonatally inoculated with 10^6 trypsinized chicken Rous tumor cells 20 days before.[5] Although three cycles of cloning were made in the course of passages, employing thymectomized newborn mice by means of the limiting dilution technique, the line arrived at its 60th generation at the longest, more than 500 days of passages. When inoculated with 5×10^6 cells per mouse into the peritoneal cavity of adult mice, $2–3 \times 10^8$ cells per mouse are easily obtainable within 7 days or so. The mean survival time of the inoculated mice is 12.6 ± 2.3 days. Since one of the subclones, Th-11-D, at its 16th generation and 115 days after the second cloning, i.e. at its more than 10^{-40} dilution, still showed the positive wing web test, this line of ascites sarcoma seems to keep fairly stable characteristics.

In case of SR-DDD, ascitic conversion occurred rather later and gradually at its 9th to 11th generation, and the line still shows the positive wing web test at its 24th generation.

In case of SR-C57BL/6, strain #2008, the cells processed rather in curious way. Original solid tumor appeared very late, 115 days after the neonatal inoculation with a cell-free extract of SR-RSV-infected chorio-allantoic membrane containing 10^4 pfu per inoculum, but the subsequent transplants grew very well in C57BL/6 adult mice. The line, however, had shown a positive wing web test until its 11th generation and the test suddenly turned out negative at its 12th generation and thereafter. Clonings were also made, but all the clones remained negative in wing web test.

The wing web test itself is very sensitive in our experience; when inoculated with 1 pfu of SR-RSV into a wing web within 7 days after hatching, chickens killed showed typical Rous sarcoma including multiple metastases in lungs, liver, heart, etc., after 30 days or so. However, in case of back-inoculations of mouse sarcoma cells, more than 10^5 whole cells were necessary to produce such a chicken tumor. The chicken sarcoma thus obtained always contained numerous active SR-RSV infectious to chorio-allantoic membranes of embryonated chicken eggs. Neither cell-free extracts nor purified nuclei preparations could induce such a tumor. Therefore SR-RSV has been considered to exist in those mouse sarcoma cells in so-called masked form.

In the course of serial passages in mice, 2×10^6 mouse sarcoma cells were usually back-inoculated bilaterally into chicken wing web. For the most part bilateral local tumor formation was observed, but sometimes only unilateral tumor formation occurred and sometimes neither. There was no regression of local tumors once developed; the chickens succumbed after 30 days or so. Every 10 generations' changes in percentage tumor take per wing web in the case of SR-C3H/He are shown in Table 2. In an earlier stage of the passages a higher percentage of tumor take is observable. Since the time of the first appearance of local tumors and the mean survival time of tumored chickens remain fairly constant throughout the course of passages, some changes might have occurred in the early stages of passages, suggesting a possible occurrence of a kind of tumor progression probably in the mode of virus localization in the cell.

TABLE 2. EVERY 10 GENERATIONS' CHANGE IN WING WEB TEST IN CASE OF SR-C3H/HE ASCITES SARCOMA

	1–10 gen.	11–20 gen.	21–30 gen.	31–37 gen.
Wing web loci tested	56	34	34	28
Positive wing web test	51	27	17	16
Percent take	91.1	79.4	50.0	57.4
First appearance of local tumor (days)	12.5 ± 3.8	15.0 ± 2.6	20.3 ± 4.2	16.3 ± 3.5
Survival time (days)	29.8 ± 6.8	30.0 ± 6.8	33.7 ± 9.3	31.0 ± 5.1

As mentioned above in the case of SR-C57BL/6, strain #2008, the transfer from positive to negative wing web test may constitute another kind of tumor progression. All three other solid types of SR-RSV-induced sarcomas in C57BL/6 mice showed similar tendencies as indicated in Table 3. Possible segregation of the masked virus leaving the malignant cellular properties

TABLE 3. DISAPPEARANCE OF SR-RSV IN THE COURSE
OF PASSAGES IN C57BL/6 TUMORS

Tumor strain	Wing web test	
	(+)	(−)
# 2001	3 gen.	4–7 gen.
# 2003	2	3–5
# 2006	6	8–
# 2008	11	12–

awaits further investigation, particularly in comparison with Macpherson's experiment in which the normalized revertants were shown to be separable from SR-RSV-induced transformed colonies of BHK 21/C 13 line cells.[6]

Chromosome Analysis

Chromosome analysis was undertaken along with the cloning experiments. Clonings were made by inoculating the limiting diluted cell suspension into thymectomized newborn mice.

In the case of SR-C3H/He, as shown in Fig. 1, the chromosome number was already diversely distributed at its 8th generation, just three passages after ascitic conversion, but after the first cloning at least two types of clones were separated according to the modal number, one with hyperdiploidy and the other with hypotetraploidy, however with minute modifications. After the second cloning quasi-pure clones in respect to chromosome number were obtained. All the clones including subclones except one showed positive wing web test, indicating almost every cell in the ascites contains the virus in masked form (Table 4).

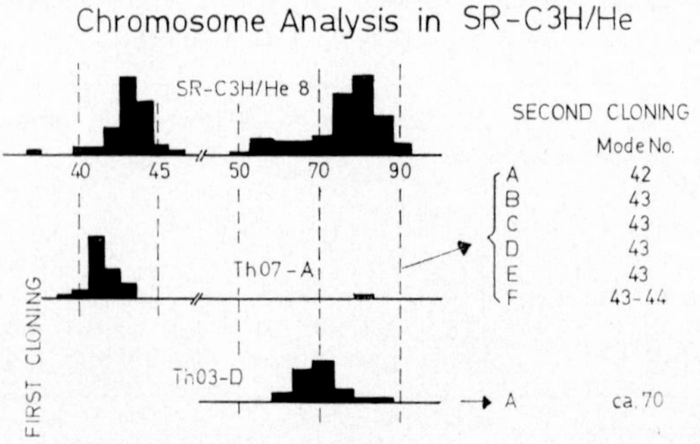

Chromosome Analysis in SR-C3H/He

FIG. 1

Figure 2 shows the results of chromosome analysis in the case of SR-C57BL/6 and SR-DDD. Both were of the hypotetraploidy type. The former with one marker chromosome of metacentric, however, showed negative wing web

TABLE 4. WING WEB TEST OF CLONED CELL LINES
OF SR-C3H/HE ASCITES SARCOMA

	Number of clones tested	Number of (+) wing web test
First cloning (21–36 gen.)	15	14
Second cloning (34–42 gen.)	9	9
Third cloning (39–41 gen.)	3	3

test at the same time as it turned out non-specific in histocompatibility. The latter with three rabbit ear- and several minute chromosomes as markers was still specific in tumor take and contained the virus in masked form. Therefore,

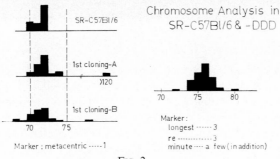

FIG. 2

as far as these observations are concerned, there was no direct correlation between the existence of masked virus and the particular modal number of chromosomes.

Immunological Studies

Possible correlation between the existing masked virus and a tumor cell nucleus was suggested from immunological studies. Figure 3 is a microphotographic picture of SR-C3H/He cells at the 10th generation, with fluorescent antibodies. FITC-labelled antibodies were derived from the γ-globulin fractions of immune ascitic fluid accumulated in syngeneic C3H/He mice. Immunization was made with living cells of allogeneic SR-C57BL/6 tumor at their early stage of passages together with a complete Freund adjuvant and the ascitic fluid was obtained by Munoz's method.[7] The nucleus was exclusively fluoresced. The suggested nuclear antigen(s) could be extracted from purified nuclei preparations in collaboration with Dr. Amano and Dr. Izawa of the National Cancer Center Institute, Tokyo. Table 5 shows

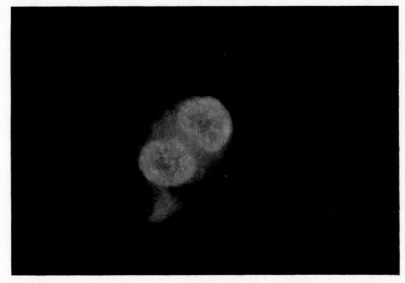

FIG. 3

a rough flowsheet of our extraction procedure. Pulse washing with citric acid was shown to effectively eliminate the protoplasmic contamination without the loss of antigenicity. A final pellet of 3×10^7 nuclei was suspended in 0.25 M sucrose, distributed in test tubes and centrifuged. The RNA/DNA ratio was usually between 0.20 and 0.22 and the DNA content per nucleus was 2×10^{-11} g. 3×10^7 sedimented nuclei were then frozen and thawed three times with 1 ml of distilled water containing 1 ml of 2-fold concentrated gelatine veronal buffer containing Ca and Mg ions (GVB^{++}).[8] The supernatant after centrifugation at 2000 rpm for 10 min was used as the comple-

TABLE 5. PREPARATION PROCEDURE OF NUCLEAR ANTIGEN(S)

—Wash the ascitic tumor cells by centrifugation at 500 rpm for 5 min with 0.02% EDTA-PBS (pH 7.4) and 0.25 M sucrose several times in the cold to eliminate erythrocytes
—Add 10 volumes of Mg^{++} (0.001 M) and Ca^{++} (0.0002 M) for osmotic shock of the cells
—Pass through Chaikoff-type press (clearance: 15 μ) into equal volumes of 0.5 M sucrose containing Mg^{++} and Ca^{++}
—Wash by centrifugation at 1500 rpm for 10 min successively with the mixture of 0.25 M sucrose and 2% citric acid (1:1 and 3:1) and with 0.25 M sucrose
—Centrifuge at 22,000 rpm for 60 min with 2.2 M sucrose
—Suspend the pellet in 0.25 M sucrose and distribute in small test tube about 3×10 nuclei
—Freeze stock the centrifuged pellet at $-70°C$
—3 times of freezing and thawing with 1 ml of aq. dist. and add 1 ml of $2 \times$ GVB^{++} per tube
—Centrifuge at 2000 rpm for 20 min. ---- Supernatant

ment fixing antigen. The micromethod was adopted according to Huebner *et al.*[9,10] for the complement fixing reaction. The antibodies were the above-mentioned γ-globulin fractions of immune ascitic fluid obtained from syngeneic C3H/He mice and they were proved not to contain neutralizing antibodies for SR-RSV. Table 6 shows one of the results obtained in comparison with the whole cell antigen(s). The two antigens reacted quite similarly, indi-

TABLE 6. COMPLEMENT FIXATION TEST COMPARING THE WHOLE CELL ANTIGEN
AND NUCLEAR ANTIGEN

Antigen		Immune ascites (conc)							
		2×	4×	8×	16×	32×	64×	128×	C
Cells	1×	4	4	4′	0	0	0	0	0
	2×	4	4	4	1	0	0	0	0
	4×	4	4	4	3	0	0	0	0
	8×	4	4	4	4′	0	0	0	0
	16×	4	4′	0	0	0	0	0	0
Nuclei	1×	4	4	0	0	0	0	0	0
	2×	4	4	4	4	2′	0	0	0
	4×	4	4	4	4	3′	0	0	0
	8×	4	4	4	4	4′	3	0	0
	16×	4	4′	0	0	0	0	0	0
	CE	4	4′	0	0	0	0	0	0

cating that the antigen(s) were localized exclusively in the nucleus. In order to eliminate any anticomplementary action, mouse ascitic amboceptors were employed in these experiments to sensitize the sheep erythrocytes, but some questions yet remained. Several efforts are being made to purify the mouse antibodies.

TABLE 7. NATURE OF NUCLEAR ANTIGEN(S)

Treatment	Antigen titer
None	8×
Trypsin (0.5 mg/ml 37°C 60′)	2×
DNase (0.25 mg/ml 37°C 60′)	16×
RNase (0.25 mg/ml 37°C 60′)	8×
Antibody titer:	16×

The nature of the antigen(s) was partially studied. As shown in Table 7, their protein nature was strongly suggested. Whether the antigen(s) are the same as "gs" antigen[11-13] or something like the T antigen of DNA-type tumor viruses[14-18] should await further analyses.

REFERENCES

1. P.K.Vogt, *Adv. in Virus Res.*, **11**, 293, 1965.
2. H.Hanafusa, T.Hanafusa and H.Rubin, *Proc. Nat. Acad. Sci.*, **49**, 572, 1963; *ibid.*, **51**, 41, 1964; *Virology*, **22**, 591, 1964.
3. H.Hanafusa, *Nat. Cancer Inst. Monogr.*, **17**, 543, 1964.
4. H.Hanafusa and T.Hanafusa, *Proc. Nat. Acad. Sci.*, **55**, 532, 1966.
5. T. Yamamoto, *Igaku no Ayumi*, **53**, 139, 1965.
6. I.Macpherson, *Science*, **148**, 1731, 1965.
7. J.Munoz, *Proc. Soc. Exp. Biol. Med.*, **95**, 757, 1957.
8. E.A.Kabat and M.M.Mayer, *Experimental Immunochemistry*. Springfield, Illinois, C.C.Thomas, 1961.
9. R.J.Huebner, D.Armstrong, M.Okuyan, P.Sarma and H.C.Turner, *Proc. Nat. Acad. Sci.*, **51**, 721, 1964.
10. P.Sarma, H.C.Turner and R.J.Huebner, *Virology*, **23**, 314, 1964.
11. G.Kelloff and P.K.Vogt, *Virology*, **29**, 377, 1966.
12. H.Bauer and W.Schäffer, *Ibid.*, **29**, 494, 1966.
13. R.M.Dougherty and H.S.DiStefano, *Ibid.*, **29**, 586, 1966.
14. R.J.Huebner, W.P.Rowe, H.C.Turner and W.T.Lane, *Proc. Nat. Acad. Sci.*, **50**, 379, 1963.
15. P.H.Black, W.P.Rowe, H.C.Turner and R.J.Huebner, *Ibid.*, **50**, 1148, 1963.
16. J.H.Pope and W.P.Rowe, *J. Exp. Med.*, **120**, 121, 1964.
17. F.Rapp, J.S.Butel and J.L.Melnick, *Proc. Soc. Exp. Biol. Med.*, **116**, 1131, 1964.
18. F.Rapp, T.Kitahara, J.S.Butel and J.L.Melnick, *Proc. Nat. Acad. Sci.*, **52**, 1138, 1964.

CLOSING REMARKS

James D. Ebert

President Akabori, Professor Seno, Professor Miura, members of the Organizing Committee, colleagues:

My valediction will not—in fact, it cannot—differ in *spirit* from my impromptu remarks at last night's festive occasion.

Put directly and simply, we thank you. You have welcomed us graciously and warmly. You have at once initiated us into the beauties and mysteries of a culture new to us, and yet made us feel at home.

The fare at the Symposium has been rich and varied. I refer not just to the culinary delights, but to the intellectual feast: a smorgasbord of ideas, accompanied by the heady wine of uninhibited discussion. All of us have welcomed the opportunity to take part in the examination of different approaches and different points of view. In a very real sense, discussion of the kind we have enjoyed makes us look at *ourselves*. Are we objective and critical? Claiming the right to criticize, are we perpared to accept criticism, to be self-critical? We are—all of us—*interdependent*. Communication and interaction give greater scope to our individual endeavors.

As Professor Spiegelman said in his opening address, with all of us "joining in the chorus" in succeeding lectures, over the years we have shared the joys (and occasionally the disappointments) of research with many able Japanese colleagues. Thus we count it a special privilege to have been invited to join in your Society's deliberations.

We hope that our contributions to the Symposium have in some small measure expressed our joy and pride in being here.

Thank you very much.

SUBJECT INDEX

471